Fundamentals and Applications of Heavy Ion Collisions

Experiments with accelerated heavy ion beams of energy less than 10 MeV per nucleon require an understanding of the basic concepts of reaction dynamics as well as of experimental technique and methods of preparing samples for irradiation, accurate measurement of sample thickness, sample irradiation, selection and calibration of detectors etc. In depth discussion of models like, promptly emitted particles model (PEPs), breakup fusion model (BUF), hot spot model, Harp Miller and Berne (HMB) and geometry dependent hybrid model etc., offer a comprehensive and up-to-date discussion to the theory and applications of heavy ion collisions at lower energies.

Experimental details of sample preparation, irradiation by HI beams, post irradiation analysis, measurement of recoil range distribution (RRD) and angular distribution of heavy residues in off-beam experiments and of spin distribution in the in-beam experiments are all covered in detail in this monograph. The application of heavy ion interactions including the study of highly rotating neutron deficient nuclei, production of super heavy nuclei and production of specific isotopes of medical applications etc. is presented for the benefit of readers.

R. Prasad is a retired Emeritus Professor of Physics from Aligarh Muslim University, Aligarh, India. He has more than 40 years experience in teaching courses on nuclear physics, electronics, thermal physics, quantum mechanics and modern physics at undergraduate and graduate levels. His field of specialization is experimental nuclear physics in which he has published more than hundred research papers. He has also authored the book *Classical and Quantum Thermal Physics*.

B. P. Singh is a Professor of Physics at Aligarh Muslim University, Aligarh, India. He has more than 25 years teaching and research experience. He has taught courses on nuclear physics, heavy-ion physics, mechanics, electronics, electricity and magnetism, thermal and statistical physics at undergraduate and graduate levels. His areas of research include experimental nuclear physics with special interest in pre-equilibrium emission in nuclear reactions and incomplete fusion reactions in heavy ion interactions, on which he has published more than hundred research papers.

Fundamentals and Applications of Heavy Ion Collisions

Below 10 MeV/ Nucleon Energies

R. Prasad

B. P. Singh

CAMBRIDGE UNIVERSITY PRESS

CAMBRIDGE
UNIVERSITY PRESS

University Printing House, Cambridge CB2 8BS, United Kingdom

One Liberty Plaza, 20th Floor, New York, NY 10006, USA

477 Williamstown Road, Port Melbourne, VIC 3207, Australia

314 to 321, 3rd Floor, Plot No.3, Splendor Forum, Jasola District Centre, New Delhi 110025, India

79 Anson Road, #06–04/06, Singapore 079906

Cambridge University Press is part of the University of Cambridge.

It furthers the University's mission by disseminating knowledge in the pursuit of
education, learning and research at the highest international levels of excellence.

www.cambridge.org
Information on this title: www.cambridge.org/9781108499118

© R. Prasad and B. P Singh 2018

Printed in India by Rajkamal Electric Press, Kundli, Haryana.

A catalogue record for this publication is available from the British Library

ISBN 978-1-108-49911-8 Hardback

Additional resources for this publication at www.cambridge.org/9781108499118

Contents

Figures

Tables

Preface

The study of incomplete fusion (ICF) reactions in heavy ion (HI) interactions at energies below 10 MeV per nucleon is a topic of resurgent interest. At such low energies, near and/or just above the fusion barrier, the complete fusion (CF) of the interacting ions is expected to be the most dominant process; however, experiments carried out during the last decade or so have indicated that a significant part of the interaction proceeds through ICF process. Some theories have been proposed to explain the process of incomplete fusion but none of them could successfully reproduce the experimental data at energies < 10 MeV/A. In order to understand the dynamics of such low energy ICF processes and to develop a viable theoretical frame work, our group carried out extensive and complementary experiments on the topic during the last decade or so. The monograph presents the details of these experiments and the analysis of the data.

The presentation has five chapters; Chapter-1 gives a historical background of the subject and discusses the motivation for the work. Chapter-2, entitled 'Theoretical Tools, Reaction Mechanism and Computer Codes' is intended to develop a sound theoretical background of the subject. Important features of computer codes available in the market for theoretical simulation are discussed in this chapter. All experimental details, including the methodology, experimental setups, formulations used for data reduction etc., are given in Chapter-3. The Chapter-4, entitled 'Measurements' contains the details of the measurements of Excitation Functions (EFs), Recoil Range Distributions (RRDs), Angular Distributions (ADs), Spin Distributions (SDs) and Feeding Intensity Profiles (FIPs) of reaction residues. Each measurement is discussed in detail and the recorded experimental data is presented both in tabular form as well as in graphical form. Chapter-5, is 'Results and Conclusions' which provides a detailed discussion of the results obtained from the critical analysis and evaluation of the data obtained in the present set of experiments. Conclusions regarding the dependence of ICF component on various entrance channel parameters, presented in this chapter may be of considerable value in developing a theoretical frame work for HI reactions at energies below 10 MeV per nucleon. The experiments detailed in this document were carried out by our research group at the Physics Department, Aligarh Muslim University, Aligarh, India, in collaboration with members of the Nuclear Physics Group of the Inter University Accelerator Centre (IUAC), New Delhi, India. The Appendix provides a list of some of the important research publications on the subject published by our research group.

During our interaction with fresh graduates desirous of having a career in accelerator based physics in general and experimental nuclear physics in particular, it was realized that they need a document that may spell out most details for carrying out experiments using accelerated beams. These details, such as, designing an experiment, preparation of samples for irradiation, their thickness measurement, choice of detectors, calibration of detectors, data acquisition and analysis etc., are generally available only in research publications and that too in brief. The present document is written with the view to provide young entrants a detailed description for carrying out experiments with accelerated beams. Details of four different types of experiments mentioned above are provided in this document. As such, the monograph is expected to serve as a handbook, a ready reference for beginners in the field. It is hoped that the monograph will be of interest both to new entrants as well as to experienced researchers in the field of low energy heavy ion interactions.

Acknowledgements

The research work reported in this book is based on the experiments carried out by our research group at the Physics Department, Aligarh Muslim University, Aligarh, India in collaboration with members of the Nuclear Physics Group, Inter University Accelerator Centre (IUAC), New Delhi, India. We wish to put on record our sincere thanks to our collaborators from IUAC, Dr Ranjan K. Bhowmik, Dr R. P. Singh, Dr S. Muralithar, Dr Rakesh Kumar, Mrs K. S. Golda, Mrs Indu Bala and Dr Ajit K. Sinha (presently, the Director General, Inter University Centre for Department of Atomic Energy Facilities, Indore, India). We appreciate, very much, the cooperation extended by you all, individually and collectively. Thank you very much.

Thanks are due to the members of our research group as well; in particular to Dr H. D. Bhardwaj, Dr (Mrs) Sunita Gupta, Dr M. M. Musthafa, Dr (Mrs) Unnati, Dr Manoj Kumar Sharma, Dr Pushpendra P. Singh, Dr Devendra P. Singh, Dr Abhishek Yadav, Dr Vijay Raj Sharma and Mr Mohd. Shuaib. Each member of the group participated, designed, and successfully carried out some part of the experimental work. We thank them all from the bottom of our hearts.

All experiments reported in this document were carried out at the IUAC, New Delhi, India using the accelerated heavy ion beams provided by the 15 UD Pelletron accelerator of the centre. We wish to put on record our heartfelt thanks to Professor Amit Roy, ex-Director and Dr Dinakar Kanjilal, the present Director of IUAC, New Delhi, for their kind cooperation and for extending all facilities required during these experiments. We thank the Pelletron crew, who provided the beams of desired ions of required energy and fluence.

Authors also wish to thank the Department of Physics and the Aligarh Muslim University, respectively for extending departmental facilities and administrative support required to successfully complete the research projects that facilitated these investigations.

Financial support in the form of research projects sanctioned by the University Grants Commission (UGC), Council for Scientific and Industrial Research (CSIR) and Department of Science and Technology (DST), Government of India is thankfully acknowledged.

We both wish to say a big thank you to Mr Gauravjeet Singh Reen, the young, smart, prompt and extremely polite but firm commissioning editor at Cambridge University Press, India for his continued cooperation in communicating with the publishers.

Though last but not the least, we wish to thank our families for their continued support.

Introduction

1.1 Background

Though the concept of the nucleus and the subsequent evolution of nuclear physics are credited to Rutherford, the earlier discovery of radioactivity by A. Henri Becquerel, Pierre and Marie Curie (1896–1898) played the most crucial role in these developments. The discovery of radioactivity opened up the way to new techniques of exploring subatomic systems – for example, by bombarding them with fast moving charged particles, a technique which is still in use, and used more vigorously now, even after hundred years.

In 1898, Pierre and Marie Curie succeeded in isolating significant amounts of two new elements from pitchblende, a uranium ore. They named the two elements polonium and radium. These new elements were found to undergo spontaneous self-destruction by emitting mysterious radiations. Passing of the collimated beam of these radiations through electric and magnetic fields revealed that they are made up of three components: negatively charged components, called beta particles; neutral components of electromagnetic waves of very short wavelength or gamma rays and a third component of positively charged particles. The negatively charged beta particles were identified as electrons, while the Curies established that the positively charged particles were doubly-ionized helium atoms, called alpha particles. The average kinetic energies of these alpha particles, beta particles and neutral gamma rays had different values for different radioactive sources. Radium and polonium, the two natural radioactive sources, emit alpha particles of energies in the range of 5 to 7 MeV. Rutherford, in his famous alpha scattering experiments, actually carried out by Geiger and Marsden[1], bombarded thin metallic foils by a collimated beam of alpha particles obtained from radium. In these experiments, it was observed that, on an average, one to five alpha particles out of about 20,000 particles, get scattered by more than 90°. Rutherford[2] concluded that this is possible only if the target atoms have very small volumes at their centres where total positive charge and almost all mass of the atom are concentrated. Rutherford named this small volume as the nucleus of the atom, a term he borrowed from biological science. The layout of the experimental setup used by Rutherford is shown in Figure 1.1. The alpha particle source (radium) was kept in a lead box with a small hole to get the collimated beam. A ZnS painted screen placed in front of the microscope served as the detector. The target metallic foil was placed normal to the alpha beam. The scattered alpha particles, on hitting the ZnS painted screen, produced tiny flashes of light, which were measured by looking through the microscope. The microscope was capable of rotating in the horizontal plane.

The technique of bombarding a specimen with charged particles in order to explore the charge structure of the specimen's nuclei and nucleons has been used by R. Hofstadter et al.[3] Earlier experiments using neutron scattering have indicated that the radius R of an atom's nucleus $^{A}_{Z}X$, where A is the atomic mass number and Z is the atomic number, is proportional to $A^{1/3}$ or $R = r_0 A^{1/3}$, where r_0 is a constant independent of A. Hofstadter used fast electrons of around 200 MeV energy and bombarded gaseous targets of many elements. He measured the angular distribution of scattered electrons with the help of a double-focusing electron spectrometer. Since electrons of energy less than 1 GeV do not split the target nucleus and interact with the target nucleus only through electromagnetic interaction, Hofstadter's experiments were best suited for studying the charge distribution of the nucleus. The analysis of the data showed that nuclei have a Fermi-type charge distribution described by two parameters – the radius c, which depends on the number of nucleons A in the nuclei and a surface thickness t, which is nearly

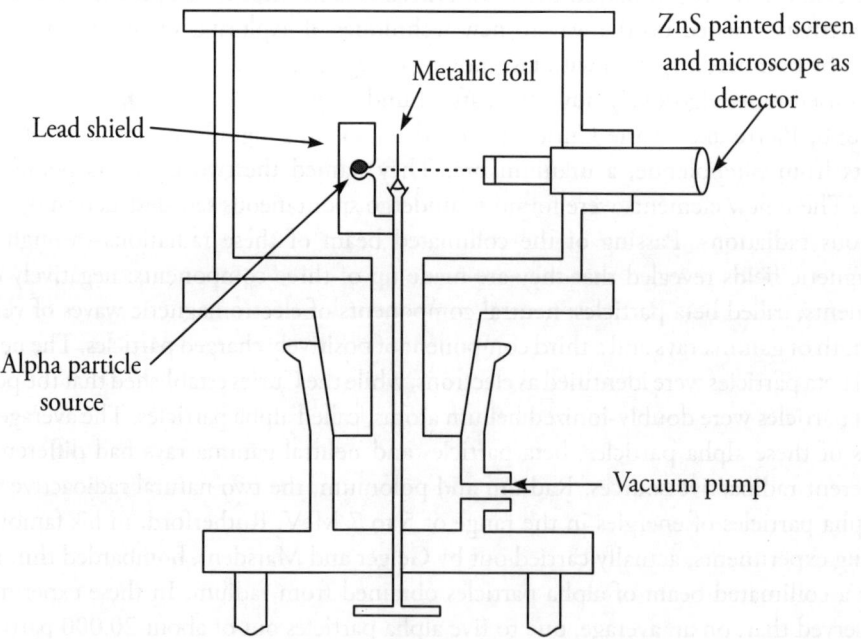

Figure 1.1 Side view of the experimental setup used by Rutherford for alpha scattering

constant for all the nuclei. Accordingly, the charge density $\rho(r)$ at a distance r from the centre of the nucleus is related to its value $\rho(0)$ at the centre by the relation,

$$\rho(r) = \frac{\rho(0)}{e^{\left(\frac{r-c}{0.23\,t}\right)} + 1} \tag{1.1}$$

where, $c = \left(1.07 \mp 0.02\right) A^{1/3}$ fm and $t = \left(2.4 \mp 0.10\right)$ fm $\tag{1.2}$

This means that all nuclei, big or small, have the same charge density at the centre and the charge density falls off smoothly for all nuclei within the same distance, but they differ in the radius of the central core of uniform charge density as shown in Figure 1.2. It may be observed from Figure 1.2 that the y-axis scale is highly magnified.

In this technique of exploring a given target using highly charged particles, the resolving power of the incident charged particle increases with the decrease of the de Broglie wavelength of the particle. The same incident particle will distinguish finer details of the target if it has a smaller wavelength, i.e., relatively higher energies. Hofstadter also estimated the size of protons and neutrons using electrons of higher energies. Since both protons and the neutrons are spin-½ particles, and the incident electron is also a spin-½ particle, the electron interacts with nucleons through its charge as well through its magnetic moments. Hofstadter obtained the values of 0.75 fm and 0.00 fm for the charge radii of proton and neutron respectively; the corresponding magnetic radii were estimated to be 0.97 fm and 0.76 fm respectively. Hofstadter got the 1961 Nobel Prize in Physics for his work on electron scattering.

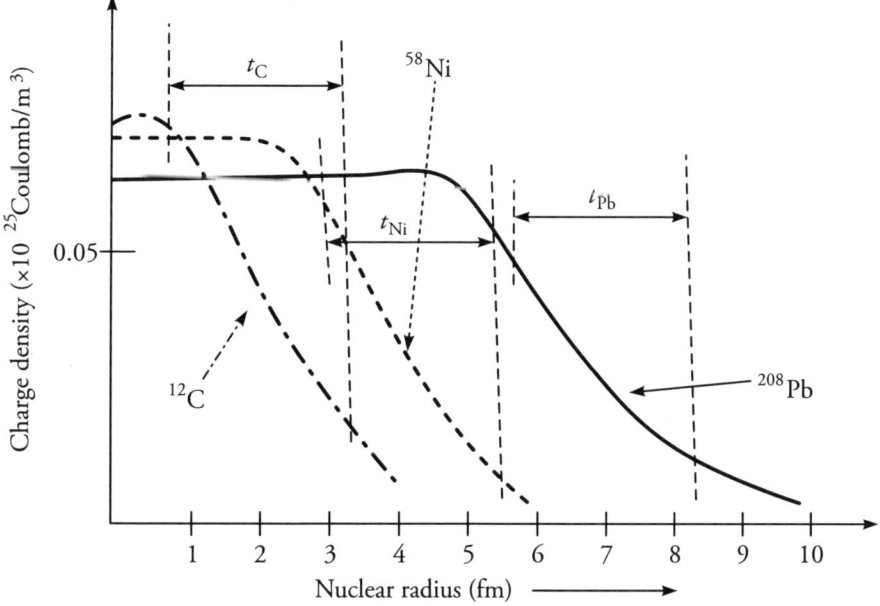

Figure 1.2 Nuclear Charge Distribution and surface thickness t for some nuclei

1.1.1 Artificial radioactivity

The alpha (α) particles provided by naturally radioactive sources like polonium and radium have energies of the order of 6 MeV, which is enough to overcome the Coulomb barrier between the alpha particle and a light nucleus. These alpha particles were extensively used to investigate nuclear reactions initiated by them in light nuclei. In January 1934, Irene Joliot-Curie, daughter of the Curies, and her husband, Frederic Joliot[4] carried out an experiment in which they bombarded a thin aluminium (Al) foil with a collimated beam of 5.3 MeV α particles obtained from a ^{210}Po source. A thin window Geiger–Müller (GM) counter was used to detect the charged particles, if any, emitted from the Al foil in the process. It was found that positively charged particles are emitted when the foil was bombarded by α particles and these particles keep coming from the foil even when the polonium source is taken away from the foil. It was further observed that the number of counts per unit time in the GM counter decreased exponentially with time, like that in the case of natural radioactivity, with a half life of 2.34 min. They attributed these observations to the formation of a new unknown isotope ^{30}P formed according to the following reaction,

$$\begin{array}{l}^{4}_{2}\text{He} + ^{27}_{13}\text{Al} \rightarrow ^{30}_{15}\text{P} + ^{1}_{0}n\end{array} \qquad (1.3)$$

Later, Curie and Frederic confirmed it by chemically separating phosphorous from the irradiated aluminium foil within a few minutes of irradiation and observing that the activity comes from the separated phosphorous. The isotope ^{30}P was found to undergo the following radioactive decay:

$$^{30}_{15}\text{P} \rightarrow ^{30}_{14}\text{Si} + ^{0}_{1}e + \nu \qquad (1.4)$$

Curie and Frederic's experiment demonstrated for the first time that radioactive isotopes may be produced artificially by nuclear reactions. At the same time, it also showed that the isotope ^{30}P decays via a new mode, called β^+ decay, which was unknown till then. The experiment led to the search of new radioactive isotopes for medical and other applications.

1.1.2 Neutron era

After establishing the presence of a nucleus at the centre of each atom, Rutherford focussed his attention on the composition of the nucleus. Since only the proton and the electron were known at that time, it was assumed that the nucleus is also made up of only these two particles. However, in 1920, Rutherford, based on the fact that the actual mass of a nucleus is much larger than what is expected from the 'proton + electron' theory, proposed that there are heavy neutral particles of the type (P + e⁻) present in the nucleus. He called them neutrons. The task of tracking and identifying these heavy and neutral 'Rutherford' particles was assigned to Chadwick.

In an experiment carried out in 1930, it was observed that when a sheet of beryllium was bombarded by α particles from a polonium source, a stream of highly penetrating radiations is emitted. Since the stream was made up of electrically neutral particles, it was assumed to be made up of gamma rays. However, unlike gamma rays, these penetrating radiations did not discharge a charged electroscope or produce ionization. It was obvious that these radiations were different from gamma rays. Frederic and Irene Curie used these mysterious radiations to hit a layer of paraffin wax (Figure 1.3), which is rich in protons, and found that a large number of protons is ejected from the wax, which would not have been possible if the radiations were gamma rays. In 1932, Chadwick, using reaction kinematics and the principle of linear momentum conservation, calculated the mass of mysterious particles emitted from the beryllium foil and found it almost equal to the mass of a proton. Thus, Chadwick established the presence of neutrons, the neutral particle (P + e⁻) suggested by Rutherford. However, from quantum mechanical reasoning, it was concluded that the electron cannot be a constituent of the nucleus. Thus, the neutron could not have the (P + e⁻) structure as suggested by Rutherford. However, Heisenberg, using the quantum mechanical reasoning, had already predicted a neutral particle of a mass nearly equal to the mass of a proton. This gave rise to the proton–neutron model of the nucleus. Chadwick explained the production of neutrons in his experiment by the following equation;

$$_2^4\text{He} + {}_4^9\text{Be} \rightarrow {}_6^{12}C + {}_0^1n \qquad (1.5)$$

It may be seen that the technique of hitting the target with energetic α particles and analyzing the emitted radiations led to the establishment of the presence of neutrons.

Neutrons, having no electrical charge, face no Coulomb barrier when impinging on a nucleus. Therefore, they enter the target nucleus with relative ease. Even very low energy neutrons, like thermal neutrons with energies $\approx 0.025\text{eV}$, can initiate nuclear reactions on entering a target nucleus. Taking advantage of this fact and taking a cue from Frederic and Irene Curie's experiment that new elements produced in a nuclear reaction may be identified through their radioactive properties or by the method of chemical separation, Fermi[5] carried out a large number of experiments in which he bombarded elements of successively higher atomic numbers with neutrons and produced hitherto unknown isotopes. The nuclear reaction

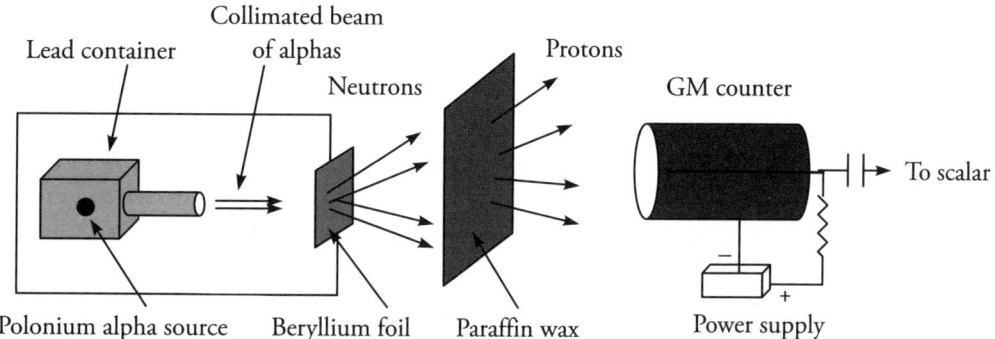

Figure 1.3   Frederic and Irene Curie's experimental setup to discover the nature of 'Rutherford particles' – neutrons

initiated by the absorption of an incident neutron by a target nucleus $_{Z}^{A}X$ may be written as follows:

$$_{Z}^{A}X + _{0}^{1}n \rightarrow {}_{Z}^{A+1}X + _{0}^{0}\gamma \tag{1.6}$$

The new isotope $_{Z}^{A+1}X$ is, generally, unstable and undergoes β^{-} decay leading to the formation of the isotope of the next higher element Y of atomic number (Z+1);

$$_{Z}^{A+1}X \rightarrow {}_{Z+1}^{A+1}Y + \beta^{-} + \overline{\nu} \tag{1.7}$$

With this method, Fermi's group in Italy produced more than 40 new isotopes in a few months.

Using heavier and heavier targets, Fermi's group bombarded the heaviest nucleus, uranium, with neutrons in May 1934. The result of their experiment surprised them – instead of producing the nucleus of the heavier element, radio nuclides of elements having mass almost half of the uranium nucleus were found in the reaction products. Fermi and his collaborators could not unravel the mystery and it took three years for other European laboratories to understand the process. Finally, in 1938, German chemists Otto Hahn and Fritz Stresemann, discovered the phenomena of nuclear fission and published their findings in 1939[6]. The discovery of fission not only proved the mass energy equivalence proposed by Albert Einstein for the first time but also led to a new method of producing huge amounts of energy both for peaceful uses and for nuclear armament.

1.2   Classification of Ions and Research with Accelerated Light Ions

As far back as the late 1920s, the need for machines that may impart higher kinetic energies to charged particles, much more than the energies available with natural radioactive sources, was felt. Higher energy charged particles can overcome larger Coulomb barriers of nuclei of heavy elements, making it possible to initiate more nuclear reactions. Moreover, they have shorter de Broglie wavelengths, and therefore, higher power of resolution to distinguish the finer details of the target. Fusion of accelerated charged nuclear particles with the target nucleus may also produce new elements. In 1927, Rutherford asked scientists to develop systems or machines that may give 'copious supply' of particles more energetic than the α and β particles available from natural radioactive sources. Several scientists responded to his call and undertook projects to build machines that may accelerate light ions. In general, charged particles with atomic mass number A < 4, like $_{1}^{1}H, _{1}^{2}D, _{1}^{3}H$ and $_{2}^{3}He$ are classified as light ions. At Rutherford's Cavendish Laboratory, Cambridge, England, John Cockcroft and E.T.S Walton produced high energy protons by accelerating them using a high-voltage transformer coupled to a voltage multiplier. Robert Van de Graaff built an electrostatic high-voltage generator that could generate up to 1 MV potential. This machine was used to accelerate protons up to 1 MeV energy.

The first cyclotron was conceived at Berkeley (USA) by Ernest Lawrence in 1929 and accelerated protons to 80 keV energies in 1930. In spite of its very small size (its acceleration chamber could be taken in one hand), this apparatus was a very important realization, and the first of an impressive series of machines of increasing power and diameter (4 inches, 11 inches, 27 inches, 37 inches, 60 inches ...). In 1938, E. Lawrence was already building his sixth cyclotron with a diameter of 184 inches (4.70 m). It aimed to accelerate protons to 100 MeV energies. This project was stopped by the Second World War. Just after 1945, magnets were used to build the Berkeley synchrocyclotron with which mesons were artificially produced for the first time. After the Second World War, powerful cyclotrons were built at several places and considerable research, using these machines and high energy light ions, were carried out to study nuclear reactions and the nuclear structure. However, to overcome the relativistic increase in the mass of the ion, cyclotrons were replaced by synchrocyclotrons. In the late 1950s, a 150 MeV proton synchrocyclotron was built at Orsay, France and a 600 MeV one at CERN. This was when the study of high energy interactions became separated from nuclear physics and came to be known as 'particle physics'.

Availability of accelerated light ions opened an era of focused attempts to study the nuclear reaction mechanism in detail. Since nuclear reactions occur in a very short time ($\approx 10^{-22}$ to 10^{-16}s), and also because the true nature of a nuclear force is not known, it is not possible to look, in real time, at what transpires during the rearrangement of nucleons, etc., when the incident light ion enters the nuclear field of the target nucleus. A possible solution is to make some simplified assumptions regarding nuclear force and based on this, develop a theory for the reaction mechanism. The first such attempt was made by N. Bohr in 1936, and it is called the 'compound nucleus reaction mechanism'. Bohr assumed that a nuclear reaction proceeds in two steps – the formation of the compound nucleus by the absorption of the incident ion by the target nucleus that continues till a thermodynamic equilibrium is established in the compound system, and the subsequent decay of the equilibrated compound nucleus into the final products. The most important assumption of Bohr's theory, called 'the independence hypothesis' was that the two steps – the formation and the decay of the compound nucleus are independent of each other. This essentially means that a compound nucleus in a given quantum state decays to a given final state with a fixed probability that does not depend on the specific way it was formed. The first comprehensive test of the independence hypothesis was carried out by Ghoshal[7] in 1950, using accelerated beams of proton and α particles. In his experiments, Ghoshal produced the same compound nucleus ^{63}Zn at the same excitation energy using two different channels and measured the excitation functions for three different decay channels in each case. The input and decay channels in Ghoshal's experiment are shown in Figure 1.4, and the measured excitation functions for the decay channels in Figure 1.5

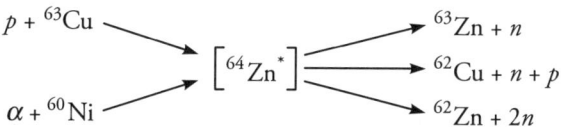

Figure 1.4 Two input and three decay channels in Ghoshal's experiment

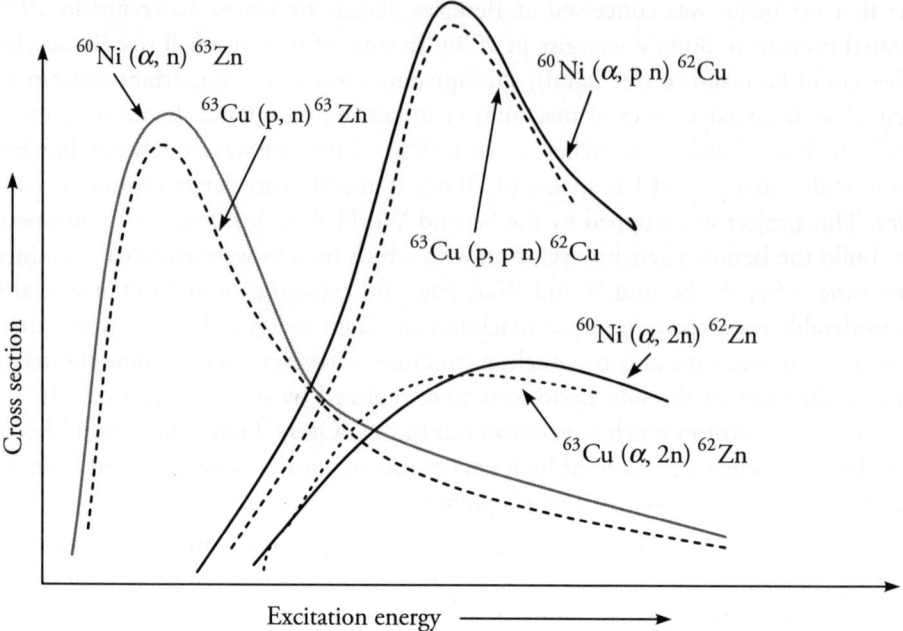

Figure 1.5 Experimental excitation functions for the three decay channels

It may be observed in Figure 1.5 that the excitation functions for the same exit channel from two different input channels are similar to each other, as expected from compound reaction mechanism, but are not exactly identical. This discrepancy may be attributed to the fact that the excited compound system ^{64}Zn produced via proton and α channels had the same excitation energy but different spin and isospin distributions. Several groups including that of J. R. Wiley[8] carried out detailed studies of the effects of isospins in compound statistical reactions. Wiley produced nine compound systems – ^{49}V, ^{52}Cr, ^{55}Mn, ^{56}Fe, ^{60}Ni, ^{63}Cu, $^{64,66}Zn$ and ^{69}Ga by 14 MeV proton bombardments of appropriate targets and also by bombarding various targets with 12–19 MeV α particles. The energy of α particles in each case were chosen to match the excitation energy of the compound system formed by proton bombardment. In the in-beam experiment, energy spectra and angular distributions of protons and α particles emitted from each compound system were measured. These spectra were used to deduce integrated cross-sections for each channel, which in turn were used to obtain the fraction of mixing of $T_<$ and $T_>$ isospin states. Neutrons cannot be accelerated in an accelerator; however, high energy neutrons may be obtained from nuclear reactions initiated by accelerated charged particles. For example, the following reaction

$$^{3}_{1}T + {}^{2}_{1}D \rightarrow {}^{4}_{2}He + {}^{1}_{0}n \tag{1.8}$$

which has a Q-value of $\approx$ 17 MeV, has often been used to generate neutrons of $\approx$ 14 MeV energy. Since the threshold energy of the reaction is low and peaks around 120–130 keV deuteron energy, small Cockcroft–Walton type accelerators, which may have 150 kV high voltage terminals, may be used for this. A Cockcroft–Walton accelerator was built at Aligarh Muslim University, India in 1960[9] for producing 14.8 MeV neutrons and has been extensively used[10] for the study of neutron-induced nuclear reactions at moderate energies. Interesting results about shell effects in statistical nuclear reactions etc., have been obtained from these studies.

The Bohr compound nuclear reaction theory assumes that the composite excited system formed by the absorption of the incident particle by the target nucleus, lives long enough till a thermodynamic equilibrium is established in the composite system, i.e., the composite system forgets all about the history of its formation. Both intuition and experimental evidences suggest that light particle like n, p, α etc., may be emitted from the composite system while it is undergoing equilibration. These particles emitted during the process of equilibration are termed as pre-equilibrium or pre-compound particles and the process is termed as pre-equilibrium or pre-compound emission. Information of considerable value on the dynamics of pre-equilibrium emission has been obtained from the analysis of measured excitation functions[11] of reaction residues, and the energy and angular distributions of emitted light particles[12].

Accelerated light ion beams have also been used in medical applications, i.e., for radiation therapy as well as for producing radioisotopes.

1.3 Accelerated Ions

As has already been mentioned, atomic ions with mass number $A > 4$ are called heavy ions. It is generally not possible to remove all the electrons of atoms of heavy elements. At best, a few electrons may be removed by the ionization process. If n electrons from a neutral heavy atom are removed, an ion with positive charge $Q = ne$, where $e = 1.6 \times 10^{-19}$ Coulomb, is formed. Similarly, if by some method, n extra electrons are added to the neutral atom, a negatively charged ion with charge Q is formed. If such an ion of charge Q is accelerated by electrostatic potential V, the energy E gained by the ion will be QV eV. The energy gained by an ion of charge Q and mass number A in a cyclotron of radius R is given by $K\dfrac{Q^2}{A}$, where K is a constant that depends on the cyclotron. In order to have higher values of Q, new multiple charged ion sources, particularly, penning ionization gauge (PIG) type sources[13] were developed. Though, historically the first accelerated heavy ion beam was carbon produced by Alvarez[14] way back in 1940, the real breakthrough was achieved by Walker et al.[15] at Birmingham, who introduced a small amount of argon gas in the accelerating column. The stripping process in the accelerating chamber increased the charge state of the ion resulting in an internal beam of high energy with a continuous spectrum of lower energies. Taking a cue from this work, two accelerating stage machines, called 'heavy ion linear accelerators' (HILACs), capable of accelerating ions up to neon to energies ≈ 10 MeV per nucleon (10 MeV/A) were built at many places including Manchester (UK), Yale and Berkeley (USA)[16]. These machines used a gas stripper cell between the two accelerating stages. Later, in 1965, a variable energy heavy ion cyclotron, called 'CEVIL' was commissioned under the supervision of Lefort[17] at Orsey,

France. With the view to accelerate ions heavier than neon, R. Basile[18], on a suggestion by Irene Joliot Curie, studied the process of stripping in detail and found that heavier ions of higher charged states may be produced if they are pre-accelerated up to about 1 MeV/A energy before undergoing stripping. This opened a new era of coupled machines for accelerating heavy ions. A linear accelerator (LINAC) with PIG ion source was used to inject pre-accelerated heavy ions into the variable energy heavy ion cyclotron (CEVIL) at Orsey, which, for the first time, could accelerate ions up to krypton to energies higher than the Coulomb barrier for uranium. This coupled machine is called ALICE. Two linear accelerators – SuperHilac at Berkeley, USA and Unilac at Darmstadt, Germany – could accelerate all ions up to uranium to energies ≈ 10 MeV/A. The French GANIL uses a combination of three cyclotrons to accelerate carbon to xenon ions to 20–100 MeV/A energies. Machines at RIKEN, Japan and Lanzhou at China can also accelerate heavy ions up to uranium to about 10 MeV/A energies. With the availability of superconducting magnets, cyclotrons of larger K ≈ 1200 were assembled at some places like Michigan State University and are used for heavy ion acceleration.

The Inter University Accelerator Centre (IUAC) at New Delhi, India, established specifically for users from the university system has a 15 UD Pelletron accelerator (see Figure 1.6). The accelerator which is a tandem Van de Graaff accelerator in vertical configuration can sustain a maximum terminal voltage of 16 MV. The accelerator can provide ion beams of energies from a few MeV to several hundred MeV, depending on the ion and its charge state. The energy range of the Pelletron accelerator has been further enhanced by installing a superconducting linear accelerator (LINAC) booster made up of eight niobium quarter wave resonators, which

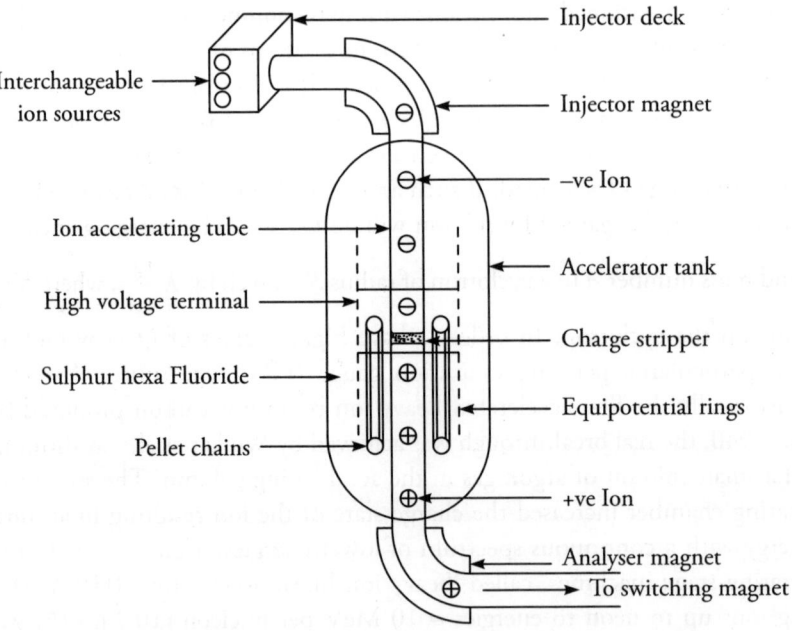

A schematic figure showing the principle of acceleration of ions in pelletron

Figure 1.6 Layout of IUAC heavy ion accelerator

has been designed and installed indigenously. In order to get high charge and high current ion beams, a new high temperature superconducting electron cyclotron resonance ion source (HTS-ECRIS) has also been installed in the IUAC Pelletron.

Another Pelletron accelerator of energy comparable to the IUAC machine is operational at Mumbai, India. This machine was setup in 1989 as a collaborative project of Bhabha Atomic Research Centre (BARC) and the Tata Institute of Fundamental Research. The accelerator is used mostly for nuclear research on advanced topics like the study of nuclear structure at high temperature, study of nuclear states of high angular momentum, transfer and fusion–fission reactions, etc. A layout of the machine is shown in Figure 1.7

Another heavy ion and radioactive ion beam (RIB) facility based on a K500 superconducting cyclotron, electron cyclotron resonance (ECR) and ISOL (on-line isotope separator) ion sources is operational at the Variable Energy Cyclotron Centre (VECC), Kolkata, India. VECC also houses a 224-inch cyclotron that may provide proton beams of 30 MeV to 60 MeV and α beams of 15 MeV to 30 MeV energies. A view of one of the external beam line of the cyclotron is shown in Figure 1.8

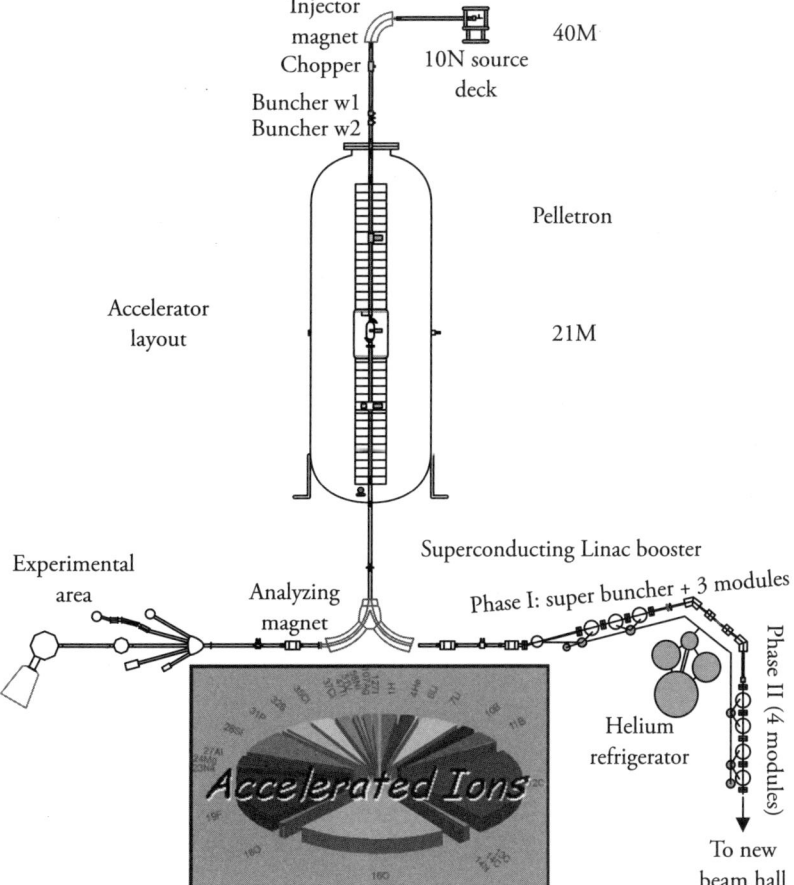

Figure 1.7 Layout of BARC–TIFR heavy ion accelerator

Figure 1.8 A view of the external beam line of the cyclotron in VECC

The Large Hadron Collider (LHC), a relativistic heavy ion machine built by the European Organization for Nuclear Research (CERN) at the border of France and Switzerland, is capable of providing up to ≈ 7 TeV energy per beam including the beams of lead. This is the highest energy to which heavy ions may be accelerated at present. The only other relativistic heavy ion collider (RHIC) at Brookhaven National Laboratory (BNL), New York can also deliver polarized protons along with heavy ions.

1.4   Special Features of Heavy Ions

The creation of heavy ion beams opened the following new fields of study:

1.   Study of high spin nuclear states,

2.   Study of the dynamics of nuclear interaction between multi-nucleon systems,

3. Study of the sequence of nuclear interactions where fusion is followed by fission,

4. Study of large amplitude collective nuclear motions that are associated with the rearrangement of large masses in nuclear reactions,

5. Synthesis and study of elements heavier than trans-uranium elements and leading to super heavy elements (SHE), etc.

Accelerated heavy ions carry large linear momentum and therefore, may impart angular momentum to the target heavy ion, which is orders of magnitude larger than that imparted by a light ion of corresponding energy. If A_P, A_T and E_{lab} respectively denote the mass numbers of the projectile, the target and the laboratory energy of the projectile in units of MeV, then the radii r_p and r_T, respectively of the projectile and the target in units of Fermi (fm) are given by

$$r_p = r_0 A_P^{1/3} \; ; r_T = r_0 A_T^{1/3} \text{ and the sum } R = r_P + r_T = r_0 \left(A_P^{1/3} + A_T^{1/3} \right) \text{fm} \tag{1.9}$$

Here, r_0 is the unit nuclear radius, which has a value ≈ 1.2 fm. The classical expression for the angular momentum of the incident heavy ion with respect to the centre of the target heavy ion in units of $\hbar$ is given as

$$\hbar R = 0.22 \sqrt{\frac{A_P A_T}{A_P + A_T} E^{CM}} \, R \; ; \text{where } E^{CM} \text{ is the centre of mass energy of the projectile} \tag{1.10}$$

$$\text{Also, } E^{CM} = \frac{A_T}{A_P + A_T} E_{lab} \tag{1.11}$$

For the case of 64 MeV (laboratory energy) ^{16}O ions incident on a ^{238}U target, $\hbar R \approx 32\hbar$. As such, the compound nucleus formed by the fusion of the incident ion with the target will have an angular momentum as high as $32\hbar$; while for a corresponding proton of laboratory energy 4 MeV (= 64/16), the value of $\hbar R \approx 2\hbar$. Thus, nuclear reactions initiated by accelerated heavy ions provide the best means of populating high spin states in reaction residues.

P.J. Twin et al.[19], in an experiment, bombarded ^{108}Pd with a 205 MeV (lab. energy) beam of ^{48}Ca ions. The compound system ^{156}Dy formed by the fusion has $\hbar R = 153\hbar$; a proton with the same energy per nucleon (= 205/48) has $\hbar R = 3\hbar$. Thus, it may be observed that very high spin rotational bands may be populated in accelerated heavy ion interactions. A compound system formed by the fusion of two heavy ions, generally, decays first by neutron emission followed by gamma decay. Since neutrons have isotropic distribution, they do not carry any angular momentum. Twin et al. observed gamma rays from the superdeformed band of ^{152}Dy with spin as high as $60\,\hbar$. Gamma ray spectrum of the superdeformed band of ^{152}Dy from the reaction ^{108}Pd + ^{48}Ca $\rightarrow$ ^{152}Dy + $4n$ is shown in Figure 1.9. The numbers written above the gamma peaks are the spins of the nuclear levels in ^{152}Dy from which that gamma ray originates.

A nucleus with charge Ze has a strong electrostatic field ε around it that decreases as $\dfrac{1}{r^2}$, where r is the distance from the surface of the nucleus. As such, the electric field $|\varepsilon|$ at the surface of the heavy ion nucleus of radius R may be written as

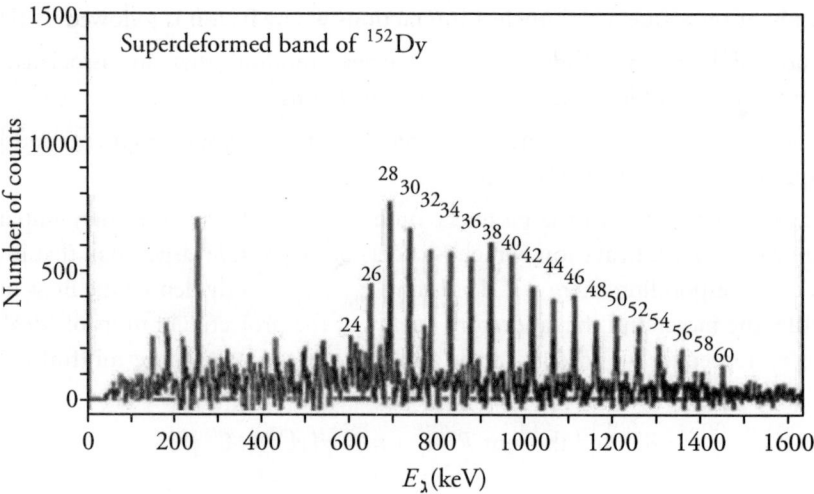

Figure 1.9 Gamma spectrum of the superdeformed band of ^{152}Dy

$$|\varepsilon| = \frac{Ze}{R^2} \cong \frac{Z}{A^{2/3}}\left(\frac{MV}{fm}\right) = ZA^{-2/3}10^{19}(V/m) \tag{1.12}$$

The electrostatic energy E stored outside the nucleus may be given as

$$E = \frac{Z^2 e^2}{R} \approx 1.2\frac{Z^2}{A^{2/3}} \quad (MeV) \tag{1.13}$$

which gives a value of 30 MeV energy to the ^{16}O nucleus and 1642 MeV to the ^{238}U nucleus. Such strong and long range electrostatic field associated with the incident heavy ion makes it possible to excite and even break or initiate nuclear reactions in the target heavy ion. These processes are called Coulomb excitations and Coulomb breakup. They are tools to study the high spin states of deformed nuclei. An accelerated charged particle emits electromagnetic waves or gamma rays if the acceleration is very large. Gamma rays emitted from very high energy heavy ions may initiate photodisintegration of the target nucleus. However, at the moment, we are interested in the interactions initiated by low and moderate energy heavy ions and will not discuss those interactions that are initiated by very high energy ions.

As both the incident heavy ion and the target ion are many-body quantum systems, their interaction is very complex and difficult to describe. However, the fact that the wavelength associated with the incident heavy ion is very small makes it possible to treat the problem semi-classically. The rationalized de Broglie wavelength of the incident ion λ is very small because of its large mass and may be written as

$$\lambda(fm) = \frac{1}{\hbar}4.55\sqrt{\left(\frac{A_P + A_T}{A_P A_T}\right)\left(\frac{1}{E^{CM}}\right)} \tag{1.14}$$

This equation gives a value of 0.012 fm for the rationalized wavelength of a 100 MeV ^{48}Ca beam incident on ^{238}U ion and a value of 0.12 fm for a 100 MeV ^{16}O beam (laboratory energy) impinging on a ^{238}U nucleus. As a consequence, one can use the methods of physical optics, i.e., one can use the trajectories obtained by solving Newton's equations of motion to describe the path taken by the wave front rays. By calculating the change in phase of each ray, one can construct the new equiphase wave front, thus taking into account the effect of the interaction.

Figure 1.10 shows the hyperbolic path of the incident ion (obtained using Newtonian mechanics) in the scattering plane due to Coulomb scattering by the target nucleus. As shown in the figure, r is the distance of the particle from the centre of the target ion, φ the scattering angle, θ the angle that r makes with the direction of incidence and b the impact parameter. The equation of the path, obtained using Newton's laws of motion, may be written as

$$\frac{1}{r} = -\frac{\hbar}{\eta}\tan^2\frac{\varphi}{2}\left\{1 - \csc\frac{\varphi}{2}\sin\left(\theta - \frac{\varphi}{2}\right)\right\} \tag{1.15}$$

The Somerfield parameter η in the aforementioned expression is given by

$$\eta = \frac{Z_P Z_T e^2}{\hbar v} \tag{1.16}$$

It is easy to show that the numerical value of the ratio $\dfrac{\eta}{\hbar}$ (in units of fm) and the centre of mass energy E^{CM} (in MeV) are related by the expression

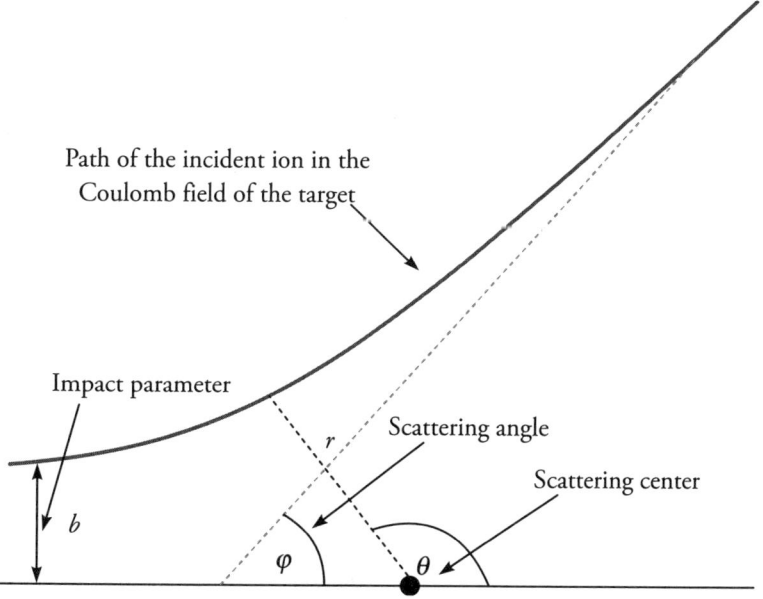

Path of the incident ion in the Coulomb field of the target

Impact parameter

Scattering angle

Scattering center

b

φ

r

θ

Figure 1.10 Coulomb scattering of the incident ion

$$\frac{\eta}{k} = \frac{Z_P Z_T}{1.37\, E^{CM}} \qquad (1.17)$$

It is obvious from the figure that r approaches ∞ both for $\theta = \pi$ and $\theta = \varphi$. The distance of closest approach d may be obtained by setting $\frac{dr}{d\theta} = 0$, that gives $\theta = \frac{\varphi}{2} + \frac{\pi}{2}$ and

$$d = \frac{\eta}{k}(1 + \csc \varphi / 2) \qquad (1.18)$$

and the impact parameter b_{cl} for closest approach has the value

$$b_{cl} = \frac{\eta}{k} \cot \frac{\varphi}{2} \qquad (1.19)$$

In the case of the grazing incidence, the distance of closest approach d_{gr} is equal to the sum of the radii of the projectile and target nuclei, i.e.,

$$d_{gr} = R = r_P + r_T = \frac{\eta}{k}\left(1 + \csc \varphi_{gr} / 2\right) \qquad (1.20)$$

The trajectory corresponding to grazing incidence just touches the surface of the target nucleus.

Figure 1.11 shows the three typical trajectories of incident ions. For all trajectories for which impact parameter b is larger than b_{gr}, the grazing impact parameter, the nuclear field of the target nucleus will not come into play; trajectories of the incident ion will be predicted by the pure Coulomb scattering or Rutherford scattering. This will lead to what is called elastic scattering. However, for the grazing incident ray, the nuclear properties of the target nucleus, whether it is fully absorbing like a black sphere, or partially absorbing, will decide the pattern of the elastically scattered nuclei in the region of the geometric shadow (forward angles), just like the diffraction of light by a spherical ball. Figure 1.12 shows a typical Fresnel scattering pattern for the scattering of ^{12}C by ^{208}Pb.

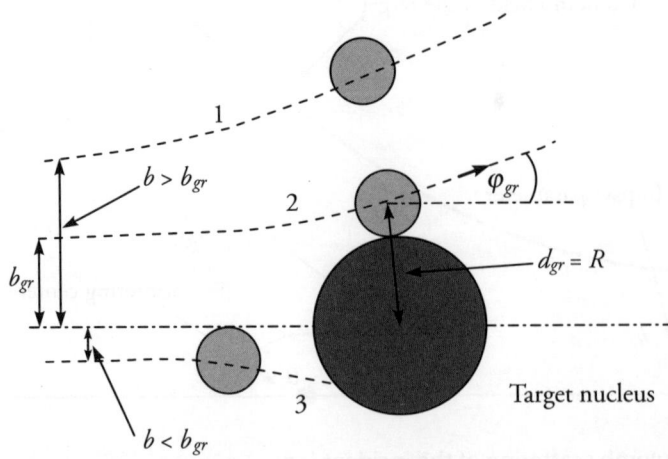

Figure 1.11 Trajectories of the incident ion for three different values of impact parameter

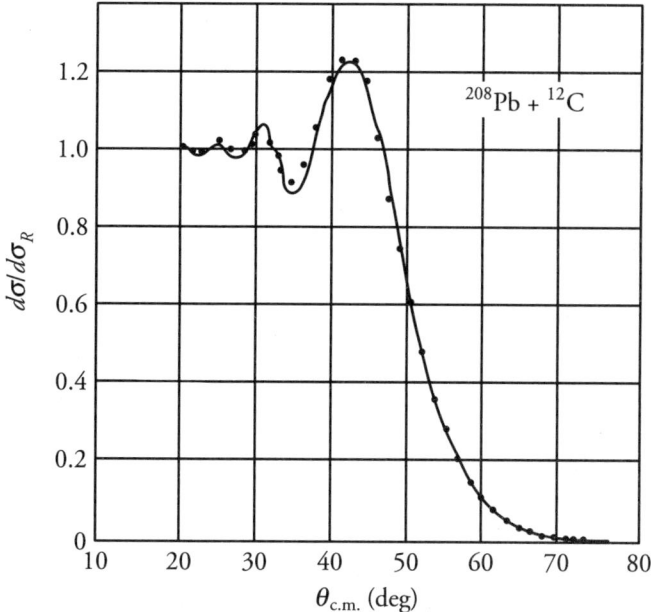

Figure 1.12 Typical Fresnel diffraction pattern

In an optical model, a nuclear potential is written as the sum of two terms, the real part V and an imaginary part W, i.e.,

$$V_{\text{nuc}} = V + iW \tag{1.21}$$

The strength of the imaginary part W of the optical potential decides the absorption. The elastically scattered ions in the forward direction show diffraction in patterns of maxima and minima, the intensity of which depends on the ratio of V/W. Assuming that there is always a reaction when $b \leq b_{\text{gr}}$ (trajectory 3 in Figure 1.11), the reaction cross-section $\sigma_{\text{rea}}^{\text{gr}}$ may be easily given as

$$\sigma_{\text{rea}}^{\text{gr}} = \pi d_g^2 \tag{1.22}$$

Experimental determination of $\sigma_{\text{rea}}^{\text{gr}}$ allows the experimental value of the sum of the radii of the projectile and target nuclei to be given by

$$d_{\text{exp}} = 1.35 \left(A_P^{1/3} + A_T^{1/3} \right) \text{fm} \tag{1.23}$$

This indicates that the magnitude of the unit radius r_0, which is normally taken as 1.2 fm is larger for these reactions.

When the impact parameter of the incident ion is close to b_{gr}, one expects nuclear reactions of short duration with no contribution to the compound nucleus formation. Such reactions are elastic and inelastic scattering and few nucleon transfer reactions, which are also called quasi-elastic scattering.

When the incident energy is moderately high and impact parameter smaller than b_{gr}, the incident nucleus may penetrate into the target nucleus. Depending on the energy and the masses involved, the reaction may end in one of the following process:

(A) Complete fusion: This occurs when the incident and the target ions are both not very heavy and the incident energy is relatively low – only slightly above the Coulomb barrier. The two ions fuse together to make a highly excited composite system with atomic mass number $A_{composite}$ and atomic number $Z_{composite}$ given by

$$A_{composite} = A_P + A_T \text{ and } Z_{composite} = Z_P + Z_T \tag{1.24}$$

Depending on the excitation energy of the composite system, the system may be stable for a sufficiently long time so that the kinetic energy of motion is distributed amongst internal degrees of freedom, and the composite system becomes a compound nucleus in thermal equilibrium. The equilibrated compound nucleus will then decay – first, by evaporating nucleons and light ions and then, by gamma emission into cold residues.

On the other hand, if the excitation energy of the composite system is large, pre-equilibrium emission of nucleons and light nuclei may take place during the process of equilibration. However, ultimately, an equilibrated compound nucleus will be formed with lesser number of nucleons and energy, which will undergo decay via nucleon, light nuclei evaporation followed by gamma emission.

(B) Incomplete fusion: When two ions of moderate masses and high energy come within the range of nuclear force, they may fuse for a while but eventually break up into two parts – one like the target but heavier than it and the other like the projectile but of less mass. This essentially means that the incident nucleus breaks into two parts, one of which fuses with the target while the other goes out with the velocity of the incident ion as a spectator. The process may be understood in terms of the large amount of angular momentum pumped into the composite system by the moderately heavy and high energy incident ion, and which could not be sustained by the system. Though incomplete fusion and pre-equilibrium emission are both processes that are likely to occur at higher incident energies and for heavier ions, strangely, both these processes have been found to occur at incident energies only a few MeV/A above the Coulomb barrier for relatively lighter heavy ions.

(C) Fission: Fission is more probable if the two interacting ions are very heavy and the incident energy is much above the Coulomb barrier. A highly excited composite system has a finite probability of undergoing fission at each stage of its decay and it is unlikely that it will end in a cold residual nucleus by the emission of pre-equilibrium or evaporation particles. The angular momentum ℓ plays a crucial role in fission. The fission barrier decreases with increasing value of ℓ and may totally vanish at a critical ℓ value denoted by ℓ^{cri}. If the angular momentum deposited by the incident nucleus, which depends on the masses of the two nuclei and the incident energy, exceeds the critical limiting value ℓ^{cri}, the composite system will undergo fission. The critical value ℓ^{cri} also hinders the formation of super heavy nuclei (SHE).

(D) Deep inelastic collision (DIC): This process occurs when the two ions are quite heavy ($A > 40$) and the incident energy is 1–3 MeV above the barrier. The target nucleus and the incident nucleus remain together for a short time; in this time, the incident nucleus shares its energy, mass and angular momentum with the target, distributing it into internal degrees of freedom and then parts from the target. In DIC, no compound nucleus is ever formed; the mutual interaction between the target and the incident nucleus lasts for a short time in which mass, energy and angular momentum are exchanged between the two. Figure 1.13 shows trajectories of elastically scattered (Rutherford scattering) incident nucleus (A) and nuclei that have undergone deep inelastic collisions (B and C). All the three nuclei have been scattered by the same scattering angle φ. It is obvious that the impact parameter d_A for particle A is larger than the grazing impact parameter d_{gr}, while d_B and d_C, impact parameters for the DIC particles B and C are both smaller than d_{gr}. Further, the kinetic energy of all elastically scattered incident nuclei, irrespective of their angle of scattering, will be the same, E_A. Both nuclei B and C are scattered by the same angle φ, but the energy E_B of B is larger than the energy E_C of nucleus C ($E_B > E_C$). This is because nucleus C interacts with the target nucleus for a longer period of time (longer path within the target) than nucleus B. Thus, a detector sitting at a particular scattering angle will receive elastically scattered nuclei all having the same energy (equal to the incident energy) and a large number of nuclei scattered through the DIC process having a spectrum of energies.

These qualitative considerations are summarized in Figure 1.14(a), where angular momentum ℓ values (or impact parameter b values) for the regions of dominance of different processes are shown with different shades. It may be realized that there are no sharp boundaries between

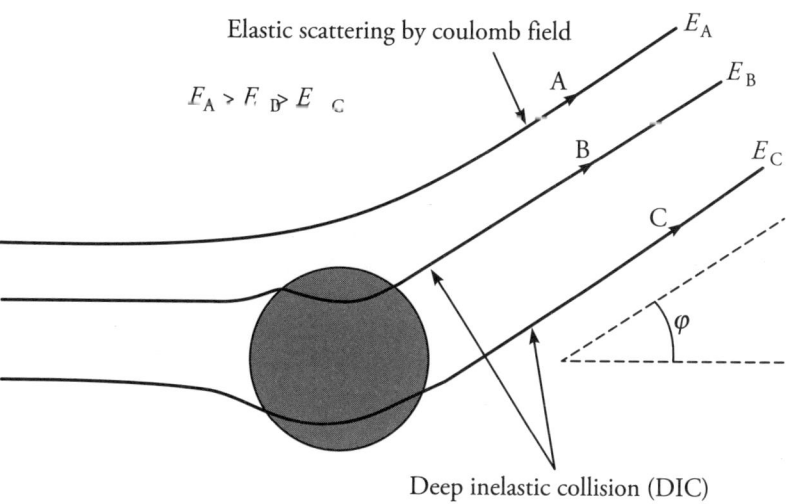

Figure 1.13 Trajectories of elastically scattered and Deep Inelastic Collision event

the different processes and there are regions where two adjacent processes overlap. As may be seen from the figure, for the higher values of angular momentum, when the incident nucleus is outside the range of the nuclear force of the target, the possible processes are elastic scattering and Coulomb excitation/photo breakup of the target nucleus. The upper limit of the graph is bound by the reaction cross-section value $\sigma(\ell) = 2\pi\lambda^2\ell^2$.

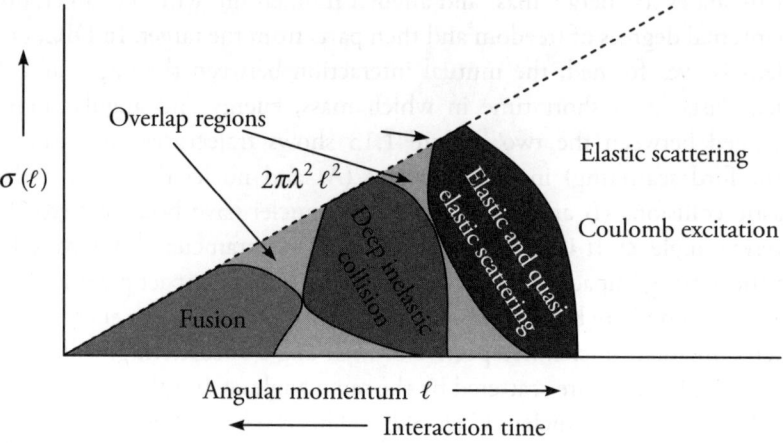

Figure 1.14(a) Qualitative division of angular momentum space into regions where different interactions dominate

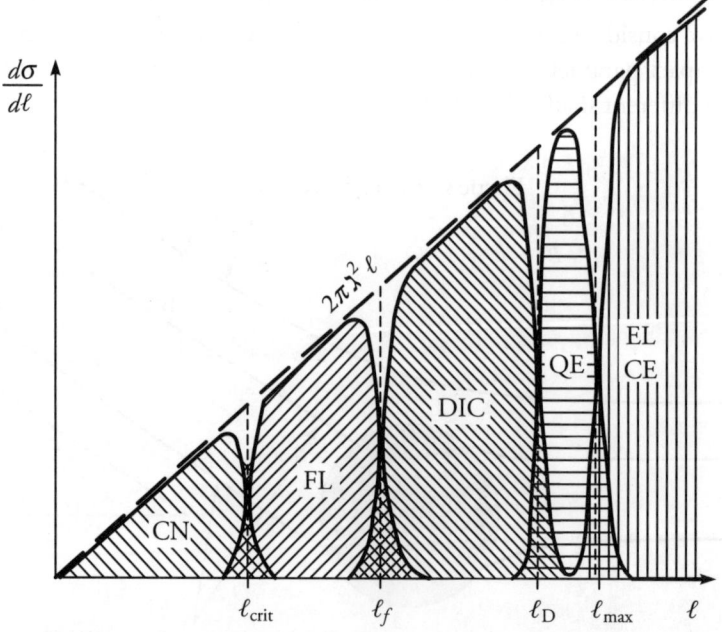

Figure 1.14(b) Domains of Compound Nuclear (CN), Fusion Like (FL) events, Deep Inelastic Collisions (DIC), Quasi Elastic Scattering (QE), Elastic Scattering (EL) and Coulomb Excitation (CE), each marked by its maximum angular momentum

Schroder and Huizenga[21] further refined Figure 1.14(a) and marked domains of compound nucleus formation (CN), fusion like events (FL), deep inelastic collisions (D), quasi-elastic events (QE), elastic collisions (EL) and Coulomb excitation (CE) events. As shown in Figure 1.14(b), each domain is bounded by the maximum limiting value of angular momentum that may contribute to the process.

A pictorial representation of heavy ion interactions is shown in Figure 1.15. As is indicated in this figure, collisions with impact parameters larger than the grazing impact parameter result in elastic scattering and/or Coulomb excitation of the target nucleus. Grazing collisions, in which the nuclear field modifies the Coulomb field by a small amount, give rise to elastic and quasi-elastic scattering. Quasi-elastic events are those in which few, maybe one or two nucleons, are transferred from the incident nucleus to the target or the vice versa. If the impact parameter or the corresponding angular momentum of the incident nucleus is still smaller and if the incident nucleus penetrates deeper into the nuclear field, it results in a deep inelastic collision. In this process, the incident nucleus remains in the nuclear field of the target nucleus for a short time, exchanges mass, kinetic energy and angular momentum with the target and then goes out of the nuclear field. The extent of the mass, angular momentum and kinetic energy transfer depends on the duration in which the two nuclei remain within the range of the nuclear force and may vary from one collision to the other; as a result, there is no direct correlation between the scattering angle and the energy of the scattered nucleus. For still smaller angular momentum (or impact parameter) values, the projectile nucleus may penetrate much deeper into the nuclear field, and stay in the nuclear field for a long time, sharing its kinetic energy, mass and angular momentum with the target nucleus; the kinetic energy of

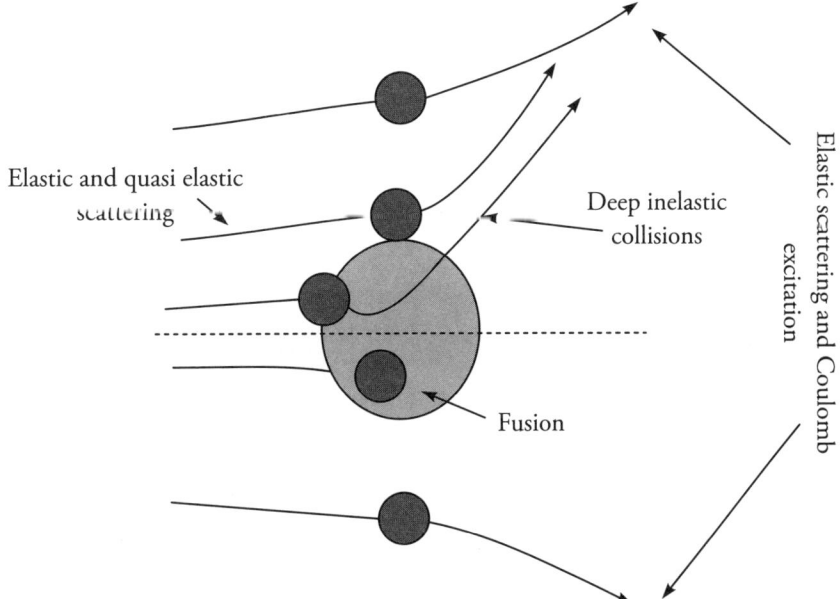

Figure 1.15 Pictorial representation of Heavy Ion Interactions

motion gets distributed in internal degrees of freedom and a highly excited composite system is formed due to the fusion of the two nuclei. The excited composite system may undergo fission before equilibration. This is called *fast fission*. On the other hand, instead of undergoing fast fission, the composite system may equilibrate to become an excited compound nucleus. In case of high excitation, pre-equilibrium emission of nucleons or light clusters may take place during the process of equilibration. An excited compound nucleus has finite probability of decay via fission at each stage of its decay, which competes with evaporation of nucleons and light ions. Fission taking place after the formation of the compound nucleus is termed *fusion–fission*.

J.W. Negele, using the time-dependent Hartree–Fock method, studied the time evolution of deep inelastic collision (DIC) and fission in the heavy ion reaction $^{16}O + {}^{40}Ca$ at around 20 MeV/A energy. The evolution of DIC is shown in sequential frames in Figure 1.16. The first four frames a, b, c and d show the formation of a di-nuclear structure as the incident nucleus enters in the mutual nuclear field. The di-nuclear structure is set in a rotatory motion by the input angular momentum of the incident nucleus. As may be seen in the figure, the neck of the di-nucleus system through which transfer of mass (nucleons), energy and angular momentum take place increases with rotation. This happens because of the joint effect of attractive nuclear and repulsive centrifugal and Coulomb forces. The length of the neck or the separation of the two nuclei increases with the degree of the rotation of the di-nuclear system and eventually after a rotation of about 90°, the two nuclei (with slightly different masses but widely different energies) separate. The incident nucleus then moves away under the influence of the long-range Coulomb force. The energy of the emerging projectile-like nucleus will depend on how long it remained attached with the target nucleus during which it dissipates its energy to the internal degrees of freedom. If the di-nuclear system separates after making one complete revolution, the process is called fast fission.

Sequential time evolution of a typical fusion event is shown in Figure 1.17. The process of fusion develops from the top left frame to the bottom right through successive stages. As may be seen in the figure, the projectile and target nuclei come in close contact forming a di-nucleus structure and its matter density profile shrinks with time, finally resulting in a deformed single nucleus.

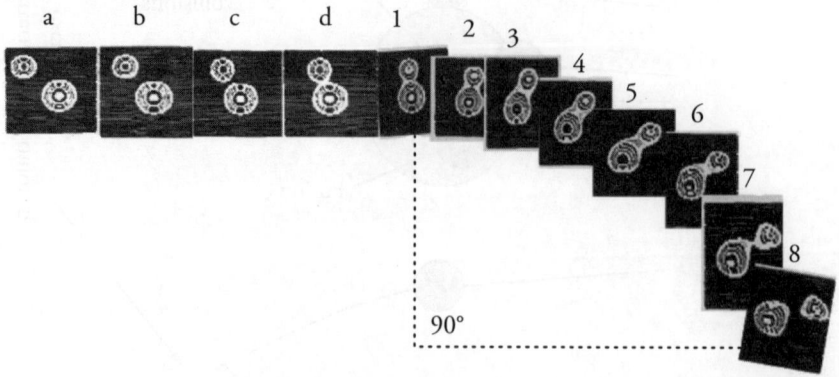

Figure 1.16 Time evolution of a Deep Inelastic event

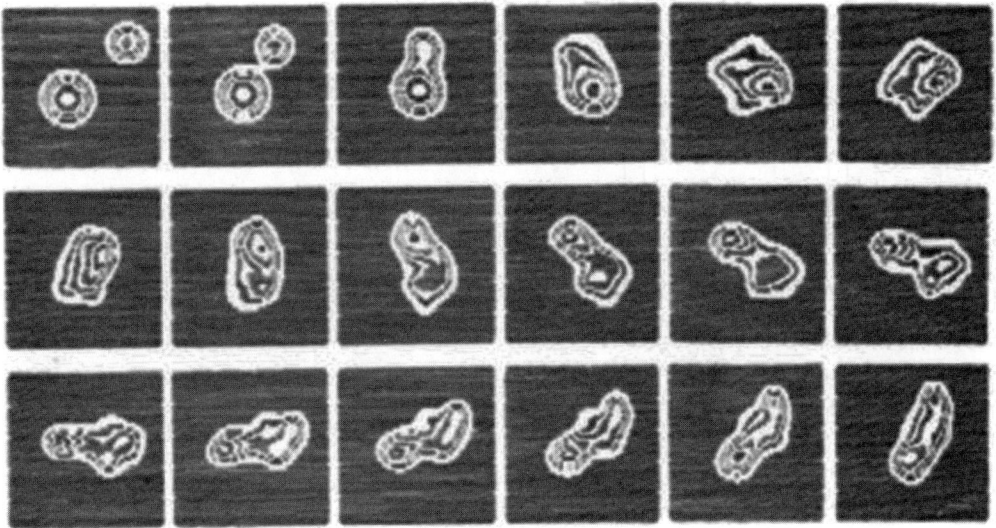

Figure 1.17 Time evolution of Fusion event

The potential for the interaction of two heavy ions, V is generally written as the sum of three components namely, a Coulomb potential V_C, a centrifugal term V_ℓ and a nuclear potential V_N. The Coulomb potential is repulsive and for two nuclei of atomic numbers Z_p and Z_T, assuming the nuclei to be point charges, is given as

$$V_C = \frac{Z_p Z_T e^2}{r} \qquad (1.25)$$

However, it is also possible to treat one of the two nuclei as a point charge and the other as having a spherical charge distribution of radius R_c In that case, Eq. (1.25) may be written[22] as

$$V_C = Z_p Z_T e^2 \left[\frac{3 - \left(\frac{r}{R_C}\right)^2}{2R_C} \right] \text{for} \, r \leq R_c \qquad (1.26)$$

The centrifugal potential may be written as

$$V_\ell = \frac{\ell(\ell+1)\hbar^2}{2\mu r^2} \qquad (1.27)$$

Here μ is the reduced mass of the system.

The nuclear potential is generally taken to be of the Woods–Saxon form defined by three parameters, the depth V_0, the radius r_0 and the diffuseness parameter a as follows,

$$V_N = -V_0 \left[\frac{1}{\left\{ 1 + e^{\left(\frac{r - r_0 \left[A_P^{1/3} + A_T^{1/3} \right]}{a} \right)} \right\}} \right] \tag{1.28}$$

Therefore, in a one-dimensional model where only the r dependence is considered, the nucleus–nucleus potential becomes

$$V = \frac{Z_P Z_T e^2}{r} + \frac{\ell(\ell+1)\hbar^2}{2\mu r^2} - V_0 \left[\frac{1}{\left\{ 1 + e^{\left(\frac{r - r_0 \left[A_P^{1/3} + A_T^{1/3} \right]}{a} \right)} \right\}} \right] \tag{1.29}$$

Here, r denotes the centre of mass distance between the incident and target nuclei of atomic mass number A_p and A_T respectively. Though several parameterizations of the Woods–Saxon potential are available in literature, the one which is frequently used is due to Akyii-Winther[23] and gives the value of V_0 of the order of few tens of MeV, $r_0 \approx 1.2$ fm and $a \approx 0.65$. Typical behaviour of potential energy, V with r and ℓ is shown in Figure 1.18(a), where it may be observed that for smaller values of ℓ, a pocket-like structure is formed due to the interplay of repulsive Coulomb and centrifugal forces and attractive nuclear force. The depth (Figure 1.18(b)) of this pocket decreases with the increasing value of ℓ and eventually, at some higher ℓ value, denoted by ℓ_{crit}, the pocket vanishes.

Since impact parameter $b = \frac{\ell}{k}$, the pocket in the potential energy V will exist only for those incident nuclei that have small impact parameters. It can be shown that the depth of the pocket decreases with the increase in the value of the product of the charges of the projectile and target nuclei $Z_p Z_T$. As such, the magnitude of the critical angular momentum ℓ_{crit} also decreases with the increase of $Z_p Z_T$. If an incident nucleus reaches the region of the pocket, it will be trapped in the pocket resulting in the fusion of the incident and target nuclei. Thus, for an impact parameter $b < \frac{\ell_{crit}}{k}$ and at a given energy, the incident and the target nuclei will have a finite probability of getting trapped in the potential minimum (pocket) for a sufficiently long time, enough to form a compound nucleus. This probability increases with the depth and the width of the pocket. The formation of the compound nucleus is accomplished by three processes: (a) conversion of kinetic energy of relative motion into the energy of mutual excitation, (b) transfer of nucleons and light clusters between the two nuclei and (c) the inter-penetration of incident and target nuclei. As a matter of fact, the aforementioned processes do not take place

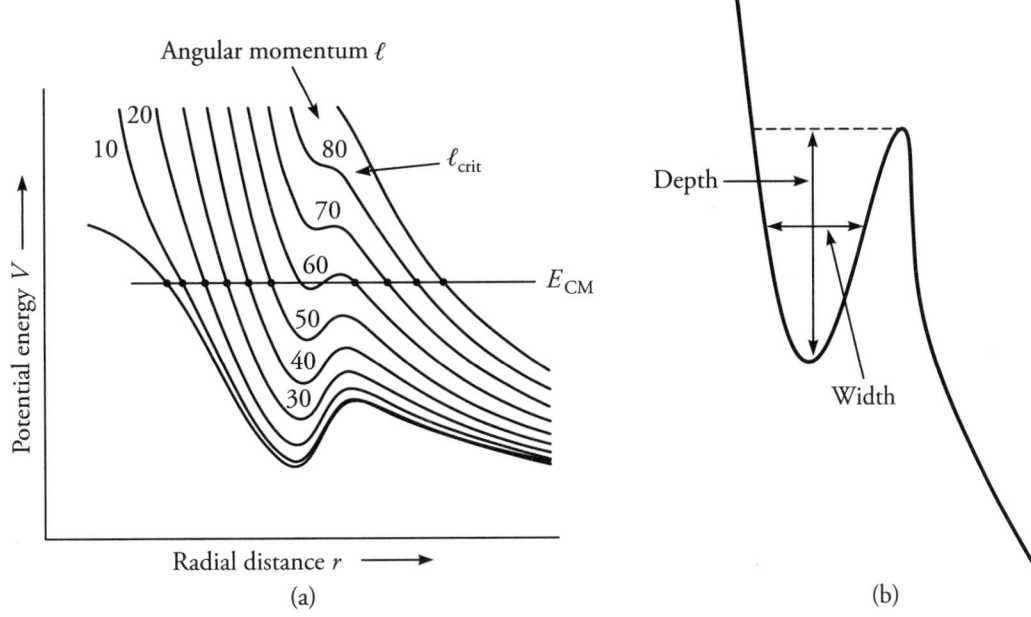

Figure 1.18 Heavy ion potential as a function of separation *r* and angular momentum *I*

only when the two nuclei are trapped in the potential minimum; they come into play before that – in fact, as soon as the two nuclei come within the range of nuclear force. As a result, before reaching the pocket, the incident nucleus is slowed down. This slowing down of the incident nucleus may be compared with motion in a viscous fluid; in theoretical calculations, it is taken into account using the viscosity of the nuclear matter. The energy loss of the incident nucleus before reaching the potential minimum means that the incident nucleus that initially had angular momentum equal to ℓ_{crit} will have angular momentum less than ℓ_{crit} when it reaches the pocket in potential. Therefore, it may fuse with the target. While trapped in the potential minimum, the composite system may emit nucleons and/or light clusters during the process of equilibration, giving rise to pre-equilibrium emission. However, if the excitation energy of the composite system is very large, even heavy clusters may be emitted. In this case, the final compound nucleus will have mass and charge less than the sum of the masses and charges of the projectile and the target. This is called incomplete fusion and is possible only at sufficiently high energies. Heavy clusters, if emitted before the formation of the compound nucleus, also carry away some angular momentum. Emission of massive clusters that may carry off angular momentum becomes necessary if angular momentum in excess to what the compound nucleus can hold is pumped in the system. This may be explained with the help of the Yrast line which restricts the population of states in the energy (*E*)–angular momentum (ℓ) plane. The Yrast line is the locus of the lowest lying energy states for a given angular momentum ℓ. Below the Yrast line, no states of the nucleus are possible. When a system reaches the Yrast line, it decays by releasing cascades of gamma rays.

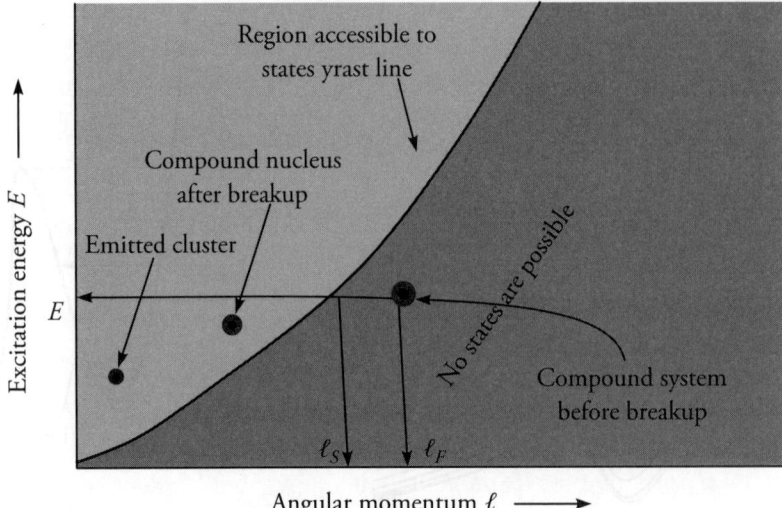

Figure 1.19 Yrast line and regions of accessible and inassessable nuclear states

As shown in Figure 1.19, a composite system with excitation energy E and angular momentum ℓ_F is formed as a result of fusion of target and projectile ions. Now according to the Yrast line, the maximum angular momentum that the composite system may hold is ℓ_S; therefore, the composite system will break into two – the compound nucleus and the emitted cluster – such that both of them fall towards the left of the Yrast line and the laws of conservation of energy and momentum hold.

Apart from the limit set by the Yrast line on the maximum angular momentum that a compound nucleus can sustain, another limit on it is set by the stability conditions of a rotating charged drop made of a non-viscous fluid. The stability of a rotating liquid drop made of non-viscous liquid against fission was studied by Cohen et al.[24] For beta stable nuclei, they derived theoretically, the maximum value of the angular momentum (say ℓ_{limit}^{Cohen}) that the nucleus may hold. If the angular momentum exceeds this value, the nucleus will undergo fission (see Figure 1.20).

When fusion occurs between two heavy ions, in the initial stage, an excited composite system with $A = A_p + A_T$; $Z = Z_p + Z_T$ and angular momentum ℓ_{max} (decided by the energy and the impact parameter of the incident ion) is formed; if the system is trapped in the potential energy pocket for sufficient time, it undergoes equilibration. However, if ℓ_{max} is larger than the ℓ_{limit}^{Cohen} value for the given A, the composite system will fission out before equilibration. This type of fission process is termed as fast fission. It may be pointed out that if $\ell_{max} < \ell_{limit}^{Cohen}$, the composite system will attain thermal equilibrium, becoming an excited compound nucleus, which at each stage of its decay will have a non-zero probability for fission. This is what is called the fusion–fission process.

M. Dasgupta[25] studied the effect of the diffuseness parameter a of the Woods–Saxon potential and the form of Coulomb potential on fusion of heavy ions. In order to reproduce the measured cross-sections for heavy ions, it has been observed that different values of both

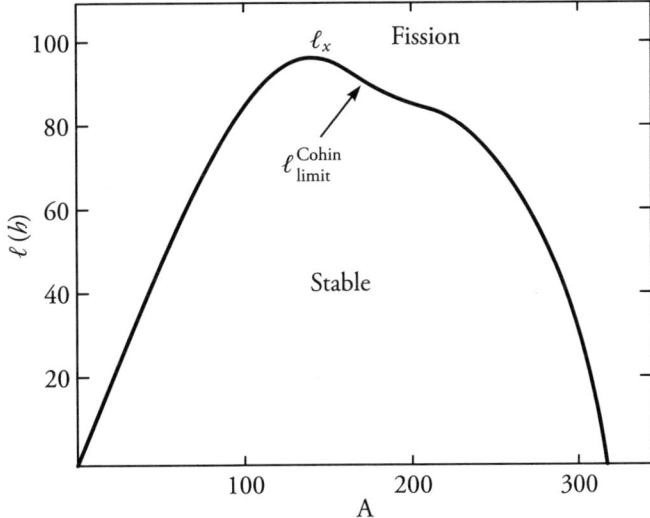

Figure 1.20 Limiting value of Angular Momentum for beta stable nuclei against fission

the depth V_0 and diffuseness a in the Woods–Saxon potential have to be taken from the usually used values given by Akyiiz et al. It has been observed that for fusion reactions initiated by light heavy ions like carbon, oxygen and fluorine, both at sub-barrier energies as well as above it, a diffuseness parameter value of ≈ 1.5 times the usually used value of 0.65 fm is required to explain the experimental data. The effect of increasing the diffuseness on the shape of the potential energy pocket in case of the $^{19}F + {}^{208}Pb$ system for $\ell = 0$ and 60 is shown in Figure 1.21(a).

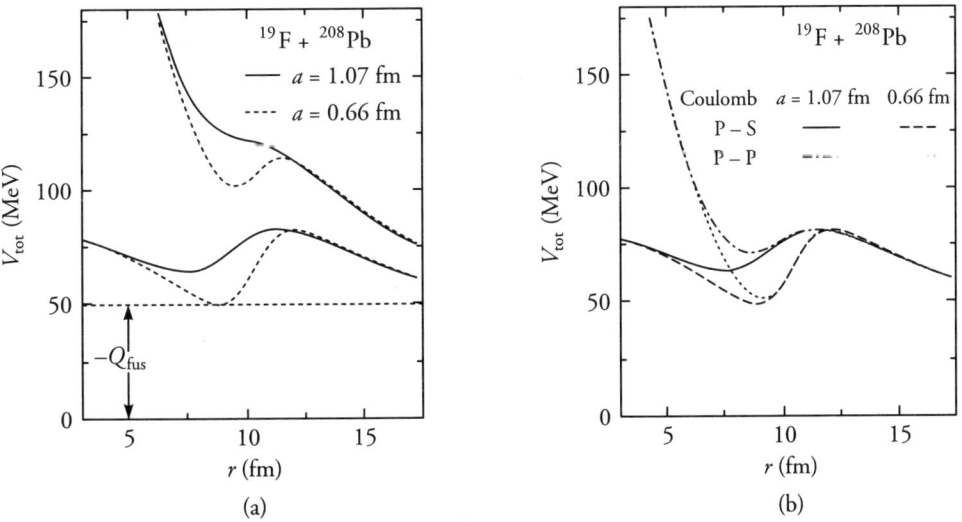

Figure 1.21 (a) Effect of diffuseness on potential energy pocket; (b) Effect of the form of Coulomb Potential on potential energy pocket

As may be seen from the figure, an increase in the value of the diffuseness parameter has the following effects: (i) the pocket becomes shallower, (ii) the fission barrier width decreases and (iii) the barrier width decreases more rapidly for higher angular momentum. A shallower potential pocket means that the nuclear force has a lesser role and the Coulomb potential has a greater role. As has already been mentioned, the Coulomb potential may be written in two forms: (a) assuming that both nuclei are point charges, denoted by P–P and (b) assuming that one of the two nuclei is a point charge and the other a spherical charge distribution of radius R_c, denoted by P–S. The effect of the two forms of Coulomb potential for the same case are shown in Figure 1.21(b). Dasgupta showed, using a lower value for diffuseness (0.66 fm) and a higher value (1.07 fm), that the fusion cross-section data could not be reproduced at higher energies. This, she opined is due to the fact that both in barrier crossing model calculations of fusion cross-section and the coupled channel method, slowing down of the nucleus before it reaches the potential pocket due to viscous effects is not taken into account. Both these methods of calculations assume the target nucleus to be like a black box and that fusion is not affected by the interior of the nucleus.

Comparison of experimental fusion cross-sections with those calculated theoretically on the basis of one-dimensional potential $V(r)$ showed that experimental cross-sections are larger than the corresponding theoretical values, particularly when the incident ion is light. Nix, Moller, Sierk and Krappe[26,27] suggested that the observed enhancement may be due to the deformation of the nuclei when they are very close to each other because of the large Coulomb force. They showed that for deformed nuclei, the Coulomb barrier is several MeV lower in some directions. For example, for the system, ^{48}Ca + ^{248}Cm, the Coulomb barrier was almost 15 MeV lower in the polar region than that in the equatorial region. Further, for spherical nuclei, the vibrational motion may reduce the Coulomb barrier which in turn reduces the fusion barrier and enhances fusion cross-section. Moller et al.[25,26] calculated the potential energy between two identical nuclei in a head-on collision ($\ell = 0$) as a function of the relative separation r between their mass centres and the deformation parameter α for the two nuclei. The deformation parameter is defined as

$$\sigma = 2\left[<Z^2> - <3^2> \right] \tag{1.30}$$

Here, Z is along the direction of the axis of symmetry. The factor of two in front of the bracket takes into account the identical deformations in the two nuclei. The graph showing these calculations for the ^{110}Pd + ^{110}Pd $\rightarrow$ ^{220}U system is shown in Figure 1.22. It turns out, from the figure, that the distance between the two spherical ^{110}Pd nuclei when they touch is $r = 1.6\ r_0$ and their deformation, $\sigma \approx 0.7 r_0$, here r_0 is the unit nuclear radius (=1.2 fm). If one now looks to the reverse process, i.e., if a compound nucleus (^{220}U in the present case) is deformed, its potential energy increases and for a particular deformation becomes maximum. The point where the potential energy has maximum value is called the saddle point for fission. If the excitation energy of the compound nucleus is more than the saddle point energy, the compound nucleus will undergo fission. Thus, if the point of contact of two spherical nuclei lies on the left of the fission saddle point, the system will have energy less than the fission barrier, and the two nuclei will fuse and form a compound nucleus. This happens in the

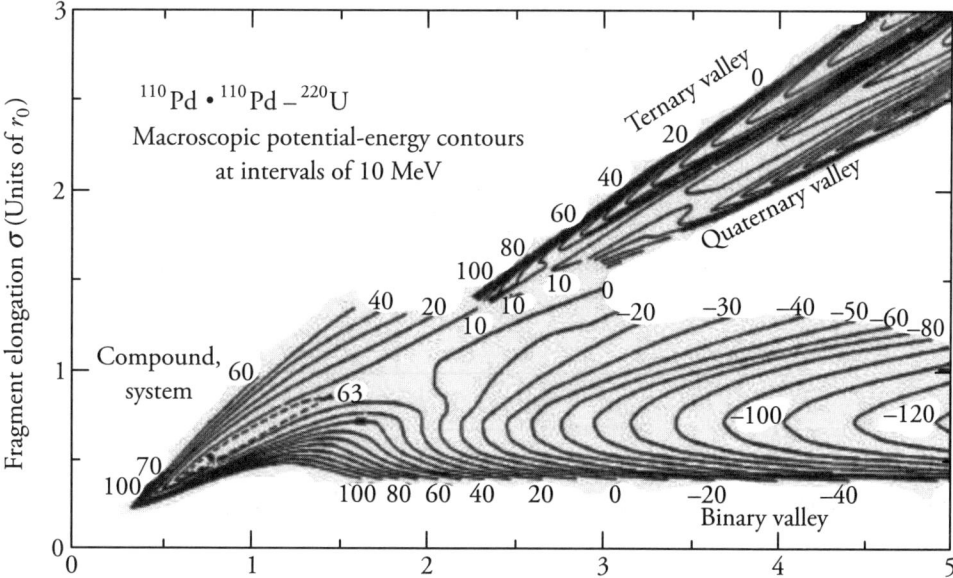

Figure 1.22 Potential energy surface for ^{220}U as a function of two shape parameters, r and σ. Whereas r directly describe the distances between the centers of mass of the two interacting nuclei, σ is a measure for the deformation of the two. The ground state of ^{220}U is normalized to zero energy; the numbers at the contours give the energies relative to it in MeV. The contact point of two spherical touching ^{110}Pd nuclei lies at $r \approx 1.6$, $\sigma \approx 0.74$

case of light nuclei. On the other hand, if the point of contact of the two nuclei falls on the right of the saddle point, the two nuclei will not be able to overcome the fusion barrier and compound nucleus formation will not take place unless additional energy is supplied to the system to overcome the fusion barrier. This additional energy is called the '*extra push*'. In the case of two very heavy nuclei, the point of contact lies on the right of the saddle point and in order to fuse and form a compound nucleus, additional energy in the form of an extra push must be supplied.

The contact point (taken from Figure 1.22) and the saddle point for the system ^{110}Pd + ^{110}Pd $\rightarrow$ ^{220}U are shown in Figure 1.23. Since the contact point falls on the right of the saddle point, an extra amount of energy Δ (extra push) will be required to fuse the two ^{110}Pd nuclei to form the composite ^{220}U. The saddle point for a given system remains fixed in the $(r - \sigma)$ plane but the position of the point of contact changes both with energy and the input angular momentum of the system. Therefore, the extra push energy Δ also depends on the target–projectile pair and the angular momentum ℓ. An empirical expression[85] for extra push energy is given as

$$\Delta E_{\text{ExtraPush}} = 200 \left(x_e - 0.7 \right)^2 \text{MeV} \tag{1.31}$$

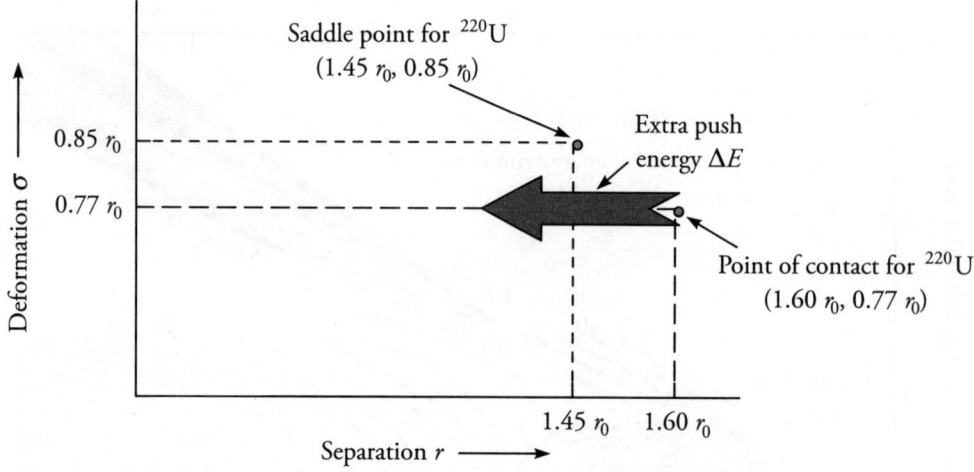

Figure 1.23 Contact point and saddle point for ^{110}Pd + ^{110}Pd = ^{220}U system in the r–σ plane

$$x_e = \left(\frac{Z^2}{A}\right)_{eff} \left(\frac{Z^2}{A}\right)_{crit}$$

where,

$$\left(\frac{Z^2}{A}\right)_{eff} = \frac{4 Z_P Z_T}{\left(A_P A_T\right)^{1/3}\left(A_P^{1/3} + A_T^{1/3}\right)} + \left(\frac{f\ell}{\ell_{ch}}\right)^2$$

and

$$\left(\frac{Z^2}{A}\right)_{crit} = 50.0\left[1 - 1.78\left(\frac{N_P + N_T - Z_P - Z_T}{A_P + A_T}\right)\right]^2$$

$$f = 0.75 \mp 0.05 \quad \text{and} \quad \ell_{ch}^2 = 0.0105\left\{\frac{\left(A_P A_T\right)^{4/3}\left(A_P^{1/3} + A_T^{1/3}\right)^2}{\left(A_P + A_T\right)}\right\}$$

We call a system formed just after fusion of the projectile–target pair a composite system. A composite system has the mass and the atomic numbers equal to the sum of the respective values for the target and the projectile, a certain angular momentum and is excited. Depending on the relative magnitudes of the relaxation times for nucleonic degrees of freedom and for fusion, there will be either no or only partial thermal equilibrium in the composite system. Once formed, the composite system may break up into fragments of nearly equal masses if the angular momentum of the system is larger than that it can sustain, resulting in fast fission or it can break up into a pair of a target-like cluster and a projectile-like cluster to shake off the excess angular momentum resulting in incomplete fusion. Fast fission fragments and/or the incomplete fusion clusters will then equilibrate to become a compound nucleus.

If the angular momentum of the composite system is sustainable, it moves towards becoming a fully equilibrated compound nucleus. During the process of equilibration, the composite system may emit light clusters or nucleons as pre-equilibrium or pre-compound emissions. Eventually, an excited compound nucleus with thermal equilibrium will be formed. The excited compound nucleus formed via fast fission, incomplete fusion or complete fusion will decay first by evaporating light clusters and/or nucleons (if energetically possible) and finally by gamma emission, below the threshold for particle emission. Proper estimation of gamma rays emitted in the decay provides a means of determining the value of the angular momentum ℓ, which is related by the following empirical relation to the gamma multiplicity M_γ;

$$\ell = 2\left(M_\gamma - 4\right) \tag{1.32}$$

Heavy residues, the end products of compound nucleus formation, may be identified, either by chemical separation or by some other specific property, like its half life, characteristic radiations etc, if radioactive. At each stage of its decay, the compound nucleus will have a finite probability of fission. In the case of the fusion of two heavier nuclei, the emission of charged particles from the compound nucleus is severely hindered because of the large Coulomb barrier and therefore, fission competes directly with neutron emission. On the other hand, in the fusion of two medium mass nuclei at higher incident energies, where a composite system with high excitation and angular momentum more than it can sustain is formed, the probability of incomplete fusion is large. In any case, when the composite system has high excitation energy, pre-equilibrium emission becomes a significant contributor to the reaction. However, in the fusion of a light weight heavy ion with another light or medium weight heavy ion at energies just above the Coulomb barrier, the sole contribution to the reaction is expected to come from complete fusion, which will result in an equilibrated compound nucleus with insufficient energy for pre-equilibrium emission.

Experimentally, the fusion cross-section may be determined by detecting the reaction residues either in an in-beam experiment or by following the activities induced in targets bombarded by projectile beams, off-line, if the residues are radioactive with measurable half lives. Information of considerable value about the reaction dynamics is contained in the energy, linear momentum and angular distributions of reaction residues. Residue excitation functions also give information regarding the reaction mechanism. On the other hand, angular and energy distributions of nucleons and light ions emitted in the reaction are very good indicators of pre-equilibrium emission.

Sometimes, the same residue is populated by the complete fusion (CF) as well as by the incomplete fusion (ICF) channels. For example, let us consider the reaction in which oxygen ion hits a target $_{Z_T}^{A_T}X$ and fuses completely, forming an excited compound nucleus $_{(Z_T+8)}^{(A_T+16)}\left[Y_{CF}\right]^*$, as shown by Eq. (1.33),

$$_{8}^{16}O + _{Z_T}^{A_T}X \rightarrow _{(Z_T+8)}^{(A_T+16)}\left[Y_{CF}\right]^* \tag{1.33}$$

Depending on the excitation energy, the compound nucleus $_{(Z_T+8)}^{(A_T+16)}\left[Y_{CF}\right]^*$ may evaporate by emitting one, two, three,... alphas (and other nucleons) leaving behind respectively the evaporation residues, $_{(Z_T+6)}^{(A_T+12)}\left[Y_{CF}'\right]^*$, $_{(Z_T+4)}^{(A_T+8)}\left[Y_{CF}''\right]^*$, $_{(Z_T+2)}^{(A_T+4)}\left[Y_{CF}'''\right]^*$,......, etc.

Next, let us consider that the projectile $^{16}_{8}O$ breaks into two clusters $^{12}_{6}C$ and $^{4}_{2}He$ and only $^{12}_{6}C$ fuses with the target; the other cluster, $^{4}_{2}He$, moves on as a spectator, without interacting. This incomplete fusion (ICF) will result in the formation of the compound system $_{(Z_T+6)}^{(A_T+12)}\left[Y_{ICF}'\right]^*$ as indicated by Eq. (1.34).

$$^{12}_{6}C + ^{A_T}_{Z_T}X \rightarrow {}_{(Z_T+6)}^{(A_T+12)}\left[Y_{ICF-I}'\right]^* \quad (^{4}_{2}He \text{ spectator}) \tag{1.34}$$

On the other hand, if $^{4}_{2}He$ fuses with the target and $^{12}_{6}C$ moves out as a spectator, then the equation of the ICF process may be written as

$$^{4}_{2}He + ^{A_T}_{Z_T}X \rightarrow {}_{(Z_T+2)}^{(A_T+4)}\left[Y''_{ICF-II}\right]^* \quad (^{12}_{6}C \text{ Spectator}) \tag{1.35}$$

However, it is also possible that the projectile $^{16}_{8}O$ breaks up into two $^{8}_{4}Be$ and one of the two ^{8}Be fuses with the target forming the compound nucleus $_{(Z_T+4)}^{(A_T+8)}\left[Y_{CF-III}''\right]^*$ via the ICF channel as given by Eq. (1.36)

$$^{8}_{4}Be + ^{A_T}_{Z_T}X \rightarrow {}_{(Z_T+4)}^{(A_T+8)}\left[Y'_{ICF-III}\right]^* \quad (^{8}_{4}Be \text{ Spectator}) \tag{1.36}$$

In an actual experiment, all the aforementioned processes take place simultaneously when the target is hit by an accelerated oxygen beam. As a result, residues, $_{(Z_T+6)}^{(A_T+12)}\left[Y'\right]^*$, $_{(Z_T+4)}^{(A_T+8)}\left[Y''\right]^*$ and $_{(Z_T+2)}^{(A_T+4)}\left[Y'''\right]^*$, produced by different routes, (CF + evaporation); (ICF of $^{12}_{6}C$); (ICF of $^{8}_{4}Be$) and (ICF of $^{4}_{2}He$), are present in reaction products. The same nuclei produced via different paths have different values for their characteristic properties, like linear momentum, excitation energy, angular distribution and spin distribution of nuclear states. The residues produced by the CF process has the largest value of linear momentum while those populated by the ICF of $^{12}_{6}C$ have linear momentum less than that of CF but larger than those produced by the ICF of $^{8}_{4}Be$. Products of the ICF of $^{4}_{2}He$ will have least value of linear momentum. Thus, the characteristic properties of residues contain a wealth of information on reaction dynamics. What is true for the characteristic properties of these residues also holds for their decay product, if these residues are unstable. Most of these characteristics for radioactive residues may be determined experimentally using the 'stacked-foil-activation' technique.

The statistical compound reaction theory is often used to describe heavy ion fusion at low and moderate excitation energies. In this theory, it is assumed that the fusion of two interacting heavy ions results in the formation of a compound nucleus in which many states of the compound system are excited and the total excitation energy is shared between all the constituents of the compound nucleus (CN) so that it becomes fully equilibrated before its decay. This implies an independence hypothesis, meaning that the decay of the compound system is totally independent of its way of formation. For a binary reaction of the type

$$A + a \rightarrow [CN]^* \rightarrow B + b \qquad (1.37)$$

where A is the target nucleus, a the projectile, $[CN]^*$ the excited compound nucleus, B the residual nucleus and b the ejectile. Since it is not possible to write the Schrödinger equation for the interaction described by Eq. (1.37), between all the nucleons in the system, it is assumed that the interaction between A and a is described by an effective optical potential. This makes it possible to numerically solve the problem by making a partial wave expansion in the centre of mass system. Thus, incident channels are defined by coupling the angular momentum ℓ_a of the projectile with its spin s_a and making the total angular momentum of the projectile $\vec{j}_a = \vec{\ell}_a + \vec{s}_a$. The angular momentum $\vec{j}_a$ is then coupled to the target spin j_A to get the compound nucleus spin j_C. The other characteristics of the system are the kinetic energy of projectile E_a, its parity π^A, and target excitation energy E_A, its parity π^A. Using conservation laws, one may write relations between the incident and exit channel parameters. In the simple case of spinless particles, the Hauser–Feshbach formalism and the diffusion theory provide a relationship between the cross-section from an incident channel α to the exit channel in terms of the scattering matrix $S_{\alpha\beta}$, which in turn may be related to the optical model transmission coefficients. However, in most practical cases, many incident and exit channels have to be considered together and summation has to be performed over all those channels that contribute to the process. Further, projectile and target spins have also to be included. A more detailed treatment of the subject will be presented later.

Pre-equilibrium emission (PE) of nucleons and light clusters may be considered as a bridge between the compound nucleus and the direct reaction mechanisms. Several semi-classical and a totally quantum mechanical treatment have been put forward to explain pre–equilibrium emission. All semi-classical theories assume two-body interactions. In the exciton model, the state of the system is classified by the number of excitons (particles and holes). Two-body interactions may derive an n-exciton state to either an $(n + 2)$ or to $(n - 2)$ exciton state. It further assumes that configurations with the same number of excitons are in equilibrium. Calculation of emission probabilities is simplified by using densities of states and transition densities. In the hybrid model, transitions and emissions are considered from each configuration individually using the Monte Carlo method. Further details of these models are provided in Chapter 2.

1.5 Motivation for this Book

As has already been mentioned, incomplete fusion (ICF) and pre-equilibrium emission (PE) in reactions between light heavy ions (^{12}C, ^{16}O, ^{19}F...) and medium weight heavy ions (HI) at energies just above the Coulomb barrier are very unlikely as in this region, complete fusion

(CF) process is expected to be the sole contributor to the reaction cross-section. However, during the last two decades or so, more and more evidence of significant contributions from both the ICF and PE processes in HI reactions at energies below 10 MeV/n is accumulating [28–41, 83, 84]. The incomplete fusion of heavy ions is generally associated with two types of particles in the exit channel, namely, the projectile-like fragments (PLFs) and the target-like fragments (TLFs). In 1961, Britt and Quinton[42] carried out some brilliant experiments and reported the observance of PLFs indicating the presence of incomplete fusion components in low energy heavy ion reactions. These findings were further confirmed by the work of Galin et al.[43]. It may be recalled that the complete fusion of two heavy ions results in a highly excited composite system with large angular momentum, while in the case of incomplete fusion, a composite system with relatively lower mass, lower linear momentum and lower excitation energy (compared to the composite system formed by CF) is formed. Further, in the semi-classical approach, the driving input angular momentum imparted to the system may be associated with the interaction trajectories[44–49] and hence, one may distinguish complete fusion and incomplete fusion events in terms of the angular momentum (ℓ) transferred to the system. According to the sharp cutoff model[50–53], the attractive pocket in the potential between the two ions vanishes at $\ell = \ell_{crit}$ and all ℓ values starting from $\ell = 0$ to $\ell = \ell_{crit}$ contribute to complete fusion events. Complete fusion corresponds to central or near central collisions (small impact parameters) and, therefore, there is almost complete overlap of the two interacting nuclei involving all nucleonic degrees of freedom that results in the formation of an equilibrated compound nucleus. On the other hand, non-central collisions with larger impact parameters (than those for the complete fusion) result in incomplete fusion. For larger impact parameters, the strength of the attractive potential is reduced because of the larger centrifugal potential, and the driving input angular momentum exceeds the value of ℓ_{crit} for complete fusion. As a result, the attractive potential between the two interacting heavy ions is not strong enough to capture the whole of the incident ion. Fusion in such cases is possible only if a part of the projectile breaks off and releases the excess driving input angular momentum. Therefore, in incomplete fusion, a part of the projectile breaks off as a promptly emitted component. If the incident beam has a well-defined cluster structure, like ^{12}C or ^{16}O beams, alpha clusters are emitted as prompt particles in incomplete fusion. With the emission of prompt particles, the remaining system (target + the remaining part of the projectile) has the sustainable input angular momentum less than or equal to its own critical limit for fusion[54, 55]. Thus, an incompletely fused composite system with relatively less charge and mass as compared to the completely fused compound nucleus is formed. It is obvious that the incomplete fusion will involve input angular momentum values higher than ℓ_{crit} for complete fusion. However, findings by Tserruya et al.[56], indicate that there is no sharp limit of input angular momenta for CF and ICF processes. Gerschel[57] observed that for HI reactions at energies below 10 MeV/n, the range of input angular momenta for CF and ICF processes also depends on target deformation. Particle–gamma coincidence measurements by Inamura et al.[58] and Zolonowski et al.[59] further advanced the understanding of HI reaction dynamics. Experiments carried out by Geoffroy et al.[60] showed that ICF originates from undamped peripheral collisions. Recently, Singh et al.[76], Chakrabarty et al.[77] and Unnati et al.[78] have shown the presence of considerable ICF contributions at energies just above the fusion barrier. Though during the last decade and a half CF and ICF dynamics have been extensively studied, no clear picture, particularly for HI reaction dynamics below 10 MeV/n has emerged so far.

On the theoretical side, models, like the exciton[69,70] model and multistep direct and multistep compound reaction model[73] have been used to treat both the pre-equilibrium emission as well the incomplete fusion of heavy ions. Dynamical models like the break-up fusion (BUF)[61,62], promptly emitted particles (PEPs)[64], hot spot [67], moving-source[68], SUMRULE[63] and fermi-jet[65,66] models have been employed to treat incomplete fusion exclusively. However, all the aforementioned models have been able to explain ICF only to some extent at energies ≥ 15 MeV/n. None of these models has been able to reproduce the experimental ICF data at energies below 10 MeV/n.

In view of the aforementioned facts, the topic of incomplete fusion of heavy ions below 10 MeV/n is still wide open and of current interest. In order to develop a theoretical framework, dependence of ICF on various entrance channel parameters is being investigated. Morgenstern et al.[74,75] investigated the projectile–target mass asymmetry dependence of ICF component while Abhishek et al.[83] showed that the ICF fraction depends strongly on the alpha Q-value. Further, incomplete fusion of heavy ions at low energies is also expected to be a promising route for the production of super heavy nuclei and states of high spin[80,81, 82].

With the view to develop a better understanding of PE emission and of ICF dynamics at energies < 10 MeV/nucleon, a detailed program of measuring linear recoil range distributions (RRDs), angular distribution (AD), excitation functions (EFs) and spin distributions (SDs) of residues populated by CF and ICF of HI interaction has been undertaken. Further, the experimental data has been analyzed to obtain the dependence of ICF on various input/output parameters. Experimental data has also been analyzed to obtain the pre-equilibrium (PE) component, if present. A new method of extracting the PE component from the experimental data on RRD and particle spectra has also been discussed.

The present monograph has been divided into five chapters and an appendix. A review of the development of the science of nuclear physics, starting from the discovery of radioactivity, the role played by the technique of bombarding a target nucleus with energetically charged and uncharged particles to decipher the target structure, development of light and heavy ion accelerators, advantages of heavy ion probes over light ion probes, complexities of heavy ion collision, simplification of using semi-classical trajectories, classification of heavy ion reactions, etc are presented in Chapter 1. Theoretical models and computer codes used in the present analysis are described in Chapter 2. Chapter 3 provides the details of the experimental techniques and formulations used for measurements. Details of the measurements are given in Chapter 4. The results and conclusions of the present measurements are given in Chapter 5. A list of important research publications related to the present study is included in the Appendix.

Theoretical Tools, Reaction Mechanism and Computer Codes

2.1 Complete Fusion of Heavy Ions

Complete fusion of heavy ions is theoretically treated in the framework of a statistical compound reaction mechanism. In heavy ion collisions, a large number of resonances is excited in the compound system, involving many degrees of freedom. A complete description of such a complex collision process is almost impossible to obtain. However, the mean value of cross-section averaged over several resonances is generally of interest, and can be estimated using the statistical approach. The statistical compound reaction model is founded on the works of Bohr[1], Bethe[2], and Weisskopf[3]. Wolfenstein[4] and Hauser and Feshbach[5] extended the model to include the conservation of total angular momentum. The statistical compound model was further refined by Moldauer[6] and Lane and Lynn[7].

Nuclear reactions may be classified in terms of different parameters, including the reaction time. Fast reactions involving reaction times of the order of the time taken by a nucleon to pass through the nucleus ($\approx 10^{-21}$ s) corresponds to direct reactions. Slower processes of reaction times of the order of 10^{-16} s or so come in the category of compound and pre-compound (or pre-equilibrium, or multistep compound and multistep direct) reactions. The compound reaction mechanism, being the slowest, assumes that the excited compound nucleus formed by the fusion of the target and the projectile lives long enough, without decay, for thorough mixing of the target and projectile nucleons to take place and a thermodynamic equilibrium be established in the compound system. Sometimes, it is convenient to call the fused system formed by the amalgamation of the projectile with the target, before the establishment of thermal equilibrium, as an excited composite system that becomes the compound nucleus (CN) when thermal equilibrium is established. Pre-compound reactions occur during the time taken by the excited composite system to transit to the compound nucleus. In this section, we consider the pure compound reaction mechanism and assume that the composite system becomes a compound nucleus without losing any nucleons or clusters. Almost all nuclear models that aim to determine reaction cross-sections make use of the optical model which enables the separation of the total cross-section into different components and provides transmission coefficients that are used in the compound nucleus model.

Consider the following binary reaction,

$$_{Z_P}^{A_P}\text{a} + _{Z_T}^{A_T}\text{A} \rightarrow \left[\text{CN}^*\right] \rightarrow _{Z_B}^{A_B}\text{B} + _{Z_b}^{A_b}\text{b} \qquad (2.1)$$

where the projectile a fuses with the target A forming an excited compound nucleus [CN*], which lives for a sufficiently long time so that thermodynamic equilibrium is established in the compound nucleus before it decays into the residual nucleus B and the ejectile b.

Here,

a is the projectile with atomic number Z_P mass number A_P, kinetic energy E_a, spin s_a and angular momentum ℓ_a, linear momentum p_a and parity π^a

A is the target nucleus with atomic number Z_T, mass number A_T, excitation energy E_A, spin J_A, linear momentum p_A and parity π^A

B is the residual nucleus with atomic number Z_B, mass number A_B, excitation energy E_B, spin J_B, linear momentum p_B and parity π^B

b is the ejectile with atomic number Z_b mass number A_b, kinetic energy E_b, spin s_b and angular momentum ℓ_b, linear momentum p_b and parity π^b.

[CN*] is the excited compound nucleus with atomic number Z_{CN}, mass number A_{CN}, excitation energy E_{CN}, spin J_{CN}, linear momentum p_{CN} and parity π^{CN}

Now from the conservation laws, it follows that:

$$Z_P + Z_T = Z_{CN} = Z_B + Z_b \qquad (2.2)$$

$$A_P + A_T = A_{CN} = A_B + A_b \qquad (2.3)$$

$$E_a + E_A = E_{CN} = E_B + E_b \qquad (2.4)$$

$$\vec{p}_a + \vec{p}_A = \vec{p}_{CN} = \vec{p}_B + \vec{p}_b \qquad (2.5)$$

$$\vec{\ell}_a + \vec{s}_a + \vec{J}_A = \vec{J}_{CN} = \vec{\ell}_b + \vec{s}_b + \vec{J}_B \qquad (2.6)$$

$$\pi^a \pi^A \left(-1\right)^{\ell_a} = \pi^{CN} = \pi^b \pi^B \left(-1\right)^{\ell_b} \tag{2.7}$$

It may, however, be noted that in Eq. (2.6), the sum is a quantum mechanical sum of three angular momentum and as such $\vec{J}_{CN}$ can have a large number of possible values for a given set of $\vec{\ell}_a, \vec{s}_a$ and $\vec{J}_A$, each having a single value. In a real experiment, compound nuclei with fixed excitation energy, definite charge and definite mass, but with a distribution of angular momentum are formed. The same is also true for the isospin of the compound nucleus.

Equation (2.6) may be implemented in two different ways. The L-S coupling scheme is the most frequently used because the optical model transmission coefficients required for further calculations in statistical compound reaction mechanisms, are mostly available with this scheme of coupling. In L-S coupling, the orbital angular momentum of the projectile ℓ_a is first coupled with its inherent spin angular momentum s_a giving the channel spin j_a of the particle. Channel spin j_a and the target spin J_A are then coupled to get the spin of the compound nucleus J_{CN}. Therefore,

$$\vec{\ell}_a + \vec{s}_a = \vec{j}_a; \qquad\qquad \vec{\ell}_b + \vec{s}_b = \vec{j}_b \tag{2.6(a)}$$

(entrance channel spin)   (exit channel spin)

and $$\vec{j}_a + \vec{J}_A = \vec{J}_{CN} = \vec{j}_b + \vec{J}_B \tag{2.6(b)}$$

Here $\vec{j}_a$ and $\vec{j}_b$ are respectively, the entrance and exit channel spins.

From the point of calculations, it is convenient to describe the reaction in terms of the input channel α and the outgoing channel β. Here, α and β are composite symbols that respectively contain all the quantum mechanical parameters defining the incident channel (like E_a, E_A, $\vec{\ell}_a, \vec{s}_a, \vec{J}_A$, etc.) and the exit channel (like E_b, E_B, $\vec{\ell}_b, \vec{s}_b, \vec{J}_B$, etc.).

2.1.1 Hauser–Feshbach formalism for spinless particles

In the case of a spinless projectile and target, it has been shown[8,9] that the diffusion theory leads to the following relation for the cross-section $\sigma_{(\alpha, \beta)}$ from the incident channel α to the exit channel β,

$$\sigma_{(\alpha,\beta)} = \frac{\pi}{k_\alpha^2} < |\delta_{(\alpha,\beta)} - S_{(\alpha,\beta)}|^2 > \tag{2.8}$$

Here, k_α and $S_{(\alpha, \beta)}$ are respectively the wave number of relative motion in the incident channel and the element of the scattering matrix. The optical model gives the value of $S_{(\alpha,\beta)}$ and also of the transmission coefficient T_α for the incident channel as

$$T_\alpha = 1 - |S_{\alpha\alpha}|^2 \tag{2.9}$$

Therefore, the cross-section for the formation of the compound nucleus via channel α may be written as

$$\sigma_{CN}^{\alpha} = \frac{\pi}{k_{\alpha}^2} T_{\alpha} \tag{2.10}$$

The assumption that the compound nucleus does not decay till a thermodynamic equilibrium is established ensures that the compound system, during the process of equilibration, forgets the history of its formation and is characterized only by its quantum numbers. This lays the foundation of the independence hypothesis which states that the decay of the compound nucleus is totally independent of its way of formation, i.e., the formation and decay of the compound nucleus are totally independent of each other. A given compound nucleus (having a set of quantum numbers) formed by diverse incident channels will decay in the same way. Hence, the cross-section for the formation of the compound nucleus by channel α and decay by channel β may be obtained by multiplying the cross-section for the formation of the compound nucleus via channel alpha σ_{CN}^{α} by the probability $G^{CN}(\beta)$ for the decay of the compound nucleus into channel β. Therefore,

$$\sigma_{(\alpha,\beta)} = \sigma_{CN}^{\alpha} G^{CN}(\beta) \tag{2.11}$$

The given compound nucleus may decay via many different exit channels, say, $\beta, \gamma_1, \gamma_2, ..., \gamma_n$ that are allowed by conservation laws. Then,

$$G^{CN}(\beta) = \frac{T_{\beta}}{\sum_i T_{\gamma i}} \tag{2.12}$$

and

$$\sigma_{(\alpha,\beta)} = \sigma_{CN}^{\alpha} \frac{T_{\beta}}{\sum_i T_{\gamma i}} - \frac{\pi}{k_{\alpha}^2} T_{\alpha} \frac{T_{\beta}}{\sum_i T_{\gamma i}} - \frac{\pi}{k_{\alpha}^2} \frac{T_{\alpha} T_{\beta}}{\sum_i T_{\gamma i}} \tag{2.13}$$

Equation (2.13) gives the Hauser–Feshbach formula for spinless particles.

Let us consider the formation of the compound nucleus by the fusion of the projectile a (Energy E_a, spin s_a, angular momentum ℓ_a) with the target A (excitation energy E_A, spin J_A) and see how many incident channels are involved in the process. Because of the quantum mechanical sum of the angular momentum, the compound nucleus will have a distribution of angular momentum values each with a certain statistical weight. Let us assume that compound nuclei with spins $J_{CN}^1, J_{CN}^2, J_{CN}^3, ..., J_{CN}^N$ are populated respectively with statistical weights $w_1, w_2, w_3, ..., w_N$. As such, N incident channels, each specified by quantum parameters $(E_a, s_a, \ell_a, E_A, J_A, J_{CN}^i, w_i)$, ($i$ going from 1 to N) will contribute to the formation of the compound nucleus. Similarly, many β channels will contribute to the decay of the compound nucleus to the ejectile b and residue B.

The first step towards writing an expression for the reaction cross-section is to determine the possible values of the compound nucleus spin J_{CN}, which is the quantum mechanical sum of three angular momentums, ℓ_a, j_a and J_A. In a given case, many incident partial waves with different values of orbital angular momentum ℓ_a will contribute to the reaction. The maximum value of the projectile orbital angular momentum ℓ_a^{max} may be determined using the classical expression $\ell = R \times p$. (see Eq. (1.10), Chapter 1). The maximum value of R in this expression may be taken as the sum of the radii of the target and the projectile, $R = r_0 \left(A_{Pz}^{1/3} + A_T^{1/3} \right)$, where $r_0 = 1.2$ fm is the unit radius. The linear momentum of relative motion p may also be calculated from the centre of mass energy of the projectile $E_a^{CM} = \dfrac{A_T}{A_T + A_p} E_a$. Now the possible values of the compound nucleus spin may have values $J_{CN} = |J_A - s_a|$ to $(\ell_a^{max} + j_a + s_a)$ in steps of unity. Further, the statistical factor of the compound nucleus with a spin J_{CN} is given by

$$\left[\frac{2J_{CN} + 1}{(2J_A + 1)(2s_a + 1)} \right].$$

Thus, for a given value of J_{CN},

The possible j_a values are: $|J_{CN} - J_A|$, $\left[|J_{CN} - J_A| + 1 \right]$, $\left[|J_{CN} - J_A| + 2 \right], \dots, (J_{CN} + J_A)$; in steps of unity.

The possible ℓ_a values are: $|j_a - s_a|$; $\left[|j_a - s_a| + 1 \right]$; $\left[|j_a - s_a| + 2 \right], \dots, (j_a + s_a)$, in steps of unity.

The possible j_b values are: $|J_{CN} - J_B|$; $\left[|J_{CN} - J_B| + 1 \right]$; $\left[|J_{CN} - J_B| + 2, \right], \dots, (J_{CN} + J_B)$, in steps of unity.

The possible ℓ_b values are: $|j_b - s_b|$; $\left[|j_b - s_b| + 1 \right]$; $\left[|j_b - s_b| + 2 \right]; \dots; (j_b + s_b)$, in steps of unity.

Thus, all channels corresponding to the possible values of j_a and ℓ_a will contribute to the formation of the compound nucleus with spin J_{CN} and all channels corresponding to the possible values of j_b and ℓ_b will contribute to the decay of the compound nucleus with spin J_{CN}.

For example, let us consider the case when a beam of 50 MeV (laboratory energy) ^{19}F ions hits the stationary target ^{45}Sc (ground state) and fuses with it, forming the compound nucleus ^{64}Zn. The spin of the projectile ^{19}F is $1/2\,\hbar$ and that of the target is $7/2\,\hbar$. Moreover, the projectile has an even parity while the target has an odd parity. The sum of the radii of the projectile and the target R, calculated using the formulations given earlier, comes out to be 7.47036 fm and the maximum value of the orbital angular momentum ℓ_{max} to be 36 $\hbar$. The minimum possible value of the compound nucleus spin J_{CN}^{min} will be given by

$$J_{CN}^{min} = |J_A - s_a| = \left| \frac{7}{2} - \frac{1}{2} \right| = 3\hbar$$

and the maximum value

$$J_{CN}^{max} = \ell_{max} + J_A + s_a = 36 + \frac{7}{2} + \frac{1}{2} = 40\hbar$$

Thus, the compound nucleus with spins 3, 4, 5,..., 40 $\hbar$ will all be populated as a result of fusion, each with its own statistical weight.

Further, for each possible value of J_{CN}, the possible values of j_a will vary from $|J_{CN} - J_A|$ to $(J_{CN} + J_A)$; and the possible values of ℓ_a will vary from $|J_{CN} - s_a|$ to $(J_{CN} + s_a)$, in steps of unity.

Suppose we chose the value of J_{CN} to be 10.

The possible values of j_a and ℓ_a will be $\frac{13}{2}, \frac{15}{2}, \frac{17}{2},...,\frac{27}{2}\hbar$, and ℓ_a from 6 to 14$\hbar$.

Similarly, one may calculate the possible values of channel spin j_b and the orbital angular momentum l_b for the decay channels.

Since the target ^{45}Sc has odd parity, odd values of orbital angular momentum will populate compound nuclei with even parity and the even ℓ values will feed to compound nuclei with odd parity. However, compound nuclei of both even and odd parities will contribute to the cross-section.

As has been mentioned earlier, when the projectile and the target have non-zero spins, the cross-section for the formation of the compound nucleus $\sigma_{(a,A,E_a)}^{compound}$ is made up of contributions from various possible values of J_{CN}.

$$\sigma_{(a,A,E_a)}^{compound} = \sum_{J_{CN}^{min}}^{J_{CN}^{max}} \sigma_{(a,A,E_a)}^{J_{CN}} \qquad (2.13(a))$$

and $\quad \sigma_{(a,A,E_a)}^{J_{CN}} = \frac{(2J_{CN}+1)}{(2J_A+1)(2s_a+1)} \sum_{\ell_a^{min}}^{\ell_a^{max}} T_{(a,A,E_a,l_a)}^{J_{CN}} \qquad (2.13(b))$

The cross-section for a given value of J_{CN} depends on two factors, the statistical factor $\left[\frac{2J_{CN}+1}{(2J_A+1)(2s_a+1)}\right]$ and the transmission coefficient $T_{(A,a,E_a,\,j_a,\ell_a)}^{J_{CN}}$, which decreases with increasing value of ℓ_a. As such, the maximum contribution to the cross-section $\sigma_{(a,A,E_a)}^{compound}$ comes from some value(s) of J_{CN} intermediate between J_{CN}^{min} and J_{CN}^{max}.

The cross-section for the reaction shown by Eq. (2.1) may, therefore, be written as,

$$\sigma(a,b) = \frac{\pi}{k_a^2} \sum_{J_{CN}=|J_A-s_a|}^{\ell_a^{max}+J_A+s_a} \sum_{\pi^{CN}=+,-} \frac{(2J_{CN}+1)}{(2J_A+1)(2s_a+1)} \sum_{j_a=|J_{CN}-J_A|}^{J_{CN}+J_A} \sum_{\ell_a=|J_A-j_a|}^{J_A+j_a} \sum_{j_b=|J_{CN}-J_B|}^{J_{CN}+J_B} \sum_{\ell_b=|j_b-s_b|}^{j_b+s_b} X \qquad (2.14)$$

where $X = \dfrac{T_{(E_a, J_A, j_a, \ell_a)} \, T_{(E_b, J_B, j_b, \ell_b)}}{\sum_\gamma T_\gamma} \, \delta\left(a, \pi^{\mathrm{CN}}\right) \delta\left(b, \pi^{\mathrm{CN}}\right)$ (2.15)

The $T_{(E_a, J_A, j_a, \ell_a)}$ and $T_{(E_b, J_B, j_b, \ell_b)}$ in Eq. (2.15) are the transmission coefficients respectively for the entrance and exit channels. The delta functions; $\delta\left(a, \pi^{\mathrm{CN}}\right)$ and $\delta\left(b, \pi^{\mathrm{CN}}\right)$ ensures the parity conservation in the process.

It is also possible to calculate the angular distribution of the ejectile in the centre of mass using the relation

$$\sigma^\vartheta(a,b) = \sum_{\text{even } L} C_L P_L(\cos\vartheta)$$ (2.16)

where ϑ is the scattering angle. Also,

$$C_L = \frac{\pi}{k_a^2} \sum_{J_{\mathrm{CN}}, \pi^{\mathrm{CN}}} \frac{(2J_{\mathrm{CN}}+1)}{(2J_A+1)(2s_a+1)} \sum_{(j_a, \ell_a j_b, \ell_b)} X \, A^{J^{\mathrm{CN}}}_{(J_A, \ell_a, j_a, J_B, \ell_b, j_b, L)}$$ (2.17)

Here, $A^{J^{\mathrm{CN}}}_{(J_A, \ell_a, j_a, J_B, \ell_b, j_b, L)} = \dfrac{(-1)^{(J_B - s_b - J_A + s_a)}}{4\pi} (2J+1)(Z_1)(Z_2)$ (2.18)

$$Z_1 = (2j_a+1)(2l_a+1)C(1100:L0)W\left(j_a j_a \ell_a \ell_a : L s_a\right)$$ (2.19)

and

$$Z_2 = (2j_b+1)(2\ell_b+1)C(1100:L0)W\left(j_b j_b \ell_b \ell_b : L s_b\right)$$ (2.20)

W and C in these equations are respectively Racah and Clebsch–Gordan coefficients.

2.1.2 Level width and level separation

Figure 2.1 shows the typical levels of a compound nucleus. At lower excitation energies, the levels are discrete and separated from each other. There are two characteristics of the level scheme, namely, the separation between two adjacent levels, called level separation, denoted by D and the energy spread of the level, called level width, denoted with Γ. The width Γ and the mean life τ of the level are related through the uncertainty principle, i.e.,

$$\Gamma . \tau \geq \hbar$$ (2.21)

Since the ground state of a stable nucleus has infinite mean life, the ground state level has the minimum width and, therefore, is the sharpest. As the excitation energy increases, the level separation D decreases and the level width increases; the levels become more and more unstable (smaller mean life). As shown in Figure 2.1, the level scheme may be divided into two parts – the lower energy end where $D \gg \Gamma$ and the higher energy side where $D \ll \Gamma$. In the higher energy side, levels overlap with each other. When the compound nucleus is excited into the overlap region, either because of the large kinetic energy of the projectile or because of the energy spread of the incident ion beam or both, the reaction is statistical, as average properties of many overlapping levels are reflected in the reaction cross-sections. On the other hand, in the lower energy region, where levels are well separated from each other, the reaction cross-section show peaks corresponding to compound nucleus levels, as shown in the extreme right in Figure 2.1. This is the region of resonance. In the resonance region, cross-section for the isolated levels is given by the well-known Breit–Wigner single level formula given by

$$\sigma\,(a,b) = \pi D^2 \left[\frac{(2J_{CN}+1)}{(2J_A+1)(2j_a+1)} \right] \left[\frac{\Gamma_{aA}\Gamma_{bB}}{(E-E_0)^2 + \left(\dfrac{\Gamma}{2}\right)^2} \right] \qquad (2.22)$$

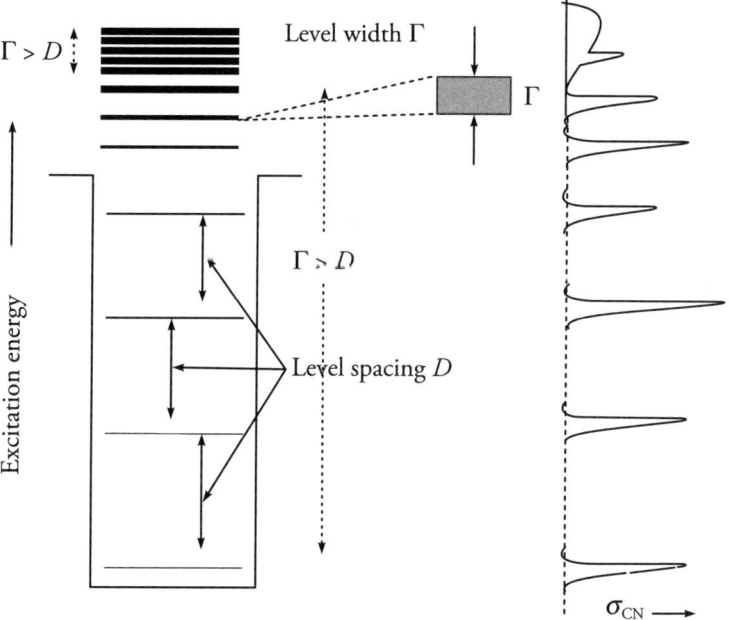

Figure 2.1 Energy levels, level width and level spacing for compound nucleus

Here, E is the centre of mass energy of the projectile, E_0 the excitation energy of the level, Γ the total width of the level, Γ_{aA} and Γ_{bB} partial widths respectively for the formation of the compound nucleus and the decay of the compound nucleus via b and B. As shown on the extreme right of Figure 2.1, cross-section sharply rises as the incident energy reaches the excitation energy of the compound nuclear level.

Further, when a nucleus is excited to the range of overlapping levels (statistical region), it is meaningless to talk about the individual levels; instead, one talks about the number of levels per unit energy interval, called the level density, denoted by $\rho(U^*)$, where U^* is the excitation energy of the nucleus. For a given nucleus, level density increases very rapidly with its excitation energy.

Coming back to the region of statistical reactions (overlapping levels, $D \ll \Gamma$), the cross-section for the reaction σ (a, b) may be written as the product of two terms, σ(CN), the cross-section for the formation of the compound nucleus by the fusion of the projectile and the target and G(C, b), the probability of the decay of the compound nucleus into ejectile b and residue B, i.e.,

$$\sigma(a,\ b) = \ \sigma(CN).G(C,\ b) \tag{2.23}$$

where,

$$\sum_{\text{all ejectiles}} G(C,\ b) = 1 \tag{2.24}$$

The excitation energy of the compound nucleus E_{CN}^* is the sum of the centre of mass energy E_a^{CM} of the projectile and the binding energy of projectile a in the compound nucleus, $(BE)_a^{CN}$. We now discuss the decay of the compound nucleus by emitting the ejectile b leaving the residual nucleus B. The maximum energy ε_b^{max} with which the ejectile b may be emitted is given by $\varepsilon_b^{max} = (E_{CN}^* - S_b)$, where S_b is the separation energy of ejectile b from the compound nucleus. However, ejectile b may be emitted with many possible values of energy ε_b up to ε_b^{max}, leaving the residual nucleus B in corresponding excited states. Suppose the ejectile is emitted with some energy $\varepsilon_b (< \varepsilon_b^{max})$ leaving the residual nucleus B with excitation energy E_B^*. Then, invoking the theory of detailed balance, it may be shown that the probability of the emission of ejectile b with energy ε_b, $P(\varepsilon_b)d\varepsilon_b$, is given by;

$$P(\varepsilon_b)d\varepsilon_b = \frac{(2j_b+1)\mu}{\pi^2\hbar^3}\varepsilon_b\ \sigma_{\text{inverse}}\ \frac{\rho(E_B^*)}{\rho(E_{CN}^*)}\ d\varepsilon_b \tag{2.25}$$

In Eq. (2.25), μ is the reduced mass and σ_{inverse} is the cross-section for the reverse process, that is, the process in which a compound nucleus with excitation energy E_{CN}^* is formed by the fusion of the ejectile b of energy ε_b and the residual nucleus B at excitation energy E_B^*. $\left[B(E_B^*)+b(\varepsilon_b)\rightarrow[CN]^*\ (E_{CN}^*)\right]$. The inverse cross-section σ_{inverse} may be calculated using the optical model transmission coefficients for the inverse process as one calculates the cross-section for the formation of the compound nucleus by the fusion of projectile a and the target A. The other quantities: $\rho\left(E_B^*\right)$ and $\rho\left(E_{CN}^*\right)$ are respectively, the level densities in the residual nucleus B, at excitation energy E_B^* and in the compound nucleus, at excitation energy E_{CN}^*.

It is easy to understand the level density dependence on the probability $P(\varepsilon_b)$ with reference to Figure 2.2. Level density of the residual nucleus gives the number of states of the system in a unit energy interval, and each energy state means that after decay, the residual system may land there. Therefore, each level of the residual system provides a path for the decay of the compound nucleus. Larger the level density of the residual nucleus B, more will be the probability of decay. It may be observed that the level density of the residual nucleus increases with the excitation energy ($E_B{}^*$), which means that emission of ejectile b is more likely when its energy ε_b is small (so that $E_B{}^*$ is large).

In general, the excited compound nucleus may decay by emitting many different ejectiles (subject to conservation laws) like b, b′, b″..., etc, leaving the residues B, B′, B″... etc, respectively. The ejectile that leaves the residual nucleus with the largest level density will be the most likely mode of decay.

Fermi gas model of the nucleus provides the following simple expression for the level density as a function of the excitation energy U^*,

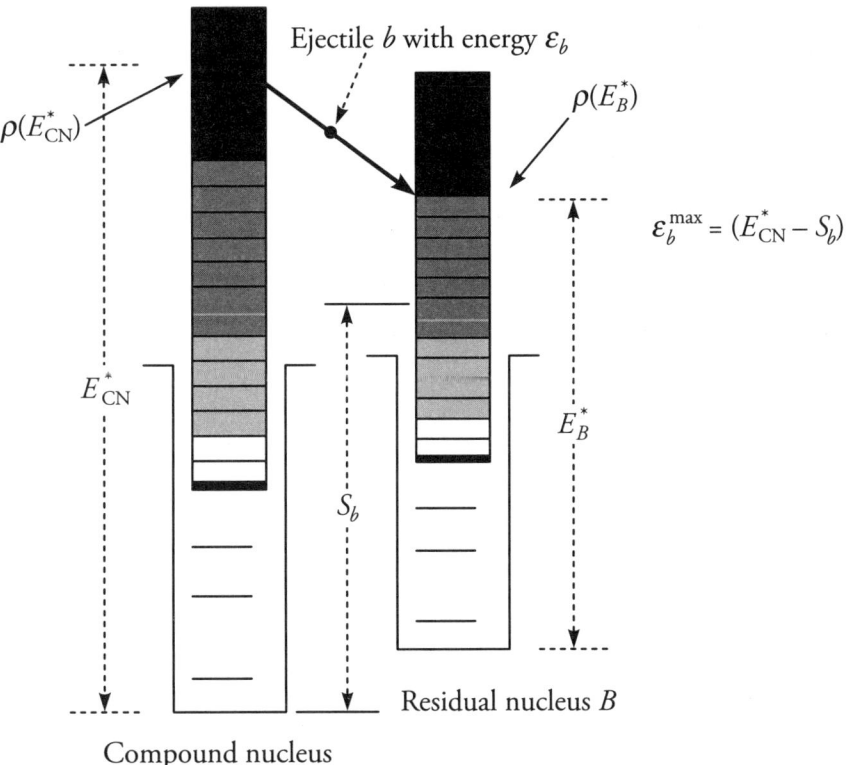

Figure 2.2 Compound nucleus decays by emitting ejectile b and residual nucleus B

$$\rho\left(U^{*}\right)=\frac{\sqrt{\pi}}{12a^{\frac{1}{4}}U^{*\frac{5}{4}}}e^{\left[2\left(aU^{*}\right)^{\frac{1}{2}}\right]}$$

(2.26(a))

A spin dependent expression for the level density is given as

$$\rho\left(U^{*},J\right)=\frac{2J+1}{2\sigma^{3}\sqrt{2\pi}}\rho\left(U^{*}\right)e^{\left[-\frac{\left[\left(J+\frac{1}{2}\right)^{2}\right]}{2\sigma^{2}}\right]}$$

(2.26(b))

Here a, called the level density parameter, is a constant for a given nucleus. The value of a, in general, varies between $A/10$ and $A/8$, A being the atomic mass number of the nucleus. As a matter of fact, level density parameter a is treated as a free parameter the value of which may be varied between $A/10$ and $A/8$ to reproduce the experimental data. However, level density parameter a is related to g, the density of states around the Fermi level,

$$a=\frac{\pi^{2}g}{6}$$

(2.26(c))

σ, appearing in Eq. (2.26(b)) is the spin cutoff parameter, often determined by fitting the experimental isomeric ratios.

2.1.3 Evaporation spectra

The emission of ejectiles from the compound nucleus may be compared to the evaporation of molecules from the surface of a hot liquid. This similarity is due to the fact that the energy spectra of emitted ejectiles follow Maxwell distribution. It can be shown that the energy spectra of ejectiles consisting of neutrons is given by

$$N\left(\varepsilon_{N}\right)d\varepsilon_{N}=\frac{\varepsilon_{N}}{T^{2}}e^{\left(\frac{\varepsilon_{N}}{T}\right)}d\varepsilon_{N}$$

(2.27)

where $N(\varepsilon_{N})$ is the number of neutrons with energy ε_{N}. In the case of emission of charged ions, Eq. (2.27) gets modified as a charged ion coming out of the compound nucleus faces a Coulomb barrier and only ions having energy larger than the Coulomb barrier may get evaporated from it. If ε_{c} is the height of the Coulomb barrier for the emitted ion, then the energy spectra of the ion may be given by,

$$N\left(\varepsilon_{\text{ion}}\right)d\varepsilon_{\text{ion}}=\frac{\left(\varepsilon_{\text{ion}}-\varepsilon_{c}\right)}{T^{2}}e^{-\left(\frac{\left[\varepsilon_{\text{ion}}-\varepsilon_{c}\right]}{T}\right)}d\varepsilon_{\text{ion}}$$

(2.28)

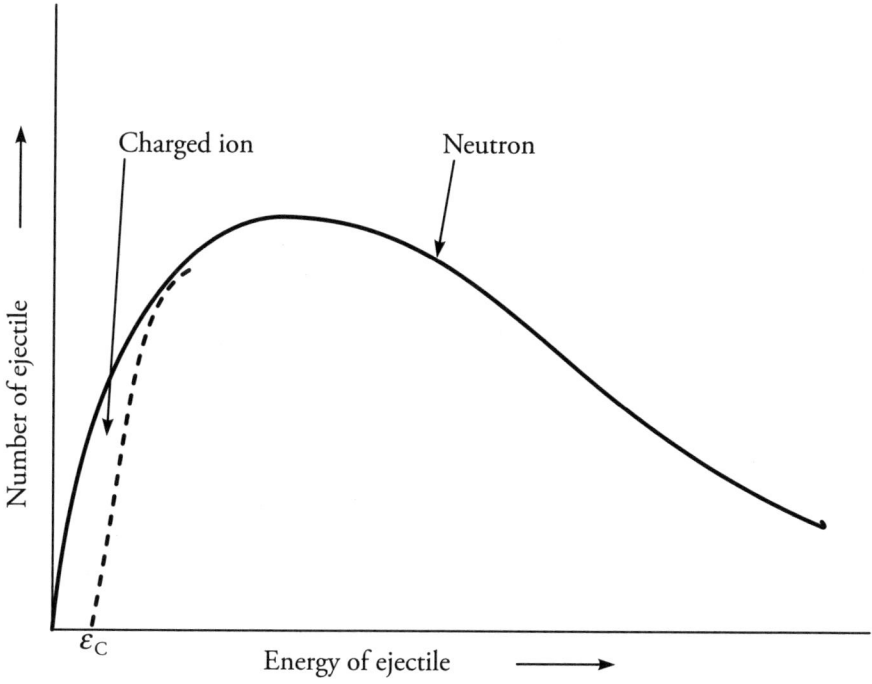

Figure 2.3 Energy spectra of ejectile evaporated from the compound nucleus

The nuclear temperature T is related to the excitation energy E^*_{CN} of the compound nucleus by the relation,

$$E^*_{\text{CN}} = aT^2 - T \tag{2.29}$$

The value of the characteristic temperature T may be obtained from the measured energy spectra using the relation,

$$T = \frac{\varepsilon_b}{\dfrac{d}{d\varepsilon}(\ln N(\varepsilon_b))} \tag{2.30}$$

The level of the compound nucleus at excitation energy E^*_{CN} has a certain width denoted by Γ_{CN}. As already mentioned, several different ejectiles, like b, b', b'',..., etc may be evaporated by the excited compound nucleus, each with its characteristic probability. In such situations, it is customary to divide the width of the compound nucleus into partial widths, Γ_b, $\Gamma_{b'}$, $\Gamma_{b''}$,...,. each proportional to the probability of the emission (with all possible energies) of ejectile b, b', b'', ..., etc. Therefore, the ratio,

$$\frac{\Gamma_b}{\Gamma_{b''}} = \frac{\int P(\varepsilon_b)d\varepsilon_b}{\int P(\varepsilon_{b''})d\varepsilon_{b''}} = \frac{j_b \mu_b E_b^{*\min} a_{(b)}}{j_{b''} \mu_{b''} E_{b''}^{*\min} a_{(b'')}} e^{2\left[(a_{(b)}E_b^{*\min})^{1/2} - (a_{(b'')}E_{b''}^{*\min})^{1/2}\right]} \tag{2.31}$$

Here j_b and $j_{b''}$ are the spins of ejectile b and b'', μ_b and $\mu_{b''}$ their reduced masses, $E_b^{*\min}$ and $E_{b''}^{*\min}$ respectively, the minimum excitation energies of the residual nuclei left after the emission of ejectiles b and b'' with maximum energy $\varepsilon_b^{\max}$ and $\varepsilon_{b''}^{\max}$. $a_{(b)}$ and $a_{(b'')}$ are the level density parameters for the two residual nuclei.

2.1.4 Width fluctuation correction

The independent hypothesis that the formation of the compound nucleus and its decay is independent of each other holds better at higher excitations but is not so good at lower energies. To compensate for this, one uses the width fluctuation correction (WFC). It is argued that a better estimate of the cross-section $\sigma_{(\alpha,\beta)}^{J_{CN},\pi^{CN}}$ for the formation of a compound nucleus of spin J_{CN} and parity π^{CN} with entrance channel α and exit channel β, may be obtained by averaging the cross-sections for all compound nucleus states of spin J_{CN} and parity π^{CN} in the energy interval ΔE, where ΔE is the spread in the energy of the incident ion. One uses the Breit–Wigner single level formula (see Eq. (2.22)) to obtain the following expression for the cross-section,

$$\sigma_{(\alpha,\beta)}^{J_{CN},\pi^{CN}} = \frac{\pi}{k_\alpha^2} \frac{2\pi}{D\left(J_{CN},\pi^{CN}\right)} \left< \frac{\Gamma_\alpha^{\left(J_{CN},\pi^{CN}\right)} \Gamma_\beta^{\left(J_{CN},\pi^{CN}\right)}}{\Gamma^{J_{CN}\pi^{CN}}} \right> \tag{2.32}$$

Here $D(J_{CN}, \pi^{CN})$ is the average separation of compound levels with spin J_{CN} and parity π^{CN} in the energy interval ΔE_a and $\Gamma_\alpha^{\left(J_{CN},\pi^{CN}\right)}$, $\Gamma_\beta^{\left(J_{CN},\pi^{CN}\right)}$ and $\Gamma^{J_{CN}\pi^{CN}}$ are respectively the entrance channel width, exit channel width and the total width of the compound nucleus levels with spin J_{CN} and parity π^{CN}. Now transmission coefficients are related to the widths and level spacing as follows

$$T_\alpha^{\left(J_{CN},\pi^{CN}\right)} = \frac{2\pi \Gamma_\alpha^{\left(J_{CN},\pi^{CN}\right)}}{D\left(J_{CN},\pi^{CN}\right)} \tag{2.33}$$

Using this in Eq. (2.32),

$$\sigma_{(\alpha,\beta)}^{J_{CN},\pi^{CN}} = \frac{\pi}{k_\alpha^2} \frac{\left(T_\alpha^{\left(J_{CN},\pi^{CN}\right)}\right)\left(T_\beta^{\left(J_{CN},\pi^{CN}\right)}\right)}{\sum_\gamma T_\gamma^{\left(J_{CN},\pi^{CN}\right)}} \cdot WC_{(\alpha,\beta)}^{\left((J_{CN},\pi^{CN})\right)} \tag{2.34}$$

Here $WC_{(\alpha,\beta)}^{\left(\left(J_{CN},\pi^{CN}\right)\right)}$ is the width fluctuation correction that may be written as

$$WC_{(\alpha,\beta)}^{\left(\left(J_{CN},\pi^{CN}\right)\right)} = <\frac{\Gamma_{\alpha}^{\left(J_{CN},\pi^{CN}\right)}\Gamma_{\beta}^{\left(J_{CN},\pi^{CN}\right)}}{\Gamma^{\left(J_{CN},\pi^{CN}\right)}}> \frac{<\Gamma^{\left(J_{CN},\pi^{CN}\right)}>}{<\Gamma_{\alpha}^{\left(J_{CN},\pi^{CN}\right)}\Gamma_{\beta}^{\left(J_{CN},\pi^{CN}\right)}>} \qquad (2.35)$$

2.1.5 Effective transmission coefficients

The transmission coefficients for the entrance channel may be easily calculated using Eq. (2.9) as they are required only for one projectile at a single value of incident energy E_a. However, it is not possible to use Eq. (2.9), for exit channels, because several ejectiles (like b, b′, b″, etc) may be emitted from the compound nucleus and with many possible values of energies. Hence, for the exit channels, one uses the concept of effective transmission coefficient, <T>, which is obtained by summing over all the possible decay channels that can be found in a small energy bin – the bin may be chosen in an arbitrary way. This essentially means that all decay channels within the bin contribute equally to the decay, an assumption that is valid only if the bin size is small. The effective transmission coefficient is then written as,

$$<T_{(b,\ell_b,j_b)}^{\left(J_{CN}\pi^{CN}\right)}> = \left\{\int\rho\left(E_B,J_B,\pi^B\right)\delta\left(b,\Pi^{CN}\right)\left[T_{(b,\ell_b,j_b)}^{\left(J_{CN}\pi^{CN}\right)}\right]dE_B\right\}_{\text{integral over the bin}} \qquad (2.36)$$

where, $\rho\left(E_B,J_B,\pi^B\right)$ is the level density of the residual nucleus B at excitation energy E_B and the term $\delta\left(b,\Pi^{CN}\right)$ takes care of the parity conservation.

Transmission coefficients for gamma decay of the compound nucleus after particle emission may be obtained using the strength function $f(x,\lambda)$, that is related to the gamma transmission coefficient by the following expression,

$$T_{(x,\lambda)}^{\left(J_{CN}\pi^{CN}\right)}\left(\varepsilon_\gamma\right) = 2\pi\, f\left(x,\lambda\right)\varepsilon_\gamma^{(2\lambda+1)} \qquad (2.37)$$

Here, x is the type of the radiation, i.e., electrical or magnetic, λ the multipolarity and ε_γ the energy of the gamma ray emitted from the level (J_{CN}, π^{CN}) of the compound nucleus. It may, however, be mentioned that, in general, the number of final accessible states of the compound nucleus is so large that instead of considering each decay channel individually, one performs a summation over all accessible levels. This leads to what is called the global gamma transmission coefficient defined as

$$\left(T_{\varepsilon_\gamma}^{\left(J_{CN}\pi^{CN}\right)}\right)_{\text{global}} = \sum_{(x,\lambda)}\sum_{(J_{CN}-\lambda)}^{(J_{CN}+\lambda)}\sum_{\pi^f}\int_0^E\left\{T_{(x,\lambda)}^{\left(J_{CN}\pi^{CN}\right)}\left(\varepsilon_\gamma\right)\right\}\left\{\rho\left(E-\varepsilon_\gamma,J_f.\pi^f\right)\right\}\left[f\left(x,\lambda,\pi^{CN},\pi^f\right)\right] \qquad (2.38)$$

where, $\rho\left(E-\varepsilon_\gamma, J_f.\pi^f\right)$ is the compound nucleus level density at excitation energy $(E-\varepsilon_\gamma)$ and J_f and π^F are the spin and parity of the final level reached by the emission of gamma rays. The factor $\left[f\left(x,\lambda,\pi^{\mathrm{CN}},\pi^f\right)\right]$ can have values 0 or 1, depending on the parity conservation, i.e., $f=0$ if

$$\pi^f = (-1)^\lambda \pi \text{ (for electric) and } \pi^f = (-1)^{\lambda+1}\pi \text{(for magnetic);} \quad \text{otherwise } f=1 \qquad (2.39)$$

In heavy ion reactions, the compound nucleus is often formed with a large value of spin and in a hurry to deplete it, the nucleus undergoes fission. At higher excitation energy, fission directly competes with neutron emission as charged particle emission is considerably suppressed by the Coulomb barrier. Since neutron/other particle emission do not change the spin angular momentum of the compound nucleus by any appreciable amount, fission provides a faster path for reducing the angular momentum. Generally, fission is treated by the liquid drop model of the nucleus and is due to the interplay between the repulsive Coulomb and attractive surface energy terms. Invoking microscopic shell corrections to the macroscopic liquid drop model results in fission barriers that may have a single, two or multiple humps, as shown in Figure 2.4.

In the case of fission, there is only one transmission coefficient that corresponds to the transmission of the deformed nucleus over the relevant fission barrier, independent of the properties of the fission fragments. This is in contrast to the evaporation of neutron etc., where transmission coefficients depend on the energy etc., of the ejectile.

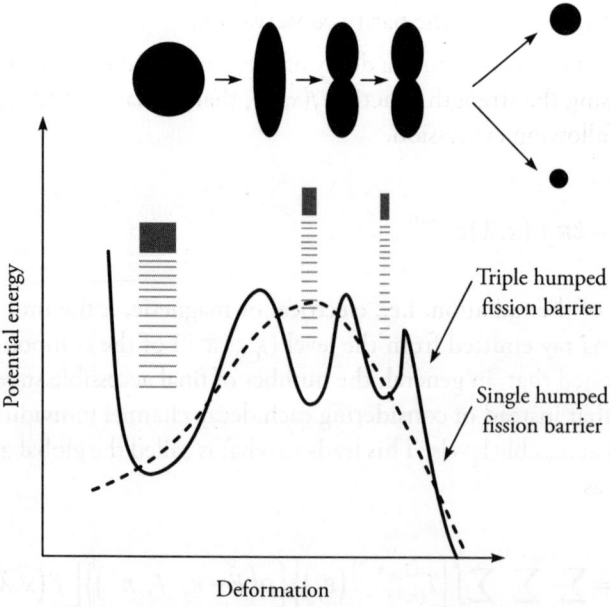

Figure 2.4 Potential energy of the nucleus as a function of deformation

In simple single fission barrier models[10], the fission barrier is taken to be an inverted parabola with barrier height B_F and barrier curvature $\omega\hbar$; the transmission coefficient is given as

$$T^{single} = \frac{1}{\left\{ 1 + e^{-\left(2\pi \frac{E - B_F}{\omega\hbar} \right)} \right\}} \tag{2.40}$$

As illustrated in Figure 2.4 and also shown in Figure 2.5, fission takes place through the transition states of the compound nucleus, the states corresponding to the deformed nucleus. Each level of the transition state, whether isolated or in continuum has its own fission barrier. Starting from the isolated compound nucleus transition level with spin J_{CN} and parity π^{CN}, all transition levels with the same values of spin and parity contribute to the fission cross-section. The averaged fission transmission coefficient is then given by

$$T^{\left(J_{CN}, \pi^{CN} \right)}(E) = \int_0^\infty \left\{ \rho\left(\varepsilon, J_{CN}, \pi^{CN} \right) \right\} \left\{ T^{sing}(E) \right\} (E - \varepsilon) \, dE \tag{2.41}$$

Here, $\rho\left(\varepsilon, J_{CN}, \pi^{CN} \right)$ is the level density of transition states at an energy ε above the top of the fission barrier.

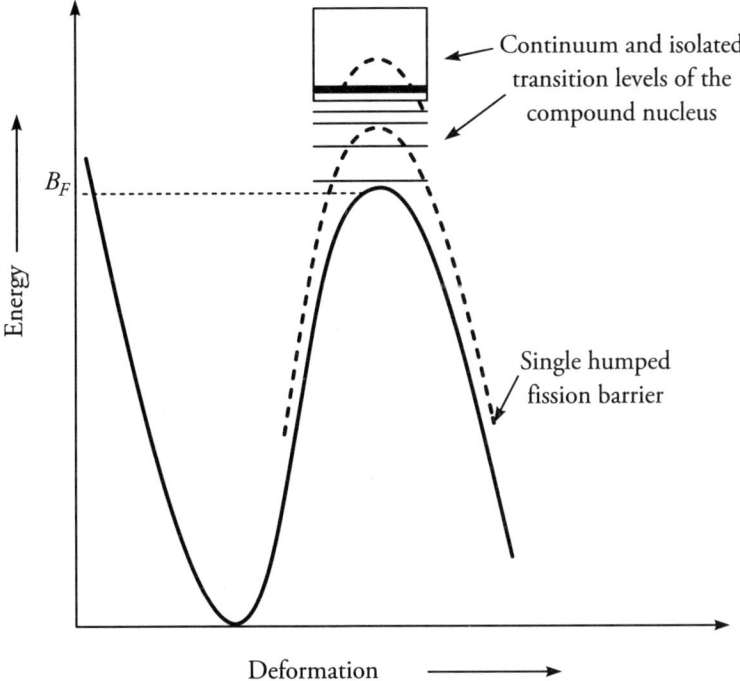

Figure 2.5 Single humped fission barrier

In the case of multiple humped fission barriers (double humped in the case of actinides), the global fission transmission coefficient may be written as

$$\left(T^{\left(J_{\text{CN}},\pi^{\text{CN}}\right)}(E)\right)_{\text{glob}} = \frac{\left[T_1^{\left(J_{\text{CN}},\pi^{\text{CN}}\right)}T_2^{\left(J_{\text{CN}},\pi^{\text{CN}}\right)}T_3^{\left(J_{\text{CN}},\pi^{\text{CN}}\right)}\cdots\cdots\right]}{\left[T_1^{\left(J_{\text{CN}},\pi^{\text{CN}}\right)} + T_2^{\left(J_{\text{CN}},\pi^{\text{CN}}\right)} + T_3^{\left(J_{\text{CN}},\pi^{\text{CN}}\right)} + ..\right]} \qquad (2.42)$$

Here T_1, T_2,.... are the transmission coefficients for barrier 1, barrier 2,... etc.

2.1.6 De-excitation sequence of the compound nucleus

It is interesting to investigate the de-excitation sequence of a compound nucleus formed by the fusion of two heavy ions. Depending on the projectile energy and the masses of the two ions, there might be two possibilities: (i) the composite excited nucleus is formed with a spin value that is more than what it can sustain. Such a nucleus A is shown in Figure 2.6, and lies on to the right-hand side of the Yrast line. The compound nucleus A undergoes fast fission to relieve itself of the excess angular momentum, leaving two fission fragments with sustainable angular momentum towards the left of the Yarst line. (ii) The other possibility is that the excited compound nucleus has an angular momentum that it can hold and lies to the left of the Yrast line (B in Figure 2.6). Apart from being unstable on account of its excitation energy, the compound nucleus B is also unstable because it is neutron deficient. The reason for it being neutron deficient is that the projectile ions, like ^{12}C, ^{16}O, ..., ^{40}Ca etc., have nearly the same number of neutrons N_1 and protons Z_1 ($N_1 = Z_1$), and when such an ion fuses with a heavy target (Z_2, N_2), the compound nucleus so formed ($Z_2 + Z_1$; $N_2 + N_1$) contains less number of neutrons than a nucleus with charge ($Z_2 + Z_1$) should have to be in the valley of stability in the N–Z graph.

The neutron deficit and excited compound nucleus B evaporates predominantly neutrons at the first stage of de-excitation, as evaporation of charged particles is greatly suppressed by the Coulomb barrier. Depending on the excitation energy, 1 to x (= 5, 6, ...) neutrons may be emitted till further emission of neutrons is energetically not possible. The compound nucleus then becomes B_1. Neutron emission depletes the compound nucleus of some amount of excitation energy but only a little angular momentum. However, it moves the compound system a little in the direction of the origin as shown in Figure 2.6. After neutron emission, the already neutron deficit nucleus becomes severely neutron deficient. It is worth noting that such excited and highly neutron deficient nuclei cannot be produced in light ion reactions. Nucleus B_1 now decays by emitting a cascade of continuum gamma rays. Gamma emission, being an electromagnetic interaction, is slower than the strong interaction processes of neutron emission, and hence, takes over only when neutron emission is no longer possible. Because of the complexity of the compound nucleus states just before gamma emission, many gamma channels become open for decay. As a result, gamma decay is characterized by a statistical 'rain cloud pattern' of continuum gammas. Copious gamma emission considerably reduces the

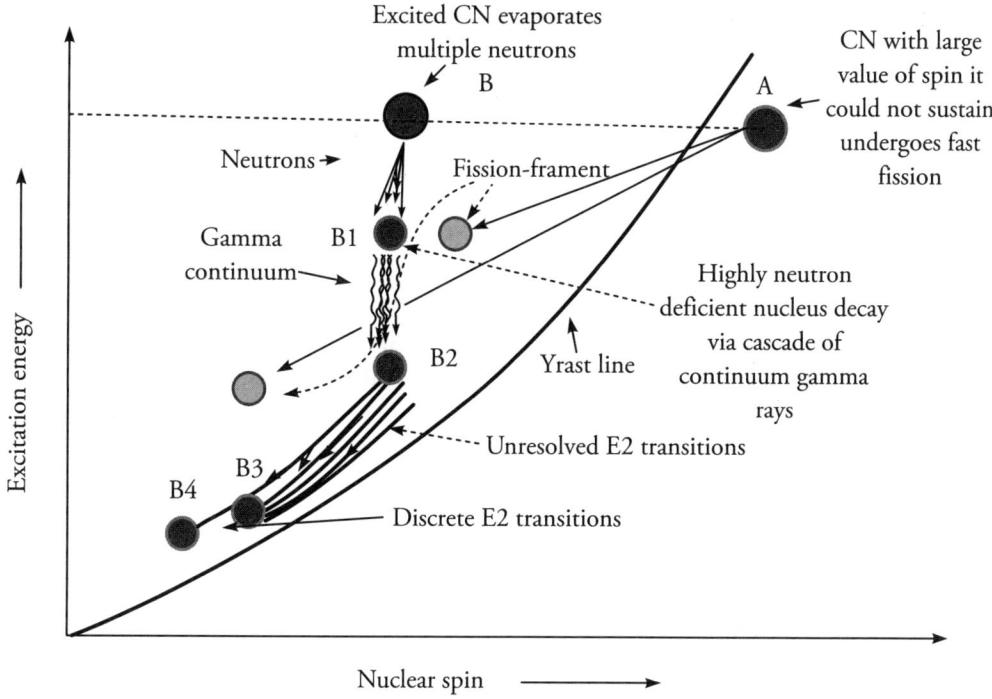

Figure 2.6 De-excitation sequence of the compound nucleus formed by the fusion of two heavy ions

excitation energy of the compound nucleus but only a negligible amount of angular momentum. The compound nucleus follows an almost vertical fall from B_1 to B_2 as shown in Figure 2.6, reaching very near the Yrast level. From here onwards, the decay pattern follows the Yrast line. Further de-excitation of the compound nucleus takes place via sequences of E2 transitions feeding down any of the several parallel quasi-rotational bands extended over large range of spins in the Yrast region. Eventually, the de-excitation of the side bands terminates at or near the band head and the intensity feeds directly into the Yrast sequence itself. From here onwards, the decay proceeds simply by stretched E2 transitions in the ground state band, B_3 to B_4.

The system of rotational bands of a nucleus may also be excited using an (n, γ) reaction with a suitable target, but there are two major differences between the rotational band structures obtained by an (n, γ) reaction and those obtained by heavy ion fusion. Firstly, the bands produced by the (n, γ) reaction are 'hot', i.e., they are at several MeV excitation. Secondly, the spread in the spin of bands is always only a few units around the target spin. On the other hand, quasi-rotational bands produced during the last stages of de-excitation of the compound nucleus in heavy ion fusion are 'cold' sitting almost at the Yrast line; secondly, the spin spread of these bands is considerably wide.

2.2   The Pre-equilibrium Emission in Statistical Nuclear Reactions

Bohr's theory of compound reaction mechanism is based on the ad hoc assumption, also called 'Bohr's amnesia', that the excited complex system formed as a result of the fusion of the projectile and the target lives long enough, without losing any energy or nucleons, till a thermodynamic equilibrium is established. Thus, the compound nucleus (CN) is an excited nucleus wherein the total excitation energy is equally distributed over all degrees of freedoms, and on an average, each nucleon has the same energy. Further, the process of equilibration is assumed to take place via two-body interactions, a multistep process that is rather slow ($\approx 10^{-16}$ s) in terms of the nuclear time scale. On the other extreme in the time scale are reactions that occur in a very short time, typically the time a nucleon takes in crossing the nucleus ($\approx 10^{-21}$ s); these are direct reactions and obviously are single or at the most a few steps, processes. Weisskopf[11] has pointed out that formation of the compound nucleus is not simple and is bound to be a complex process. Both intuition and precise measurements eventually reviled that both energy and nucleons and even clusters may be lost by the complex initial composite excited system on its way to become an equilibrated compound nucleus. In general, the experimentally measured energy spectra (at a fixed angle) of particles evaporated in statistical nuclear reactions is found to have the shape shown in Figure 2.7. The two ends of the spectra, the extreme left having a broad peak, may be assigned to the compound nuclear process and the discrete sharp resonance-like peaks on to the extreme right may be assigned to single step direct processes. If only these two processes (compound nuclear and direct reaction) were involved in particle emission, than the compound nuclear peak, which is expected to have Maxwell distribution, should have dropped down much faster as indicated by the dotted curve. However, considerable number of particles with energy more than that expected from the compound process is present in experimental spectra. These particles with energy more than that expected from compound processes are called pre-equilibrium or pre-compound particles and the process of their emission is called the pre-equilibrium (PE) (or pre-compound) process. Attempts have been made to explain PE emission using a semi-classical approach, however, the nucleus being a quantum system, it has been argued that a totally quantum mechanical theory for the PE process is required. In the total quantum mechanical theory, the PE component is further divided into two parts – the multistep compound and the multistep direct.

Experimentally measured angular distributions $\left(\dfrac{d^2\sigma}{d\Omega dE} \right)$ of emitted particles for different energy ranges give information about the validity of Bohr's independence hypothesis. Figure 2.8(a) shows the angular distribution of particles corresponding to the compound peak in Figure 2.7.

As may be observed in Figure 2.8(a), the angular distribution is almost isotropic, meaning that the compound system has forgotten the history of its formation confirming the validity of the independence hypothesis. Figure 2.8(b) shows the angular distribution of particles corresponding to the PE region of energies. A highly forward peaked angular distribution is a reflection of the fact that the emitting system remembers the direction of the incident ion

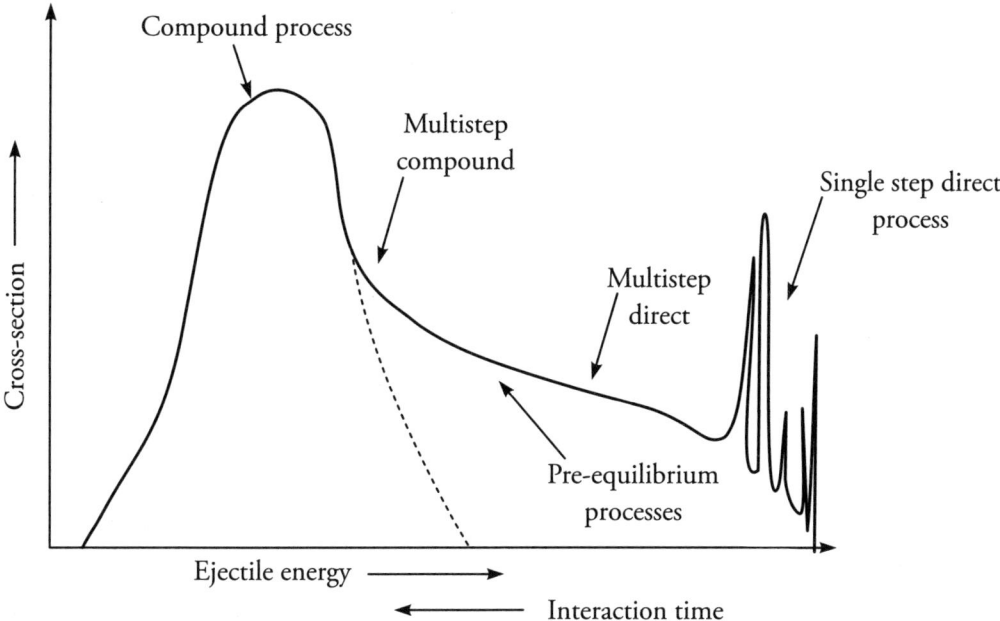

Figure 2.7 Typical ejectile energy spectra at a fixed angle

and has not yet forgotten the history of its formation, i.e., the breakdown of the independence hypothesis.

The angular distribution of particles corresponding to the direct process (Figure 2.9) has an oscillatory nature, somewhat like a diffraction pattern. Analysis of the slopes of the oscillatory curves gives information about the spin and the parity of the levels of the residual nucleus.

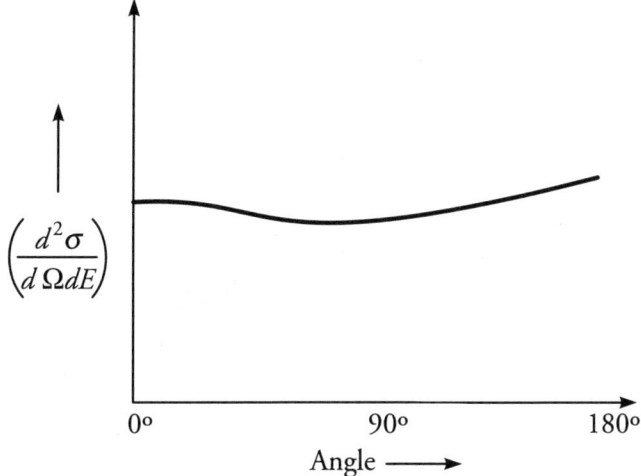

Figure 2.8a Angular distribution of ejectiles corresponding to CN emission

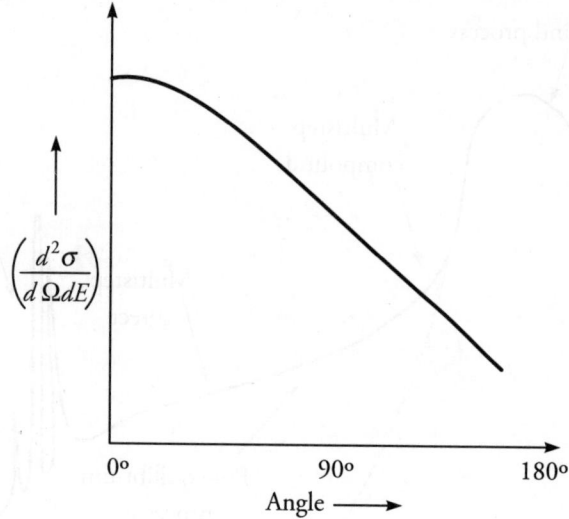

Figure 2.8b Angular distribution of ejectiles corresponding to PCN emission

Since the interaction time increases when going from the process described in Figure 2.9 to that described in Figure 2.8(a), it is obvious that thermal equilibrium is established a considerable time after the initial interaction of the projectile with the target; particle emission from the equilibrating compound system takes place during this time.

Since the initial work of Griffin[12], many classical/phenomenological models have been developed to describe the PE emission. However, the most frequently used model is the exciton model. We shall therefore, discuss the exciton model in detail and briefly touch upon the other models for PE emission later.

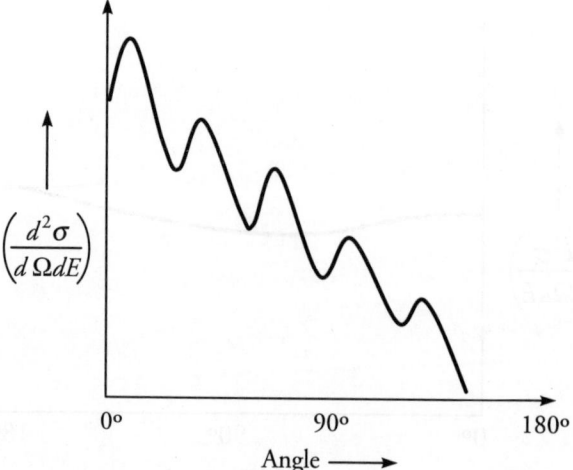

Figure 2.9 Angular distribution of ejectiles corresponding to direct processes

2.2.1 The exciton model

The exciton model[12,15] is one of the most widely used model of classical nature. Like other similar models, it assumes stepwise creation and annihilation of particle–hole pairs. The important parameter of the model is the exciton number n, which is the sum of the numbers of particles (p) above the Fermi level and the number of holes (h) below it. In this model, a nuclear state is characterized by the value of n, the exciton number and the excitation energy E^*. There may be several different states that have the same values of n and E^*, but they differ in how the excitation energy E^* is distributed between the particles and the holes. A basic assumption of the model is that all such states that have the same values for n and E^* have a priori equal probability. Since many particle and hole states are involved at each step, the model is statistical in nature. The model further assumes that state transitions are energy conserving, an assumption that makes it a classical model. Bertini et al.[13] have shown that the exciton model in the limit of thermal equilibrium reduces to the Hauser–Feshbach/Weisskopf–Ewing model. In the exciton model, it is assumed that on fusion with the target, the nucleons of the projectile fill the empty level of the composite system. For an incident nucleon, the initial state of the composite system is shown in Figure 2.10.

As shown in the figure, the incident nucleon with energy E_p enters from the left and occupies the vacant energy level at an excitation energy $E^* = (E_p + B_e)$. All energy levels of the complex system are filled up to the Fermi level, denoted by E_F. The initial state of the system is a state with exciton number $n = 1$, having one excited particle and no hole, $n = 1(1p + 0h)$. It is assumed that two-body interactions now drive the initial state to the next with exciton number $n = 3$ (2p + 1h); then to a state with $n = 5$ and so on. Except from the initial state, it

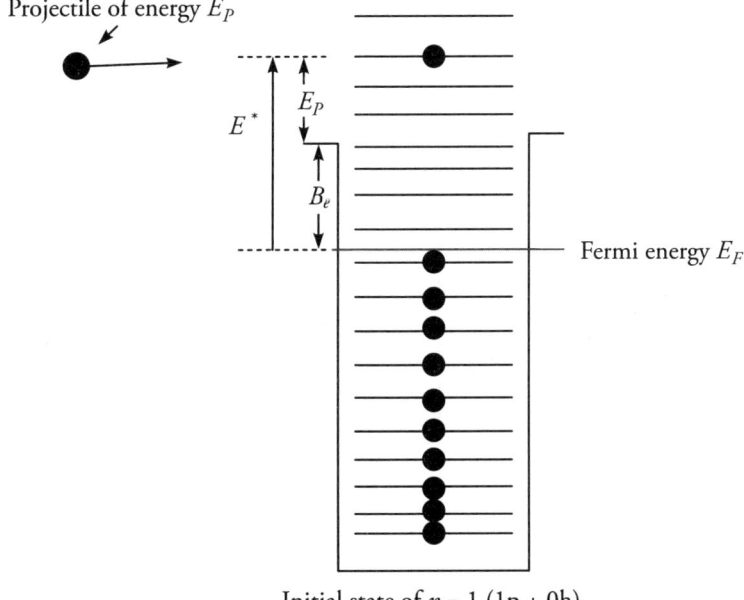

Initial state of $n = 1$ (1p + 0h)

Figure 2.10 Initial configuration of the complex system when the projectile is a nucleon

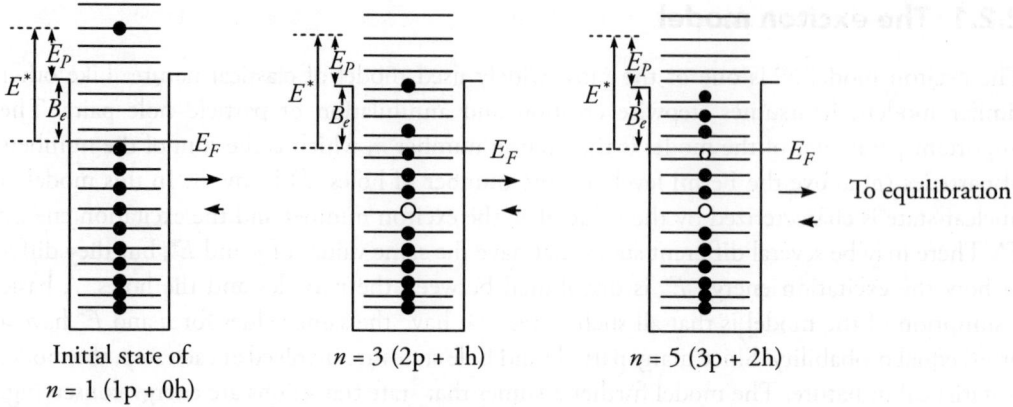

Figure 2.11 Two body interactions drive the initial state to state of higher exciton numbers

is possible for the system to return to the previous state; but the probability of moving up in exciton number is larger than reverting back down. Essentially, the two-body interactions may change the exciton number by $\Delta n = +2, -2$ or 0, which means that a state with exciton number n may become a state of exciton number $(n + 2)$ or may go back to the state $(n - 2)$.

For a given set of the exciton number n and excitation energy E^*, there may be several possible states (or configurations) in which the particles and holes occupy different levels subject to the condition that the sum of the energies of all holes and of all particles for each configuration is equal to E^*.

Figure 2.12 shows the three different sub-states corresponding to $n = 3$ (2p+1h) with excitation energy E^*. Transition from one sub-state to another sub-state of the same n corresponds to $\Delta n = 0$. It is obvious that the number of possible sub-states for a given value of

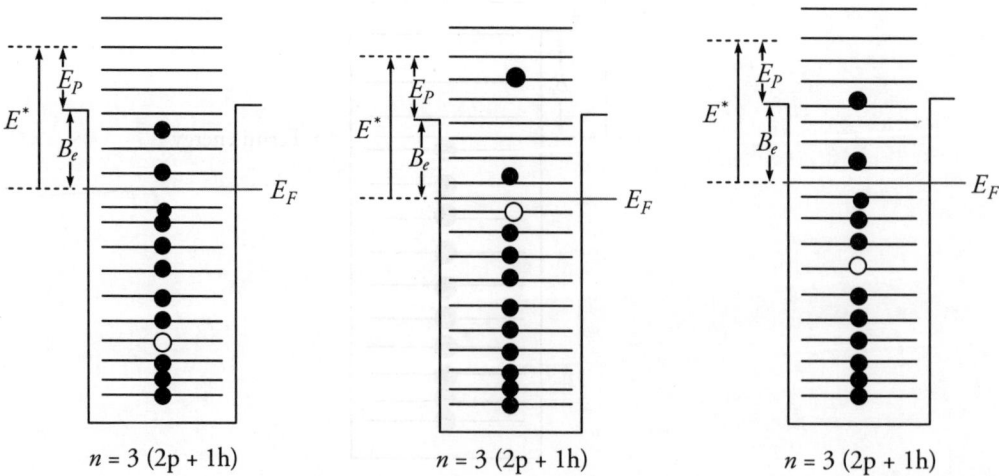

Figure 2.12 Three different configurations for $n = 3$ (2p + 1h) state

n increases rapidly with the increase in the value of n. This is so because with the increase of n, the possible ways of sharing the excitation energy E^* between particles and holes multiply, giving many possibilities. It may be observed that in these three configurations, the two particles occupy levels of different energies (energy of levels occupied by particles are measured from the Fermi level in the positive Y direction, i.e., upwards, while energies of levels occupied by holes are measured from the Fermi level in the downward direction) and so also the holes. As is shown in Figure 2.12, when particles fill levels of lower energy, the holes occupy a level of higher energy deeper inside the potential well such that

$$\sum_1^{n^p} \varepsilon_i^{\text{particle}} + \sum_1^{n^h} \varepsilon_k^{\text{hole}} = E^*; \ n^p + n^h = n \tag{2.43}$$

In Eq. (2.43), n^p, $\varepsilon_i^{\text{particle}}$, n^h and $\varepsilon_k^{\text{hole}}$ are respectively the number of particles, energy of the ith particle, number of holes and energy of the kth hole. Equation (2.43) holds for all sub-states corresponding to a given value of n, i.e., for those transitions in which $\Delta n = 0$. It may be worth mentioning that Eq. (2.43) holds not only for transitions in which $\Delta n = 0$ but also for those in which $\Delta n = \pm 2$. This is because the excitation energy deposited by the projectile in the initial interaction does not change with the increase (or decrease) of the complexity of the state and hence, for each state, the sum of energies of all particles plus the sum of energies of all holes is always equal to the initial excitation energy E^*. It is to be noted that no energy is associated with the creation or annihilation of electron–hole pairs. This accounts for the typical classical structure of the procedure; whereas, treating the nucleus as a system of quantized levels brings in the element of quantum physics. Therefore, the exciton model is semi-classical.

The probability of transition from a state of n excitons to a state of $(n + 2)$ excitons is larger than that for the reverse process because for not too large values of n, the possible number of sub-states of $(n + 2)$ are much larger than the number of sub-states of n. Moreover, from the Golden Rule of transition rates, the system moves pre-dominantly in the direction of higher number of states. When n is very large, there is not much difference between n and $(n + 2)$ and at that instant, the forward and backward transition rates become equal, indicating thermal equilibrium.

There is a finite non-zero probability of the emission of a particle from any state as the system moves from small n values towards higher n values with time. If a particle is emitted in the early stages, i.e., when n is small, the emitted particle will obviously have higher energy and will remember the direction of the projectile; hence, it is forward packed. Thus, particles (or clusters) emitted during the early phase of the reaction, i.e., within the first few transitions, will constitute the PE component. If no emission occurs during the initial few two-body interactions, the system will move towards equilibrium and be ready to evaporate particles and/or clusters. Thus, the exciton model describes both the PE emission and the evaporation from the CN as two different phases of evolution of a unified model.

Though there are many variants of the exciton model, the master equation approach most completely and quantitatively describes the equilibration process.

2.2.1.1 Master equation approach to the exciton model

This approach was proposed by Cline and Blann[15]. If one defines by $q(n, t)$ the probability that the system is in exciton state n at time t, then the rate of change of this probability with time $\left(\dfrac{dq(n,t)}{dt}\right)$ may be written as

$$\left(\frac{dq(n,t)}{dt}\right) = [\text{Rate of increase in the population of state with exciton no. } n \text{ at time } t - \text{rate of}$$

$$\text{depletion in the number of state with exciton no. } n \text{ at time } t] \tag{2.44}$$

It is easy to understand that the following states will feed the state of exciton no.(n):

State with exciton number $(n + 2)$, probability $q(n + 2, t)$, will feed state (n) via $(\Delta n = -2)$ transition; state with exciton number $(n - 2)$, probability $q(n - 2, t)$, will feed state (n) via transition $(\Delta n = +2)$; and sub-state of exciton no.(n), probability $q(n)$, will feed state(n) via transition $(\Delta n = 0)$.

Similarly, the state of exciton no.(n) will get depleted because of its decay to state $(n + 2)$ via transition $(\Delta n = +2)$ decay to state $(n - 2)$ via transition $(\Delta n = -2)$; and to sub-state (n) via transition $(\Delta n = 0)$.

In addition to the aforementioned decays, the probability of state (n) will also be reduced if some particles of different types and/or clusters are emitted from the state. Thus, (PE) emission from a state works like a sink so far as the depletion of the state is concerned. If one defines $w(n)$ as the total emission rate from state (n) summed over all particles/clusters and energies, then the total depletion at time t due to this sink term will be $w(n)q(n,t)$, where $q(n,t)$ is the probability of state (n).

If one associates $\lambda^+(n)$, $\lambda^-(n)$ and $\lambda^0(n)$ with the interstate transition rates respectively for $(\Delta n = +2)$, $(\Delta n = -2)$ and $(\Delta n = 0)$, then Eq. (2.44) becomes

$$\left(\frac{dq(n,t)}{dt}\right) = [\{\lambda^+(n-2)q(n-2,t) + \lambda^-(n+2)q(n+2,t) + \lambda^0(n)q(n,t)\}$$

$$-\{\lambda^+(n)q(n,t) + \lambda^-(n)q(n,t) + W(n)q(n,t) + \lambda^0(n)\ q(n,t) \tag{2.45}$$

A flow chart for Eq. (2.45) is shown in Figure 2.13

In Eq. (2.45), the term $\lambda^0(n)q(n, t)$ gets cancelled. However, if one is interested in angular distribution or energy spectra of the emitted PE particles, this term has to be retained.

Initial exciton number n_0 plays an important role and for nucleon-induced reactions, n_0 is taken as 3, while for alpha rays–induced reaction, n_0 is 5 (= 4p+1h). However, for heavy ion–induced reactions, a larger value depending on the projectile is adopted. Further, the relation $q_0(n) = q(n, t = 0) = \delta_{n,n_0}$ must also hold.

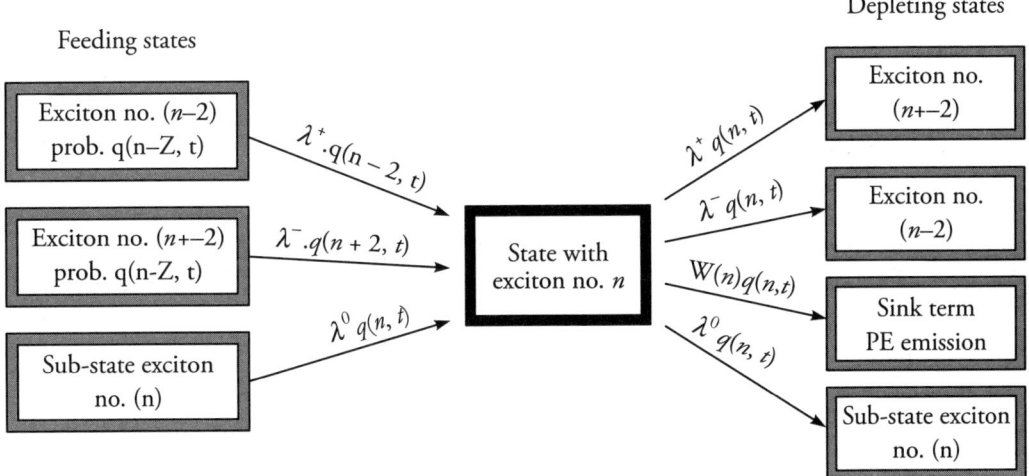

Figure 2.13 Transitions that feed and deplete the state with exciton no. *n*

Cline and Blann[15] calculated the PE emission rates using the theory of detailed balance. If a particle b of reduced mass μ_b, energy ε_b and spin s_b is emitted from a state of excitation energy E^*, of n^p particles and n^h holes leaving a residual nucleus with excitation energy U, with $(n^p - n_b^p)$ particles and n^h holes, where n_b^p is the number of particles in particle b, then the emission rate is given by

$$W\left(n,\varepsilon_b\right) = \frac{\left(2s_b + 1\right)}{\pi^2 \hbar^3} \mu_b \varepsilon_b \sigma_{\text{inv}}\left(\varepsilon_b\right) \frac{\omega\left(\left(n^p - n_b^p\right), n^h, U\right)}{\omega\left(n^p, n^h, E^*\right)} Q_b(n)\phi_b \tag{2.46}$$

In Eq. (2.46), $\sigma_{\text{inv}}\left(\varepsilon_b\right)$ is the inverse cross-section, which may be calculated using the optical model, $Q_b(n)$ and ϕ_b are respectively factors that differentiate between neutrons and protons; they are also the factors that give the probability of finding a pre-formed b, if b is a complex particle or cluster. ϕ_b is unity if b is a neutron, proton or gamma ray. It may be noted that $W(n)q(n, t)$ that appears in Eq. (2.45) is obtained by integrating Eq. (2.46) over all outgoing particles of all energies.

The particle–hole state density $\omega\left(\left(n^p - n_b^p\right), n^h, U\right)$ is given by[16-19] as follows;

$$\omega\left(\left(n^p - n_b^p\right), n^h, U\right) =$$

$$\left[\frac{g^n}{n^p! n^h! (n-1)!}\right] \sum_{j=0}^{n^h} (-1)^j \binom{n^h}{j} \left\{E^* - A_{n^p, n^h} - jE_F\right\}^{(n-1)} \Theta\left(E^* - E_{pp} - jE_F\right) \tag{2.47}$$

In Eq. (2.47), the variables have following meaning: n^p and n^h are respectively the number of particles and the number of holes, $n = n^p + n^h$. E_F is the Fermi energy, E_{pp} is the minimum energy required to excite n^p particles and n^h holes satisfying Pauli's exclusion principle and is given by

$$E_{pp} = \left[\frac{\left(n^h\right)^2 + \left(n^h\right)^2 + n^p - n^h}{2g} \right] \qquad (2.48)$$

Θ is a function that takes care of the finite hole depth and g is the single particle state density at the top of the Fermi level and is related to the level density parameter a by the relation

$$g = \frac{6a}{\pi^2} \qquad (2.49)$$

A_{n^p,n^h}, the Pauli correction factor is given as

$$A_{n^p,n^h} = \left[\frac{\left\{ n^p \left(\left(n^p - 1 \right) \right) + n^h \left(n^h - 1 \right) \right\}}{4g} \right] \qquad (2.50)$$

Williams[18] and Betak et al.[19] have derived the following expressions for interstate transition rates:

$$\lambda^0(n) = \frac{2\pi}{\hbar} < M^2 > \frac{g}{2n} \left[gE^* - 1/2 \left\{ \left(n^p \right)^2 + (n^h)^2 \right\} \right] \left[n^p \left(n^p - 1 \right) + n^h \left(n^h - 1 \right) + 4n^p n^h \right] \quad (2.51)$$

$$\lambda^+(n) = \frac{2\pi}{\hbar} < M^2 > \frac{g}{2(n+1)} \left[\frac{\left\{ gE^* - A\left(n^p + 1, n^h + 1 \right) \right\}^{(n+1)} - \left\{ C\left(n^p + 1, n^h + 1 \right) \right\}^{(n+1)}}{\left\{ gE^* - A_{n^p,n^h} \right\}^{(n+1)}} \right] \qquad (2.52)$$

and

$$\lambda^-(n) = \frac{2\pi}{\hbar} < M^2 > \frac{gn^p n^h}{2}(n-2) \left[1 - \left\{ \frac{C\left(n^p - 1, n^h - 1 \right)}{gE^* - A\left(n^p - 1, n^h - 1 \right)} \right\}^{(n-3)} \right] \qquad (2.53)$$

Here, <M^2> is the average squared matrix element for two-body residual interaction. Moreover, $C(n^p, n^h)$, the finite depth correction is given by

$$C\left(n^p, n^h \right) = gn^h \left(E^* - E_F \right) - n^h A\left(n^p, n^h \right) \qquad (2.54)$$

The averaged squared matrix element is often treated as a free parameter and the following expression is often used to represent it.

$$< M^2 >= \frac{c}{A^3 E^*} \tag{2.55}$$

Here, A is the atomic mass number of the compound nucleus, E^* the excitation energy and c a free parameter, which may be varied to match the experimental data.

The approximate values of $\lambda^+(n)$ and $\lambda^-(n)$ may be obtained from Eqs (2.52) and (2.53) as follows

$$\lambda^+(n) \sim \frac{g^3 E^{*2}}{n} \quad \text{and} \quad \lambda^-(n) \sim gn^3 \tag{2.56}$$

Therefore, for small values of n, $\lambda^+(n) \gg \lambda^-(n)$; hence, in the initial stages, the system proceeds predominantly in the direction of increasing n or towards states of higher complexities. But for larger n, $\lambda^+(n) \ll \lambda^-(n)$; and therefore, at later times, an equilibrium is established with exciton concentration centred around an equilibrium exciton number $\bar{n} \sim \sqrt{gE^*}$; where $\lambda^+(\bar{n}) = \lambda^-(\bar{n})$.

The master equation (2.45) may be solved for the exciton probability distribution $q(n,t)$ by substituting the values of various parameters using Eq. (2.46) in Eq. (2.55). On integration, one gets:

$$-q_0(n) = [\lambda^+(n-2)t(n-2) + \lambda^-(n+2)t(n+2) - \{\lambda^+(n) + \lambda^-(n) + w(n)\}t(n-2) \tag{2.57}$$

While deriving Eq. (2.57), it has been assumed that at $t = \infty, q(n, t = \infty) = 0$. The mean life time $\tau(n)$ of an exciton state is defined as

$$\tau(n) = \int_0^\tau q(n,t)\,dt \tag{2.58}$$

Once mean life time $\tau(n)$ is known, it is easy to calculate the average differential cross-section for the emission of particle b with energy ε_b as

$$\frac{d\sigma^{PE}}{d\varepsilon_b}(a,b) = \sigma(a) \sum_n w_b(n,\varepsilon_b)\tau(n) \tag{2.59}$$

Equation (2.59), denotes the cross-section for the formation of the complex composite nucleus by projectile a.

Different methods of calculating the mean life time $\tau(n)$ resulted in different prescriptions of the exciton model. However, the simplest approach, called the 'never-come-back approximation' assumes that in the beginning of the cascade, the system moves only towards the more complex states, i.e., $\lambda^-(n) = 0$. Under this assumption, the average PE emission cross-section may be written as

$$\frac{d\sigma^{PE}}{d\varepsilon_b}(a,b) = \sigma(a)\left[\frac{w_b(n,\varepsilon_b)}{w(n)+\lambda^+(n)} + \sum_{n=n_0+2,\Delta n=2}^{\bar{n}}\left\{\prod_{j=n_0,\Delta j=2}^{n-2}\frac{\lambda^+(j+2)}{w(j)+\lambda^+(j+2)}\right\}\frac{w_b(n,\varepsilon_b)}{w(n)+\lambda^+(n)}\right] \quad (2.60)$$

An exact solution for this master equation has been given by Akkermans[20]. In this approach, one first obtains an expression for the mean life time $\tau(N)$ for the state with maximum number of excitons N,

$$\tau(N) = T_N H_N \sum_{j=n_0,\Delta j=2}^{N-2} q_0(j)\left(\prod_{i=j,\Delta i=2}^{N-2}\lambda^+(i)T_i H_i\right) \quad (2.61)$$

where,

$$T_i = \frac{1}{\left[\lambda^+(i)+\lambda^-(i)+w(i)\right]}; \text{ and } H_{(i+1)} = \frac{1}{\left[1-H_i\lambda^+(i)T_i\lambda^-(i+1)T_{(i+1)}\right]} \text{ with } H_1 = 1 \quad (2.62)$$

In Eq. (2.61), for $j = N$, the term in the bracket is replaced by 1. The remaining mean life times may be successively calculated using the following recursion formula:

$$\lambda_n^+\tau(n) = \frac{1}{T_{(n+2)}}\tau(n+2) - \lambda^-(n+4)\tau(n+4) - q_0(n+2) \quad (2.63)$$

where, $n_0 \leq n \leq N-2$ and $\tau(N+2) = 0$ \quad (2.64)

For primary pre-equilibrium emission, the common initial situation of the projectile and the target is specified by $q_0(n) = \delta_{n,n_0}$. In this case, the mean life time of the nth exciton state is given as

$$\tau(n) = T_n H_n\left(\prod_{m=n_0,\Delta m=2}^{n-2}\lambda^+(m)T_m H_m\right)\left\{1 + \sum_{s=n+2,\Delta s=2}\left[\prod_{k=n,\Delta k=2}^{s-2}\lambda^+(k)T_k H_k\lambda^-(k+2)T_k + 2H_{k+2}\right]\right\} \quad (2.65)$$

In Eq. (2.65), when $n = n_0$, the term $\left(\prod_{m=n_0,\Delta m=2}^{n-2}\lambda^+(m)T_m H_m\right)$ is equal to 1; when $n = N$, the term

$$\sum_{s=n+2,\Delta s=2}\left[\prod_{k=n,\Delta k=2}^{s-2}\lambda^+(k)T_k H_k\lambda^-(k+2)T_k + 2H_{k+2}\right] = 0.$$

2.2.2 The Harp–Miller–Berne (HMB) model

The underlying concept of the HMB model is shown in Figure 2.14. As is indicated in the figure, the nuclear single particle states are grouped in bins of energy of some convenient size, say $\Delta E = 1$ MeV.

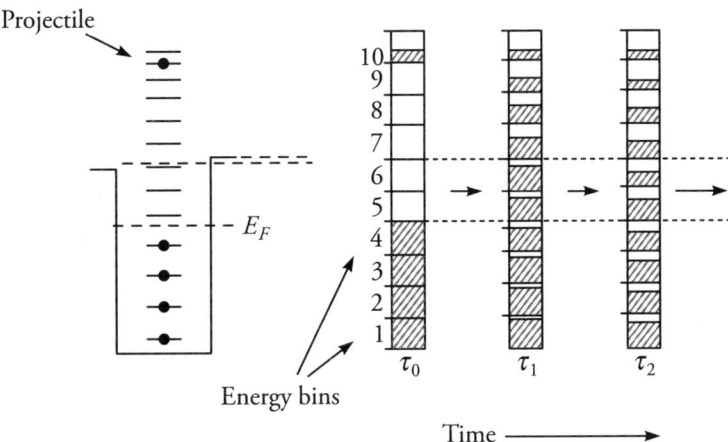

Figure 2.14 Pictorial representation of HMB model

The number of single particle levels in each bin are calculated and stored as the bin occupation number. Generally, a Fermi gas distribution is employed in calculations. In the beginning, at time = τ_0, all bins below the Fermi energy (E_F) are completely filled, while bins at energy higher than E_F are empty, except the one corresponding to the incident ion. Two-body interactions then change the bin population with time. One then calculates the average probability of a state in the ith bin as a function of time. This is done by calculating the relative probabilities of scattering into a particular bin from all other bins, the relative probability of scattering out from this particular bin and the relative probability of emission into the continuum above the binding energy of the emitted particle. Thus, the occupation numbers of bins change with time and by proper bookkeeping, the number of particles emitted along with their energy may be recorded at each step. Free nucleon–nucleon scattering cross-sections are used to obtain the revised occupation numbers of energy bins by solving coupled differential equations. A master equation of the following type describes the time evolution of bin populations,

$$\frac{d}{dt}\left(n_i, g_i\right) = \sum_{j,k,l} \omega_{(kl,ij)}\, g_k n_k g_l n_l \left(1 - n_i\right)\left(1 - n_j\right) g_k g_j -$$

$$\sum_{j,k,l} \omega_{(i,j,kl)}\, g_i n_i g_j n_j \left(1 - n_k\right)\left(1 - n_l\right) g_i g_l - n_i g_i \lambda_c\left(i'\right) \tag{2.66}$$

Equation (2.66) physically gives the net rate of change of the number of nucleons in the ith bin with g_i as the number of single particle states per MeV and n_i the number of nucleons in the ith state. It has three terms – the first represents the gain in the number of particles in the ith bin due to scattering from other bins; the second term gives the loss of particles due to scattering into other bins and the third term corresponds to the loss of particles due to emission of particles into the continuum with the emission rate $\lambda_c(i')$. Here, $\omega_{(kl,ij)}$ represents the transition rate for a nucleon in the ith bin to collide with a nucleon in bin jth such that the scattered nucleons go in those bins k and l that conserve the total energy. Free nucleon–nucleon cross-sections for collision at 90° are generally used in calculations.

2.2.3 The hybrid model

As the name suggests, this model, proposed by Blann[21], combines the HMB model and the exciton model in such a way that while the evolution of the reaction process inside the nucleus follows the exciton model procedure, the counting of levels is done using the HMB model approach. The hybrid model has the simplicity and transparency of the exciton model but spectral yields are calculated following the HMB prescription. In the exciton model, the integrated transition and emission rates are proportional respectively to the probabilities of internal scattering within the nucleus and the emission of particles into the continuum and therefore, in the exciton model, the competition between internal scattering and emission into the continuum is determined by these integral transmission and emission rates. On the other hand, in the hybrid model, the probability of finding a hole or a particle with energy ε is calculated first using the hole and particle densities from the exciton model; then, the transition rates for internal scattering and emission into the continuum are calculated for a particle of energy ε to get the relative probabilities for internal scattering and emission into the continuum.

As has been mentioned already, the hybrid model is similar to the exciton model where the internal transitions are associated explicitly with nucleon–nucleon scattering rather than the two-body matrix element. The emission spectrum in hybrid model is given by

$$\frac{d\sigma_x^{\text{pre}}}{d\varepsilon} = p_x(\varepsilon)\sigma_{\text{abs}} \tag{2.67}$$

Here $p_x(\varepsilon)$ is the probability of the emission of particle x with energy ε and σ_{abs} is the absorption cross-section. Further,

$$p_x(\varepsilon) = \sum_{n=n_0,\Delta n=2}^{\bar{n}} R_x(n) g \frac{\omega(n-1, E^* - \varepsilon - B_e)}{\omega(n.E^*)} \frac{\lambda_e(\varepsilon)}{\lambda_e(\varepsilon) + \lambda_+(\varepsilon)} \tag{2.68}$$

where $R_x(n)$ is the number of particles of type x in the n exciton state and B_e is the binding energy of the particle x. The transition rate $\lambda_+(\varepsilon)$ is given by

$$\lambda_+(\varepsilon) = \frac{\upsilon}{\text{Mean free path at energy } \varepsilon} = \rho_0 <\sigma(\varepsilon)> \sqrt{\frac{2(\varepsilon + V)}{m}} \tag{2.69}$$

and the emission width $\lambda_{e(\varepsilon)}$ is

$$\lambda_{e(\varepsilon)} = (2s+1)\frac{\mu\varepsilon\sigma_x(\varepsilon)}{\pi^2\hbar^3 g} \tag{2.70}$$

2.2.3.1 The geometry-dependent hybrid model

The geometry-dependent hybrid model improves the agreement of the hybrid model with experimental data by taking into account the effects of the nuclear surface. It first associates a value of the angular momentum ℓ with the impact parameter b using the standard rule

$$kb = \ell + 1/2 \tag{2.71}$$

Then, the average density $\overline{\rho_0(i)}$ is calculated in terms of the integral of a Woods–Saxon density on a straight line trajectory with impact parameter b as follows

$$\overline{\rho_0(i)} = \left\langle \frac{\rho_0}{1+\exp\left[\left(\sqrt{b^2+z^2}-R_0\right)/a\right]} \right\rangle \tag{2.72}$$

The average density is used to determine the mean free path as well as the Fermi energy, which determines the depth V of the potential and the limit to hole excitation, which enters into the density of states.

2.2.3.2 The Kalbach systematic

Kalbach observed that the pre-equilibrium/equilibrium double differential spectrum consists of two parts: (1) a forward-peaked continuum contribution and (2) a bound-state contribution that is symmetric about 90°. Combining the two gives

$$\frac{d^2\sigma}{d\varepsilon d\Omega} = \frac{1}{4\pi}\frac{d\sigma}{d\varepsilon}\frac{a}{\sinh a}\left[\cosh(a\cos\theta)+f_{\text{MSD}}\sinh(a\cos\theta)\right] \tag{2.73}$$

where the continuum fraction f_{MSD} must be determined separately. Kalbach obtained an excellent description of many pre-equilibrium angular distributions by parameterizing the coefficient a in terms of the incident and exit particle energies.

2.2.4 The intra-nuclear cascade model

Some 40 years back, the intra-nuclear cascade model[13,14] was the only model to theoretically obtain the angular and energy distributions of PE nucleons in nuclear reactions initiated by light ions in the energy range > 100 MeV or so. In this model, the classical trajectories of nucleons inside the nucleus are followed, one at a time, in three-dimensional coordinate space using the Monte Carlo method. At incident energies $\geq$ 100 MeV, the wavelength of the incident particle is smaller than the average distance between two adjacent nucleons inside the nucleus and therefore, the collision between the incident particle and the target nucleon may be assumed to be a quasi-free scattering event (Figure 2.15). The model assumes two-body scattering at each step and for numerical calculations, one uses experimental values of free nucleon scattering cross-sections and angular distributions. Though the model gives the correct order of magnitude for forward emissions, cross-sections in the backward hemisphere are substantially underestimated.

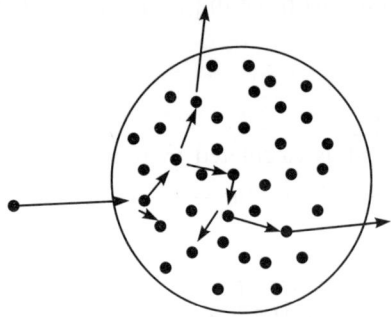

Figure 2.15 Pictorial representation of intra nuclear cascade model

2.2.5 The totally quantum mechanical model of pre-equilibrium emission

The semi-classical models for PE emission described in the previous sections were criticized on the ground that the nucleus is a quantum mechanical system and all processes involving a quantum system must be described using the quantum approach. Therefore, during the last few decades, considerable efforts have been made to develop a totally quantum mechanical theory for the compound reaction mechanism. The widely used Feshbach–Kerman–Koonin (FKK) model[22] is one such model. In this model, it is assumed that the matrix elements of the Hamiltonian that brings about transitions between intermediate nuclear states of different complexities and coupling to the PE channels are randomly distributed. If one compares the totally quantum mechanical and the semi-classical models, it will be observed that while the calculations using quantum mechanical pre-equilibrium models are more involved, lengthy and may require a fast computer, the calculations of semi-classical models are simple and may be performed on a desk computer. Further, the angle integrated particle spectra of PE particles predicted by quantum mechanical models has almost the same accuracy as that obtained by the semi-classical approach. Feshbach, Kerman and Koonin developed a statistical theory for the entire energy spectrum of particles emitted in statistical reactions (see Figure 2.7). According to the FKK theory, two types of intermediate steps, namely, multistep compound (MSC) and multistep direct (MSD) processes, build up the entire spectrum. The two types of processes are differentiated on the basis of the angular distribution of emitted particles – the multistep compound processes give a particle spectrum that is symmetrical about 90° in the CM frame; while the multistep direct processes produce a spectrum that is forward peaked. It may be mentioned that similar features were also observed by Kalbach in experimental spectra on which he based his systematic model. As one moves away from the regime of the compound nucleus towards the high energy side, the angular distribution of emitted particles first become anisotropic although it remains symmetrical around 90°. Later on, moving to still higher energies, the angular distributions become forward packed. Like the exciton model, the FKK theory assumes that the reaction proceeds through stages that become successively complex. The excitation of the nth stage is more complex than the excitation of the $(n-1)$th stage and less complex than the $(n+1)$th stage. Both types of processes, MSC and MSD, are involved at each complex stage. In the MSD process, it is assumed that at least one particle is definitely in continuum, while in the MSC process, it is assumed that all particles are bound. The building up of the reaction may be terminated at any stage by a transition to the final stage, a process that competes with (1) the excitation of a stage of higher complexity and (2) de-excitation back to a stage of less complexity. A schematic representation of the FKK model is shown in Figure 2.16, where the P chain represents stages of increasing complexity of MSD processes and the Q chain represents stages of increasing complexity for MSC the process.

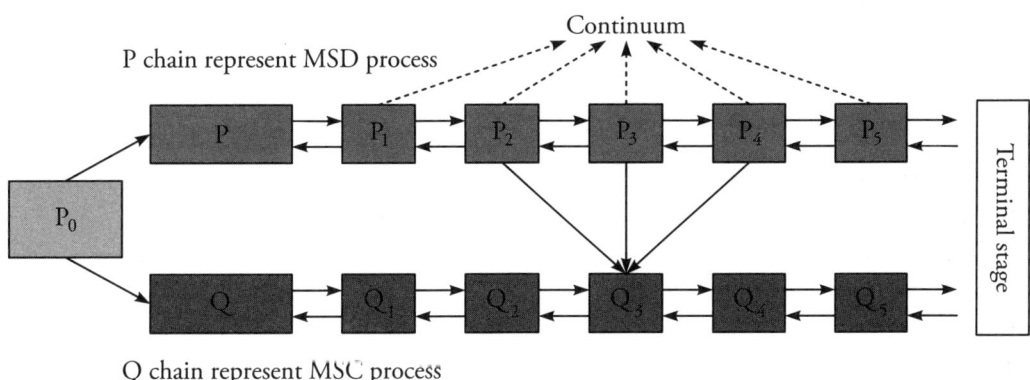

Figure 2.16 Schematic representation of MSD and MSC processes in the FKK model

In this theory, it is also assumed that the two-body residual interactions may change from the nth stage to $(n \mp 2)$th stage and this assumption is known as chaining hypothesis. The wave functions in the MSC matrix are assumed to have random phases so that on taking average, no interference term is left. The angular distribution generated from the MSC matrix, therefore, is symmetrical about 90° in the CM frame. On the other hand, in MSD processes, only those wave functions that interfere constructively on averaging accounts for the emission in continuum. The particle spectra generated by MSD, therefore, retain the memory of the incident channel and the spectrum is forward packed. The cross-section for MSC processes strongly depend on the level density ρ_n, which rapidly increases with n till the chain is terminated. The terminal stage contains all those configurations that have not been included till the $(n - 1)$th stage; it is assumed that the component of the wave function is statistically distributed over all these configurations.

The cross-section from an initial channel i to a final channel f may be given as

$$\sigma_{i,f} = \frac{4\pi^2}{k^2} \left(|T_{if}| \right)^2$$

(2.74)

The transition matrix element T_{if} in Eq. (2.74) is made up of two terms corresponding to the MSD and MSC processes, i.e.,

$$T_{if} = T_{if}^{MSC} + T_{if}^{MSD}$$

(2.75)

The appropriate transition matrix elements may be obtained from the solution of the appropriate Schrödinger equation for the optical model Hamiltonian.

2.3 The Incomplete Fusion of Heavy Ions

Incomplete fusion is a well-known phenomenon at higher energies and specifically for massive ions. As has been mentioned, if the excited composite nucleus carries an angular momentum larger than what it can sustain, the composite nucleus breaks into two parts – a target-like part forming the incompletely fused system and a projectile-like part being the spectator. During the brief period when the target and the projectile remain within the range of the attractive nuclear potential, some nucleons or clusters may be exchanged between them, after which they fly apart.

The division of the heavy ion reaction products into two types, namely, the target-like and projectile-like components may also be interpreted as the outcome of incomplete fusion of the incident and target ions. Target-like parts are heavier than the target as they are those parts who have captured a fraction of the incident ion; while the lighter projectile-like components are the residual part of the projectile.

However, recently, several group of workers[53–58] have observed incomplete fusion and pre-equilibrium emission in heavy ion reactions below 10 MeV/n energies, which is quite unlikely as complete fusion and compound reaction mechanisms are the most likely processes at such low energies. These findings have once again brought the reaction dynamics of incomplete fusion and pre-equilibrium emission in heavy ion reactions at energies below 10 MeV/n to the forefront of research.

Many models for incomplete fusion (ICF) of heavy ions have been proposed. Some authors treat ICF as a continuation of pre-equilibrium emission, where clusters rather than nucleons are emitted during the time of equilibration of the composite system. One such example is the hot spot model proposed by Weiner and Westrom[23], details of which are outlined in the following section.

2.3.1 The hot spot model

Weiner and Westrom developed a method of deriving information regarding heat conductivity in nuclear matter by looking into asymmetry effects in the angular distribution of evaporated products. They established a link between the transport properties of the nuclear matter and the pre-equilibrium emission from a composite excited nucleus. Their argument is that before the nuclear matter reaches thermal equilibrium, a (heat or temperature) diffusion phenomenon takes place in the nuclear matter. The initial excitation, which is concentrated in a small region where the projectile has hit the target, called the hot spot, diffuses from the hot spot into the nuclear matter. The diffusion of heat affects the energy and the angular distribution of emitted pre-equilibrium particles or clusters, in case of heavy ions.

In the hot spot model, it is assumed that the incoming projectile creates a hot spot in the peripheral region of the target nucleus, whose spatial spread is much smaller than the radius R of the target nucleus. It is further assumed that the four momentum q deposited by the projectile is absorbed by the target essentially before the heat diffusion process begins. Since the distribution of the excitation energy between nucleons via two-body residual interactions is a diffusion-like process, it may be assumed that one may apply local thermodynamics to the system of the hot spot and write the following diffusion equation:

$$\rho c_p \frac{\partial T}{\partial t} = \mathrm{div}\left(k \ \mathrm{grad} \ T\right) \tag{2.76}$$

Here, ρ is the density of the nuclear matter, c_p the thermal capacity of the nuclear matter at constant pressure and k the thermal conductivity of the nuclear matter. Equation (2.76) is applicable to the energy dissipation by the hot spot only if the hot spot has lost its memory of q, i.e., after a minimum time, $\tau_0 \approx \frac{2r_0}{v_F} \approx 10^{-23}\,\mathrm{s}$, where v_F is the Fermi velocity. Pre-equilibrium particles or clusters are those that are emitted within a time t, such that $\tau_0 < t < \tau_R$, where τ_R is the relaxation time. By solving the diffusion equation, Eq. (2.76), with suitable boundary conditions, it is possible to obtain the integrated energy spectra of the pre-equilibrium component. Further, it has been shown that the values of constants like the thermal conductivity and heat capacity at constant pressure for the nuclear matter may be obtained from the observed characteristics of the pre-equilibrium particles. However, the hot spot model does not give cross-sections for incomplete fusion; it only provides information regarding the mechanism of incomplete fusion and characteristics of the spectrum of light projectile-like pre-equilibrium component.

2.3.2 The promptly-emitted particles (PEPs) model

This model, like the hot spot model, describes a different mechanism of the emission of prompt particles and/or light clusters in heavy ion interaction and, from that point of view, a different mechanism for incomplete fusion. As such, at present, the model is not in a form from which cross-sections for incomplete fusion residues may be obtained. However, for the sake of completeness, a brief discussion of the mechanism of prompt-particle emission is included here. Further, as will be seen, the model is applicable at relatively higher energies > 10 MeV/n.

Light ions and light clusters in heavy ion interactions may be emitted from (i) the excited fused composite nucleus, (ii) the di-nuclear structure formed before complete fusion, or (iii) from the fragments flying apart after the deep inelastic collision (DIC). The energy spectrum of the emitted ions show two components – one with temperatures ≈ 1 – 2 MeV, may be attributed to the particle evaporation from the compound nucleus; the other, harder component, corresponding to a nuclear temperature ≈ 6 – 7 MeV, is generally called the pre-equilibrium component. Bondorf et al.[24] developed the promptly-emitted particles (PEPs) model for the hard component of the emitted light ions. The authors argued that application of the exciton model to heavy ion fusion requires some adjustable free parameters, like the number of excitons and their energies in the initial state, which may not be determined precisely. Therefore, such models are incomplete.

The central argument of the PEPs model is that as the two interacting nuclei approach/penetrate each other, the interaction barrier between the nuclei decreases and when it drops to a value that allows the free flow of nucleons, the coupling of the relative velocity of nucleons with the Fermi velocity in the donor nucleus will give a big boost of energy to the transferred nucleons as compared to the nucleons in the recipient nucleus. These nucleons with boosted energy will now pass through the recipient nucleus and may either be absorbed in the nucleus or be emitted if allowed by the energy and angular momentum considerations. These particles with boosted energy are called quasi-free PEPs. Figure 2.17 shows the geometry of the emission of PEPs.

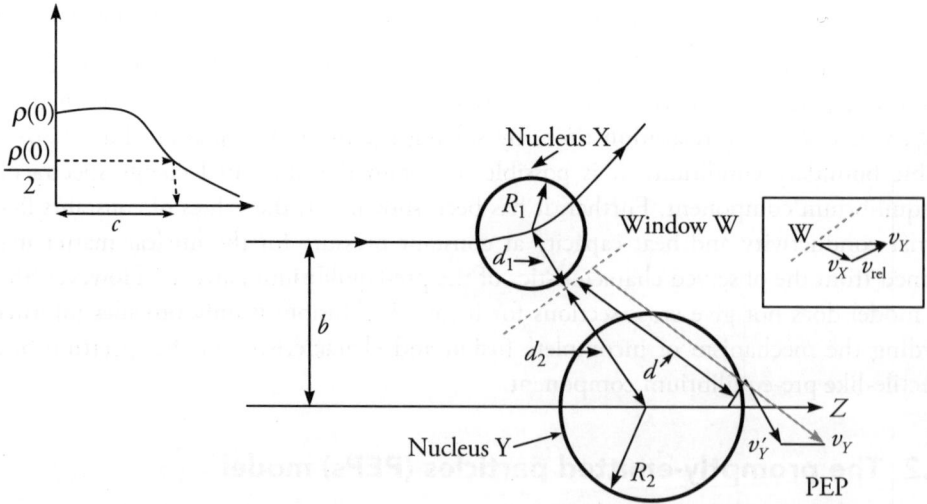

Figure 2.17 Geometry for the emission of PEPs, b is the impact parameter, Z is the beam axis, position of the window W is given by $c_1/c_2 = d_1/d_2$, where c_1 and c_2 are the half density radii of the two nuclei. $d_1 + d_2$ is the distance between the centers of the two nuclei. The distance travelled by the PEP in nucleus Y is d. R_1 and R_2 are the sharp surface radii of the nuclei X and Y

As mentioned earlier, close approach/overlapping of the two heavy ions results in the lowering of the interaction barrier between the two and when the barrier is so low that the nucleons from both the incident and target ions freely move from one to the other, a common interface area between the ions is established. Let us assume that nucleons from the nucleus X are transferred to the nucleus Y. In this common interface, the nucleons of the donor nucleus X will have their intrinsic velocity equal to the Fermi velocity (say v_X) of the donor nucleus X. However, the velocity v_Y of these transferred nucleons in the recipient nucleus Y will be different from v_X because of the relative velocity v_{rel}. If the transferred nucleons are treated as point particles, then

$$v_Y = v_{\text{rel}} + v_X \tag{2.77}$$

Although the nuclei X and Y are different and therefore, the symmetry and Coulomb terms of the liquid drop model will be different for X and Y, it is assumed that the depth of the single particle potential is the same within the volume of the two nuclei. The kinetic energy ε_Y of the transferred nucleons in the framework of the recipient nucleus Y is then given by

$$\varepsilon_Y = \frac{1}{2}mv_Y^2 = \varepsilon_X + \varepsilon_{\text{rel}} + 2\cos\varphi\sqrt{\varepsilon_X \varepsilon_{\text{rel}}} \tag{2.78}$$

Here φ is the angle between v_X and v_{rel} and m, the nucleon mass. It becomes obvious from Eq. (2.78) that the term $2\cos\varphi\sqrt{\varepsilon_X \varepsilon_{\text{rel}}}$ may have large positive values for some selected values of φ and, therefore, ε_Y may become substantially large for these selected φ values. One may ask a question as to what provides for this increase in the energy of transferred nucleons. The increase is at the cost of the reduction in the relative velocity $v_{\text{rel}}^{\text{before}}$ of the remaining part of the donor nucleus left after the transfer of the nucleons. For example, if one nucleon is transferred from X to Y, then the relative velocity $v_{\text{rel}}^{\text{after}}$ of the remaining part $(A_X - 1)$ of the donor nucleus is given by

$$v_{\text{rel}}^{\text{after}} = v_{\text{rel}}^{\text{before}} - \frac{v_X}{(A_X - 1)} \tag{2.79}$$

This follows from the conservation of linear momentum before and after the transfer of nucleon. In Eq. (2.79), A_X is the number of nucleons in the donor nucleus X before the transfer of nucleons.

Some of the transferred nucleons get scattered by the other nucleons in the recipient nucleus and are thus absorbed. A few nucleons, however, reach the surface of the recipient nucleus and may be emitted as PEP if they have sufficient energy to overcome the barrier at the surface. Classically, the condition for the emission of transferred nucleons as PEP may be written as:

$$\varepsilon_Y > (V_{\text{nuc}} + V_{\text{Coul}}) + V_{\text{escape}} \tag{2.80}$$

Here V_{nuc}, V_{Coul} and V_{escape} are respectively, the nuclear, Coulomb and escape potentials for the emitted particles. It is reasonable to assume that the sum $(V_{nuc} + V_{Coul})$ remains constant over the nuclear volume and that V_{escape} vanishes for charge-less particles like neutrons.

The distribution of velocity v_X, denoted by $p(v_X)$, may be given by the zero temperature Fermi distribution as

$$p(v_X) = \frac{3}{4\pi v_F^3} \vartheta(v_X - v_F)$$

(2.81)

In Eq. (2.81), ϑ is the theta function. The kinetic energy of the transferred nucleon just after their emission is given by

$$\varepsilon_Y^{emission} = \varepsilon_Y - (V_{nuc} + V_{Coul} + V_{escape})$$

(2.82)

In general, $\varepsilon_Y^{emission}$ and ε_Y may point in different directions and have different magnitudes.

Since the transferred nucleons move in the spherically symmetric nuclear field of the recipient nucleus Y, their angular momentum about the centre of nucleus Y is conserved. In order to conserve the angular momentum, the magnitude of the radial momentum, mv_Y^r, (v_Y^r being the radial velocity) of the transferred nucleons decreases with increase in r. This puts a more realistic limit on energy of the emitted nucleons as

$$\frac{1}{2}mv_{Y,sur}^r > (V_{nuc} + V_{Coul} + V_{escape})$$

(2.83)

Here $v_{Y,sur}^r$ denotes the radial velocity of the transferred nucleons at the surface of nucleus Y. It may be observed that the condition put by Eq. (2.83) overrides the condition given by Eq. (2.80).

For a fixed value of the impact parameter b, the probability P_{PEP} for the emission of PEP particles may be written as

$$P_{PEP} = \int dt \int dW \int dv_X \, p(v_X) N(v_Y) e^{-\left(\frac{d}{\lambda}\right)}$$

(2.84)

Here W is the area of the window at the interface, $N(v_b)$ the flux of nucleons passing from X to Y through the window W, d the distance travelled in recipient nucleus Y and λ the mean free path of transferred nucleons in nucleus Y. $e^{-\left(\frac{d}{\lambda}\right)}$ is the correction factor for the absorption of transferred nucleons in the recipient nucleus Y. The total PEP emission cross-section may be given by

$$\sigma_{total} = 2\pi \int_0^{b_z} b \, db \, P_{PEP}$$

(2.85)

Here b_z is the grazing impact parameter.

PEP may also be emitted when nucleons transfer from Y to nucleus X. This emission may be treated as done for transfer from X to Y. It may be noted that all the equations will hold if subscripts X and Y are interchanged. In the overall CM frame, the nucleon velocities after escaping the recipient are

$$\left(v'_Y\right)_{CM} = v'_Y + \left(V_Y\right)_{CM} \text{ and } \left(v'_X\right)_{CM} = v'_X + \left(V_X\right)_{CM} \tag{2.86}$$

Here $(V_Y)_{CM}$ and $(V_X)_{CM}$ are respectively the velocities of nucleus Y and nucleus X in CM. Since the ratio $(V_Y)_{CM} / (V_X)_{CM} \propto \dfrac{\text{Mass of X}}{\text{Mass of Y}}$, it follows from the formula given in Eq. (2.86) that when the projectile and the target ions have very different masses, the PEP spectra will be asymmetric in the CM frame. However, the evaporation spectra in the CM frame will always be symmetrical about 90° irrespective of the masses of the interacting ions.

Typical PEP and evaporation spectra for emitted neutrons at a backward angle of 160° and a forward angle of 15°, in the collision of a heavy target and light projectile are shown respectively in Figure 2.18 (a) and (b). As may be observed, the PEP component is totally masked by the evaporation spectra at backward angles. The nature of the PEP spectrum for charged particles will be quite different from that for neutrons because of the non-zero value of V_{escape}.

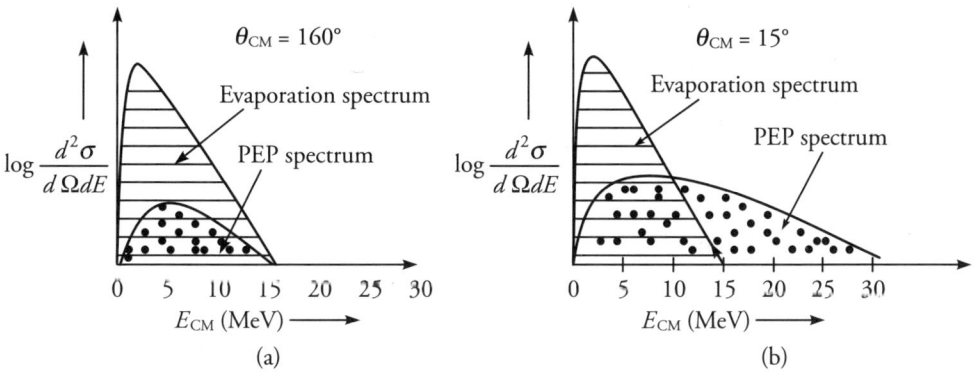

Figure 2.18 Typical PEP and evaporation spectra for emitted neutrons (a) at backward CM-angle (b) at forward CM-angle for a heavy target and light projectile. At backward angles, the PEP contribution is totally masked by the evaporation spectrum.

2.3.3 The sum rule model

The sum rule model assumes that ICF takes place in peripheral collisions and therefore, angular momentum values larger than ℓ_{crt} (for complete fusion) contribute to ICF. Before discussing the sum rule model put forward by Wilezynski et al.[25], let us outline the reactions to which the model is applicable. The sum rule model is applicable only to binary reactions, i.e., reactions in which there are only two interacting nuclei, both in the input as well as the exit channels. In heavy ion reactions initiated particularly by alpha cluster projectiles, like ^{12}C, ^{16}O etc, experimental evidence suggests that often a component of the incident ion, say ^8Be of ^{12}C fuses with the target and in the exit channel, one observes the ejectile and the residue corresponding to the incompletely fused system (target + ^8Be). Though the unfused part of the projectile ^4He in the present example is present both in the entrance and the exit channels, it is assumed that the projectile breaks into two parts (^4He and ^8Be in the present example) before the fusion and that the unfused part does not take part in the interaction and proceeds with the same initial velocity of the projectile as a spectator. Bondorf et al.[26] have shown that if the cross-section for a reaction depends exponentially on the ground state Q value, denoted by $Q_{g.g}$, then the reaction is binary in nature and that a partial statistical equilibrium has been setup in the composite system before its decay into the two components of the exit channel.

The sum rule model considers the formation of a di-nucleus when the projectile heavy ion collides with the target heavy ion. This may happen only for central and hard grazing collisions that are specified by a certain range of impact parameters. In both the central and hard grazing collisions, the two interacting nuclei penetrate so far into each other that the nuclear density in the overlapping region becomes saturated and a strongly interacting di-nuclear structure is formed. It is reasonable to assume that a thermal equilibrium is established in the di-nuclear system from the time of the collision to the time when the nuclear density gets saturated. Thus, the di-nuclear system in thermal equilibrium may now either (i) completely fuse to make a completely fused compound nucleus or (ii) exchange some nucleons/clusters and then fly apart into two fragments. Based on the assumption that the di-nuclear system attains thermal equilibrium during collision, Bandorf et al.[26] argued that the probabilities for the aforementioned two possible final stages will be proportional to the final state level densities. Further, if it is assumed that a part of the energy associated with the transfer of nucleons is converted mostly into rotational energy and if one uses the constant temperature level density formalism, it can be shown[26] that the reaction probability $p(i)$ for a given channel i is proportional to the exponential factor $e^{\left\{ \frac{[Q_{g.g}(i) - Q_c(i)]}{T} \right\}}$ and therefore, one may write,

$$p(i) \sim e^{\left\{ \frac{[Q_{g.g}(i) - Q_c(i)]}{T} \right\}} \qquad (2.87)$$

Here T is the mean temperature taken as constant. The term $Q_c(i)$ arises because of the change in Coulomb energy when charge gets transferred from one component of the di-nuclear system to the other. It is given by

$$Q_c(i) = \frac{\left[Z_1^f Z_2^f - Z_1^{in} Z_2^{in} \right] e^2}{R_c} \qquad (2.88)$$

Z_1^f, Z_2^f, Z_1^{in} and Z_2^{in} are respectively the atomic numbers of the final fragments and the initial interacting partners. R_c is the effective distance where the transfer of charge takes place.

2.3.3.1 Assumptions of the sum rule model

The sum rule model takes the validity of Eq. (2.87) as its first assumption. It further assumes that Eq. (2.87) gives the probability of the fusion of the two nuclei.

The second assumption of the model deals with the angular momentum transfer in the reaction. Let the two interacting nuclei X and Y in the entrance channel have, respectively, atomic mass numbers A_1 and A_2. After the formation of a di-nuclear system till the complete fusion of the interacting nuclei, let n nucleons from the nucleus X be transferred to the nucleus Y and m nucleons flow in the opposite direction (See Figure 2.19).

Let us consider that n nucleons from nucleus X transfer to the nucleus Y and fuses with it to form a new nucleus Z with mass number $(A_2 + n)$. It is now assumed that the transfer of n nucleons from X and their fusion with nucleus Y will be possible only if the angular momentum ℓ_n of the relative motion of n nucleons with respect to the nucleus Y is equal or less than the critical angular momentum ℓ^Z_{crt} corresponding to nucleus Z, i.e., $\ell_n \leq \ell^Z_{crt}$. The same will be true for the transfer and fusion of m nucleons from nucleus Y to nucleus X, i.e., $\ell_m \leq \ell^{Z'}_{crt}$, where Z' represents the nucleus formed by the fusion of m nucleons with nucleus X. Or,

$$\ell_{(n \text{ or } m)} \leq \ell^{(Z \text{ or } Z')}_{crt} \tag{2.89}$$

Equation (2.89) is now required to be translated in terms of the angular momentum of the relative motion of X and Y in the entrance channel. The following relation may be used for this transformation:

$$\ell_{lim}(X \text{ versus } Y) = \frac{A_1 A_2}{nA_2 + mA_1} \ell_{crt}(Z \text{ or } Z') \tag{2.90}$$

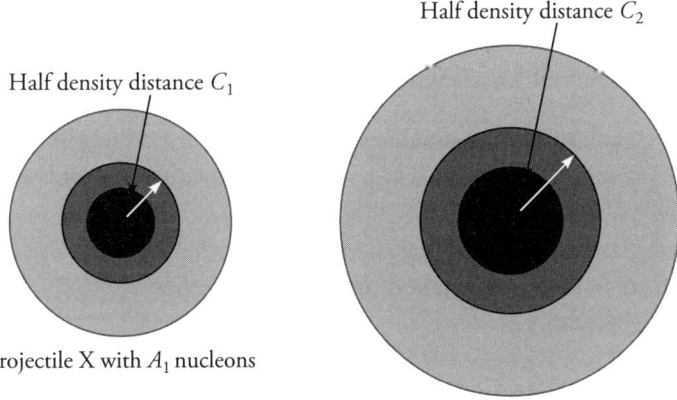

Half density distance C_2

Half density distance C_1

Projectile X with A_1 nucleons

Projectile Y with A_2 nucleons

Figure 2.19 The projectile and target nuclei with the distances C_1 and C_2 where the charges density falls to half of the central value

In the sum rule model, it is further assumed that there is a smooth cutoff for each individual reaction channel in ℓ space, i.e.,

$$T_\ell(i) = \cfrac{1}{\left[1 + e^{\left\{ \frac{(\ell - \ell_{\lim}(i))}{\Delta} \right\}} \right]} \tag{2.91}$$

Here Δ is the diffuseness parameter for the cutoff in T_ℓ space.

It is easy to show that for all channels, the transmission coefficients for small angular momentum values are unity. For larger values of angular momentum, successive reaction channels get switched off once the angular momentum value exceeds the corresponding limiting value.

Combining Eqs (2.87) and (2.91), Bondorf et al.[26] proposed the following sum rule for the reaction probabilities for a partial wave ℓ.

$$N_\ell \sum_i T_\ell(i) e^{\left\{ \frac{Q_{g.g}(i) - Q_C(i)}{T} \right\}} = 1 \tag{2.92}$$

Here N_ℓ is the normalization constant for each channel.

From the aforementioned derivation, it is clear that the sum rule can be applied only if all participating channels are counted. This includes the complete fusion channels as well. Moreover, only those partial waves that result in the formation of a di-nuclear system should be considered. With these restrictions, the absolute cross-section for a particular channel may be explicitly written as

$$\sigma(i) = \pi \lambdabar^2 \sum_{\ell=0}^{\ell_{\max}} (2\ell + 1) \frac{T_\ell(i) \, p(i)}{\sum_j T_\ell(j) \, p(j)} \tag{2.93}$$

Here $\lambdabar^2 = \dfrac{\hbar^2}{2\mu E}$ is the reduced wavelength for the entrance channel.

$\ell_{\max}$ may be defined as the largest value of ℓ for which the colliding nuclei penetrate into the region where the total nucleus–nucleus potential is attractive and /or the distance of closest approach is smaller than the sum of the half density radii (Figure 2.20)

The value of $\ell_{\max}$ can be calculated by choosing a potential for heavy ion interaction, which is generally assumed to be a combination of a Coulomb term, a Woods–Saxon type nuclear term and a centrifugal term.

$$R_0 = C_1 + C_2$$

$$V_0 = b_{\text{surf}} \left[A_1^{2/3} + A_2^{2/3} - (A_1 + A_2)^{2/3} \right] \tag{2.94}$$

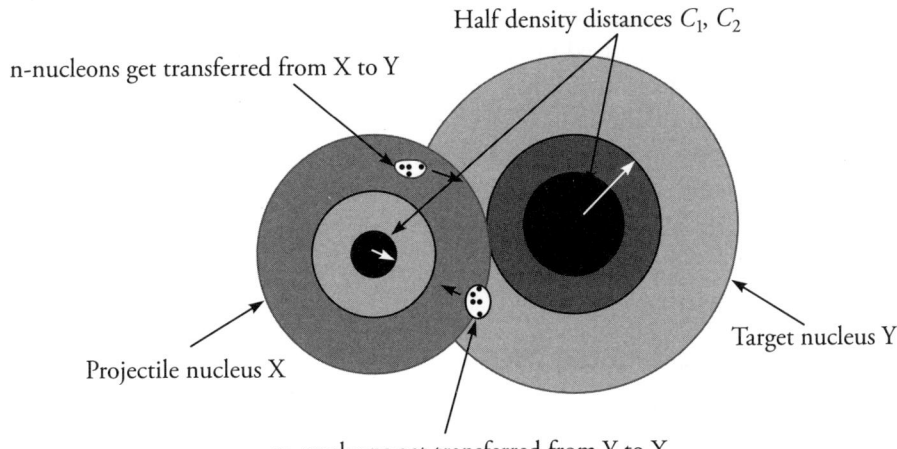

Figure 2.20 In the overlap region nuclear density saturates and *n*-nucleons from *X* transfer to *Y* while *m*-nucleons from *Y* go to *X*

and $a = \dfrac{V_0 R_0}{16\pi\gamma C_1 C_2}$

where C_1 and C_2 are the distances from the centre where the nuclear charge density falls to half of its central value respectively for the nuclei X and Y. b_{surf} = 17.9 MeV is the coefficient of the surface energy term in the semi-empirical mass formula and

$$y = 0.95(1 - 1.78I^2) \text{ MeV - fm}^{-2} \tag{2.95}$$

where y is the surface tension coefficient with the isospin correction for the combined system of the two nuclei defined as $I = (N - Z)/A$.

The following expression, given by Myers[27], may be used to calculate the value of C.

$$C = R\left[1 - \frac{b^2}{R^2} + \ldots\right] \tag{2.96}$$

where b = 1 fm and R is given by

$$R = 1.28A^{\frac{1}{3}} - 0.76 + 0.8A^{-\frac{1}{3}} \text{ fm} \tag{2.97}$$

The critical angular momentum ℓ_{crt} may be calculated from the following simplified formula,

$$\left(\ell_{crt} + \frac{1}{2}\right)^2 = \frac{\mu(C_1 + C_2)^3}{\hbar^2}\left[4\pi\gamma\frac{C_1 C_2}{(C_1 + C_2)} - \frac{Z_1 Z_2 e^2}{(C_1 + C_2)^2}\right] \tag{2.98}$$

Three free parameters, the mean temperature T, the effective Coulomb radius R_c and the diffuseness parameter Δ are involved in the sum rule model. The first two of these are purely empirical and cannot be determined theoretically. The Coulomb interaction radius may be estimated to be the sum of the radii of the interacting nuclei. Wilezynski et al.[25] in their measurements for the ^{14}N+^{159}Tb system at 140 MeV could get a good reproduction of the data with the following set of free parameters:

$$T = 3.5 \text{ MeV}; \qquad R_C = 1.5\left(A_1^{1/3} + A_2^{1/3}\right) \text{ fm}; \quad \text{and} \quad \Delta = 1.7\hbar \tag{2.99}$$

2.3.4 Breakup fusion model (BUF)

Starting from the pioneer experiments of Britt and Quinton[28], many other investigations[29,30] have shown that fast particle emission in heavy ion collisions is almost always accompanied by the fusion of the remaining part of the projectile with the target nucleus. Such reactions are often called massive mass transfer reactions or simple massive transfer reactions. The breakup fusion model was initially proposed by Kerman and McVoy[31] and has been further refined by Udagawa and Tamura[32]. The model assumes that the massive mass transfer reactions proceed in two stages: in the first stage, the projectile breaks up into two parts and in the second stage, the massive part fuses with the target nucleus.

The massive transfer reaction may be schematically represented as:

Stage I: Projectile a breaks into two parts: lighter part α and the heavier part x; i.e., $a \rightarrow \alpha + x$

Stage II: x fuses with target A

Kerman and McVoy gave the following expression for the cross-section of the massive transfer reactions

$$\frac{d^2\sigma}{dE_\alpha d\Omega_\alpha} = \left[\frac{m_a m_\alpha}{\left(2\pi\hbar^2\right)^2}\right]\frac{k_\alpha}{k_a}\sum_\ell A_\ell \sum_m |\beta_{\ell m}(k_\alpha)|^2 \tag{2.100}$$

Here a denotes the projectile that breaks into α and x, i.e., $a \rightarrow \alpha + x$. $\beta_{\ell m}\left(\vec{k_\alpha}\right)$ is the amplitude of the breakup process in which α is emitted with a momentum $\vec{k_\alpha}$ and x with an angular momentum ℓm relative to the target A.

By setting $A_\ell = 1$, all values of ℓ in Eq. (2.100) will give the singles α cross-section associated with the elastic breakup process. Kerman and McVoy further showed that putting $A_\ell = \dfrac{P_\ell}{4}$ in Eq. (2.100) gives the additional contribution coming from the breakup fusion (BUF) process to the singles alpha cross-section, where P_ℓ gives the absorption of x by A.

Kerman and McVoy used Green's function to describe the propagation of x from its creation by BUF to its absorption by the target A, and used only the on-shell part of the Green's function to obtain the relation $A_\ell = \dfrac{P_\ell}{4}$. Udagawa and Tamura, on the other hand, argued that the off-shell part of Green's function contributes as much as the on-shell part and therefore, used the relation

$$A_\ell = \frac{P_\ell}{|s_\ell|^2} \tag{2.101}$$

Here s_ℓ denotes the scattering matrix. Udagawa and Tamura showed that Eq. (2.101) evolves when one describes the relative motion of x and A by an optical potential, the complete set of states required to work with Green's function must be bi-orthogonal and that results in Eq. (2.101). In order to get the value of A_ℓ from Eq. (2.101), the following expression has been used

$$|s_\ell| = s_0 + (1 - s_0) \frac{1}{\left\{ 1 + e^{-\left(\frac{\ell - \ell_s}{\Delta} \right)} \right\}} \tag{2.102}$$

In Eq. (2.102), $s_0 = |s_{\ell=0}|$. The values for s_0 and Δ were determined by first choosing an appropriate optical potential $U_{x,A}^{opt}$ between x and A, and then calculating the elastic scattering between A and x. Further, ℓ_s was taken to be equal to ℓ_β, which is the value of ℓ at which $|\beta_{\ell m}|$ takes the maximum value. The following analytical form P_ℓ has been used;

$$P_\ell = \frac{1}{\left\{ 1 + e^{\left(\frac{\ell - \ell_{crit}}{\Delta} \right)} \right\}} \tag{2.103}$$

With these substitutions, Eq. (2.100) may be used to compute differential cross-sections for massive transfer reactions.

2.4 Computer Codes

The complete fusion of heavy ions is theoretically treated using the statistical compound reaction mechanism. In this mechanism, it is assumed that the first stage of the reaction fusion of the two ions produces an excited composite system that undergoes equilibration to become a compound nucleus with excitation energy, on an average, equally distributed amongst all degrees of freedom. Since the excitation energy is sufficiently large, the compound nucleus is excited in the region of continuum and the properties of the compound nucleus are statistically averaged over large number of levels. It is assumed that particles or clusters are not emitted during the process of thermal equilibrium and in the second stage of the reaction, particles and clusters are evaporated from the compound nucleus like the evaporation of molecules from a boiling liquid. The first and the second stages of the reaction are treated as independent of each other, assuming that the compound nucleus forgets the history of its formation and therefore, the same compound nucleus, formed by whatever way, will decay in its characteristic way.

Computer codes that may be used to calculate the complete fusion cross-sections for heavy ion reactions are available in the market. A few of these codes also include calculation of the pre-equilibrium emission either using the exciton model/hybrid model. Salient features of these codes are discussed in the following section.

2.4.1 Computer code PACE 4

The code derives its name from projection angular momentum coupled evaporation. This code is based on an earlier code named JULIAN developed by Rossner et al.[33]. In code PACE4, the decay of an excited compound nucleus is traced using the Monte Carlo method. Cross-sections are calculated using the Hauser–Feshbach method. Use of the Monte Carlo method makes it possible to obtain correlations between the evaporated particles, their energy, angular distribution and emitted gamma rays. Sequential emission of particles, one after the other, till it is energetically possible, is considered in the code. Conservation of energy and angular momentum is considered at each stage of particle evaporation. The final state to which the nucleus will decay is decided by selecting a random number. After completing the decay sequence of a selected compound nucleus, the process is repeated for the next and so on till all the compound nuclei are exhausted. An optical model sub-routine[34,35] calculates the transmission coefficients for neutron, proton and alpha particles. The competition with fission decay mode is simulated using the rotating liquid drop fission barrier sub-routine FISROT. The Bass model[36] is used to get fusion cross-sections. Since it is possible in the code to trace back the sequence of decay for each event, the angular distribution of evaporated particles, etc, may be obtained by proper bookkeeping. The level density of residual nuclei plays an important role in determining the cross-sections for statistical nuclear reactions. The code follows Fermi gas formalism for level densities, which is identical to the constant temperature level densities at low energies. The level density parameter a in the code can either be calculated or it may be obtained using the default option $a = A/K$, where A is the atomic mass number of the nucleus and K, a free parameter, the value of which may be varied to obtain a good fit to the experimental data. The rotational energy E_{rot} (J) may be calculated in two ways, viz., using the spin cutoff parameter or from the ground-state rotational energy of the finite range rotating drop model[37]. In those cases where the values of J, Z or A exceed the range of Sierk routine, they are taken from the work of Cohen at al.[38]. The Bohr and Wheeler[39] formalism is used to calculate fission probabilities. It is possible to run the code with a very large number of events, up to 50,000, to obtain better statistics.

The partial fusion cross-section σ_ℓ is given by

$$\sigma_\ell = \pi \lambda^2 (2\ell + 1) T_\ell \tag{2.104}$$

Here λ is the reduced wavelength and T_ℓ the transmission coefficient given by

$$T_\ell = \frac{1}{1 + e^{-\left(\frac{\ell - \ell_{max}}{\Delta}\right)}} \tag{2.105}$$

where Δ is the diffuseness coefficient and ℓ_{max} is the maximum angular momentum, the value of which is obtained from the total fusion cross-section σ_F;

$$\sigma_F = \sum \sigma_\ell \tag{2.106}$$

2.4.2 Computer code CASCADE

Code CASCADE developed by Puhlhofer[40] also calculates complete fusion cross-sections for heavy ions using the Hauser–Feshbach formalism. The code computes the reaction cross-sections for product nuclei, both stable and radioactive, in the ground state formed by the de-excitation of the compound nucleus. In this code, the decay probabilities of an excited compound nucleus are calculated using the level densities of the residual nuclei and the transmission coefficients for different decay channels. Similar to the code PACE4, this code also does not take into account the possibility of pre-equilibrium emission and/or incomplete fusion. However, the present version of the code includes fission competition for which the liquid drop fission barrier is assumed. Sub-routines TLCALC and MASS calculate respectively, the transmission coefficients for the emitted particles and masses of nuclei of interest, which are permanently stored on the disc. Alpha particle transmission coefficients are calculated using the optical model potential of Satchler[42] while potentials of Becchetti and Greenlees[41] are used for nucleons.

The partial cross-section for the formation of the CN of spin J_{CN} from a projectile a of spin J_a and target A of spins J_A is given by Eq. (2.13(b)) as

$$\sigma^{J_{CN}}_{(a,A,E_a)} = \frac{(2J_{CN}+1)}{(2J_A+1)(2s_a+1)} \sum_{\ell_a^{min}}^{\ell_a^{max}} T^{J_{CN}}_{(a,A,E_a,l_a)} \tag{2.107}$$

In Eq. (2.107), s_a is the channel spin given by $s_a = J_a + \ell_a$, ℓ_a being the orbital angular momentum of the projectile a. Transmission coefficients in Eq. (2.107) are approximated by Fermi distribution as a function of angular momentum. The Fermi gas model is used to calculate the level densities for the product nuclei. At low excitation energies, the parameters can be determined empirically; however, attention is required for the spin dependence of level densities in the region of high excitation. This is because of the high angular momenta involved in heavy ion reactions. Level densities in the Fermi gas model are given by Eq. (2.26(a)) and Eq. (2.26(b)), while the thermodynamic temperature is given by Eq. (2.29). The moment of inertia $\mathcal{I}$ of the nucleus is given as

$$\mathcal{I} = \frac{2}{5}mr^2; \text{ here } m \text{ is the nuclear mass and } r = r_0 A^{1/3}$$

The locus of the minimum rotational energies for different values of the spin J is called the Yrast line and is given by

$$E_{rot}(j) = \frac{j(j+1)\hbar^2}{2\mathcal{I}} + \Delta,$$

(2.108)

here Δ is the pairing energy shift.

Generally, in heavy ion reactions, the excitation energy is high. Therefore, it is required to use energy dependent parameters. In code CASCADE, the total excitation energy is divided into three parts, viz. (i) excitation energy <4 MeV, experimentally known energy levels are used, particularly for light nuclei; (ii) in the range of 4 MeV < excitation energy < 10 MeV, the analytical level density formula is used, the level density parameter a and Δ for each nucleus may be calculated individually; (ii) at still higher excitation energies, it is assumed that all nuclei behave like a liquid drop and the value of parameter a is taken as A/8 MeV^{-1}. The pairing shift Δ in this region of energy may be calculated using one of the following options: formulas given by Ramamurthy et al.[43], Dilg et al.[44] or Myers–Swiatecki[45].

2.4.3 Computer codes GNASH and McGNASH

The GNASH code calculates both the compound and the pre-equilibrium components of reaction cross-sections. The compound component is calculated using the Hauser–Feshbach formalism. The pre-equilibrium component is simulated using the exciton model PRECO-B code developed by Kalbach[47] that takes into account complete conservation of angular momentum. Transmission coefficients of particles and low-lying level schemes of intermediate nuclides required for calculations are introduced in the code as external input files. The code is also capable of calculating cross-sections for direct processes if additional data is provided via external input files.

Calculations for the PE component using the exciton model requires the matrix element for two-body residual interaction that brings about transitions from less complex to more complex exciton configurations. The matrix element has been parameterized in the code as follows:

$$|M|^2 = \frac{K}{A^3\varepsilon}\sqrt{\frac{\varepsilon}{7\text{ MeV}}}\sqrt{\frac{\varepsilon}{2\text{ MeV}}} \quad \text{for } \varepsilon < 2\text{ MeV}$$

$$= \frac{K}{A^3\varepsilon}\sqrt{\frac{\varepsilon}{7\text{ MeV}}} \quad \text{for } 2 < \varepsilon < 7\text{ MeV}$$

$$= \frac{K}{A^3\varepsilon}\sqrt{\frac{15\text{ MeV}}{\varepsilon}} \quad \text{for } \varepsilon > 15\text{ MeV}$$

(2.109)

where ε = excitation energy per nucleon of the compound nucleus = Uc/n and Uc is expressed in MeV. The value of constant K in Eq. (2.103) is generally taken between 130 and 160 MeV. The PE emission of clusters has been taken into account using the Kalbach[47] phenomenological systematic model, which does not consider angular momentum effects.

Level densities of residual nuclei play a crucial role in determining the magnitude of the CN component. In code GNASH, there are three options for calculating level densities: (i) the back shifted Fermi gas model, (ii) the Gilbert–Cameron model and (iii) the Fermi gas model with energy-dependent shell corrections proposed by Ignatyuk. The code automatically adjusts the parameters for each of the aforementioned options so that the information provided for the low lying discrete energy levels as input file is matched. The gamma widths are also adjusted in an analogous way.

McGNASH is a modular code that is being developed at Los Alamos. The code can calculate both the CN and PE cross-sections for reactions induced by photons, neutrons and protons in the energy range of a few keV to 150 MeV. Written in Fortran-95 language, the code extensively uses the concept of modular programming. In fact, the McGNASH code is really a collection of such Fortran modules, each dealing with a specific (and often independent) part of the nuclear reaction sequence calculation. The modules are written with the option of further extension and are capable of being used on many computer platforms without disturbing the inner coding structure.

In order to get default calculations running for the most common nuclear reaction data needs, it is necessary to provide the code with default input data (e.g., discrete level schemes, optical model parameters, etc). This has been achieved by directly linking the RIPL-2 database[48] to McGNASH, hence providing default data for many nuclei and nuclear reactions. Moreover, the 1996 version of the ECIS coupled-channels code[49] is still used to provide McGNASH with the transmission coefficients used in the Hauser–Feshbach equations. ECIS 96 also provides the total and reaction cross-sections, along with inelastic scattering cross-sections and angular distributions for excited discrete levels.

2.4.4 Computer code ALICE 91 and ALICE IPPE

ALICE 91 is an improved version of the code ALICE initially developed by Blann. The code calculates differential cross-sections for both the CN and PE processes. The Weisskopf–Ewing model is used for CN calculations while the hybrid model is used for simulating PE emissions. The hybrid model differential cross-section for PE emission is given by Eqs (2.67) and (2.68). Using the concept of mean free path, Blann obtained the following expression for the transition rate $\lambda_{+(\varepsilon)}$ for the transition from a state of n excitons to one of $(n + 2)$ excitons

$$\lambda_{+(\varepsilon)} = 1.4 \times 10^{21} \left(\varepsilon + B_e \right) - 6.0 \times 10^{18} \left(\varepsilon + B_e \right)^2 \text{ s}^{-1} \tag{2.110}$$

Here B_e is the nucleon binding energy in the nucleus. All energies in Eq. (2.110) are in MeV. However, comparison of experimental data with the calculations based on the hybrid model required that the strength of $\lambda_{+(\epsilon)}$ given by Eq. (2.110) be reduced by a factor of 5 for a good match between the two.

ALICE 91 is capable of calculating reaction cross-sections for all those reactions that leave the residual nuclei up to 9 units of charge and 11 units of mass away from the compound nucleus. The binding energies and Q values for all intermediate nuclides are calculated within the code using the mass formula of Myers–Swiatecki/Lysekil[45]. The Thomas[26] parabolic model is employed for the calculation of heavy ion transmission coefficients and the inverse cross-sections are generated using optical model sub-routines.

The code requires the specification of three free parameters as input. They are: (i) the level density parameter a, which is given as A/K, where K is a constant that may take values from 8 to 10; (ii) the initial exciton number n_0, that depends on the projectile; if the incident nucleus has A nucleons, it is a common practice to take $n_0 = (A + 2)$, assuming that the initial projectile–target interaction produces one hole and $(A + 1)$ particles resulting in $(A + 2)$ excitons; (iii) the mean free path multiplier COST. The mean free path in the code is calculated either using the optical model potential of Becchetti and Greenlees[41] or the free nucleon–nucleon scattering cross-sections with Pauli correction. The free parameter COST is meant to adjust the calculated mean free path so as to reproduce the experimental data.

The code has two options for calculating level densities of product nuclei: (i) using the Fermi gas model or (ii) the constant temperature approach. The level density for the Fermi gas model and the constant temperature approach are respectively given by the following expressions (see Eq. (2.26(a)))

$$\rho(U^*) = \frac{\sqrt{\pi}}{12a^{\frac{1}{4}}U^{*\frac{5}{4}}} e^{\left[2\left(aU^*\right)^{\frac{1}{2}}\right]} \tag{2.111}$$

$$\rho(U^*) \propto \frac{1}{T} e^{\frac{U^*}{T}} \tag{2.112}$$

where, U^* is the excitation energy of the nucleus.

Since in the Weisskopf–Ewing model it is assumed that the nucleus has infinite moment of inertia, no consideration is given to rotational energy and there is no cutoff for the level density at higher spin. Further, the code does not consider angular momentum in HI reactions.

The Obninsk group modified ALICE 91 in two ways: (i) they included PE emission of light clusters like deuterons, tritons and alphas using the formalism of Iwamoto and Harada[52] and (ii) they incorporated the generalized superfluid model[51] of nuclear level density that includes the collective enhancement of level densities due to energy-dependent shell effects. The modified code called ALICE–IPPE uses the ALICE 91 optical model prescription for nucleons but the optical model parameters for alphas and deuterons are slightly modified to reproduce low energy absorption data.

2.4.5 The computer code EMPIRE

EMPIRE is a code that is frequently used for practical evaluation of nuclear reaction data over a wide energy range. The code uses the Hauser–Feshbach–Moldauer model for CN calculations, the exciton model for PE emission and optical and direct reaction formalism for calculating cross-sections in case of direct reactions. In the code version EMPIRE 2.19, coupled channel code ECIS03[49] has been included to perform optical model calculations using the potentials in the RIPL 3 database. To simulate PE emission, there are two options in the code: (i) using the module PCROSS, which uses one component of the exciton model with the emission of clusters, nucleons and gamma rays; (ii) using module HSM that implements the Monte Carlo approach to the hybrid model. There are several options in the code for calculating level densities; however, the default option called EMPIRE specific uses a superfluid model below the critical temperature and the Fermi gas model above it. The code can take into account corrections to the level densities arising due to nuclear rotation, vibration, their temperature dependence, deformation-dependent collective effects, shell corrections, pairing etc.

Experimental Details and Formulations

3.1 Introduction

As has been mentioned in the introductory chapter, the initial interaction between a projectile and the target may result in the formation of an excited composite system from which nucleons or clusters may be emitted before a completely fused compound nucleus is formed. Such a process is generally referred to as the pre-compound emission (in case of nucleonic emissions) or incomplete fusion (when cluster emission takes place). Incomplete fusion/PE-emissions become more important as the incident beam energy increases; in fact, they become dominant at energies above 15 MeV/n. The measurement and analysis of excitation functions for the population of reaction residues may provide valuable information regarding the dynamics of incomplete fusion reactions. The resulting product nucleus of incomplete fusion has a momentum that is severely reduced as compared to the residues of complete fusion events. The measurement and analysis of momentum transfer via recoil range distribution is one of the most direct and irrefutable method of identifying incomplete fusion events. Details of the measurement of linear recoil range distributions (RRD) will be discussed later in the chapter. In incomplete fusion (ICF), residues recoil before the establishment of a thermodynamic equilibrium, and therefore, carry information about the initial system parameters that is reflected in the angular distribution of residues. Details of the measurement and analysis of residue angular distributions will also be presented in this chapter. In a typical experiment, residues are formed via complete fusion as well as via incomplete fusion processes. The product residues of complete fusion carry larger excitation energy and higher spin angular momentum when compared to the residues populated via incomplete fusion. This difference in their properties affects the spin distributions of their excited levels. In order to further investigate such systems and study the role of input angular momenta in ICF reactions, in-beam experiments involving particle–gamma coincidence method have been performed. Details of these experiments will be presented in the following sections. In recent years, incomplete fusion reactions have been observed even at energies as low as 3 – 7 MeV/n, where only complete fusion is likely to dominate. The present monograph deals with the description of such reactions in the low energy regime. Pre-equilibrium emission of nucleons in complete fusion reactions has been observed at energies > 15 MeV/n and is unlikely to occur at energies below it. In our experiments, at energies ≈ 3 – 7 MeV/n, we have observed pre-equilibrium emission of nucleons, particularly in those cases that are products of complete fusion. Other experiments carried out by different groups of workers have also confirmed our findings. Further, a new method of deciphering pre-equilibrium emission from the recoil range distribution data has been developed.

Before providing further details of experiments and their analysis, it is necessary to develop the necessary formulations required to deduce information of interest from experimental data. These formulations will be explained in the following sections.

3.2 Formulations for Measuring Cross-section

The production cross-section $\sigma_r(E)$ is a measure of the probability of the formation of a particular reaction product. Experimentally, the cross-section of a reaction $X(a,b)Y$ may be defined as the number of events of a given type that occur per unit area per unit target nucleus per unit time. Cross-section is generally expressed in units of barn which is equal to 10^{-28} m². As such, the reaction cross-section may be represented as,

$$\sigma_r(E) = \frac{\text{Number of events of a given type } X\left(a,b\right)Y \text{ / area}}{N_0\,\phi\,t} \tag{3.1}$$

where, N_0 is the number of nuclei in the target, ϕ is the incident beam flux and t is the irradiation time. In order to determine the cross-section of a particular reaction, the quantities given in the denominator of Eq. (3.1) have to be known and the quantities in the numerator, i.e., the number of events of a given type $X(a,b)Y$ are required to be measured. This is done either by employing off-beam or in-beam techniques.

A detailed formulation for determining the cross-section of a nuclear reaction, where the reaction products are radioactive nuclei, called activation analysis technique is given here. It may be mentioned that when a sample is irradiated by a beam of particles, then many reactions may be initiated, and many isotopes are likely to be formed by the process of nuclear transmutation. The rate of formation of a particular radioactive nuclide may be given by the expression[1],

$$N = N_0 \Phi \sigma_{X(a,b)Y} \tag{3.2}$$

where, $\sigma_{X(a,b)Y}$ is the reaction cross-section or activation cross-section for the particular channel as it is sometimes called.

If some of the isotopes so formed are radioactive, Eq. (3.2), will have to be modified to take into account their simultaneous decay. The disintegration rate of the induced activity in a sample after a time t from the stop of irradiation by the beam may be given by the expression;

$$\left[\frac{dN}{dt}\right]_t = N\frac{[1-e^{(-\lambda t_1)}]}{e^{(\lambda t)}} \tag{3.3}$$

where t_1 is the duration (time) of beam irradiation and λ is the disintegration constant of the induced activity, given as

$$\lambda = \frac{\ln 2}{t_{1/2}} \tag{3.4}$$

The factor $[1- \exp(-\lambda t_1)]$ takes care of the decay of reaction residues during the irradiation of the target by a beam of incident particles and is typically called the saturation correction. The number of decays of the induced radioactivity in a very small time dt may be given as

$$dN = N\frac{[1-e^{(-\lambda t_1)}]}{e^{(\lambda t)}}dt \tag{3.5}$$

If the radioactivity induced in the irradiated target sample is recorded/measured for a time of duration t_3, after a lapse of time t_2, then the number of nuclides decayed in the time interval between t_2 and $(t_2 + t_3)$ may be given as

$$C = N\frac{[1-e^{(-\lambda t_1)}][1-e^{(-\lambda t_3)}]}{\lambda.e^{(t_2)}} \tag{3.6}$$

In an actual experiment, the activity induced in the sample is generally observed by a suitable detector. In case the detector is a gamma ray spectrometer of efficiency G_ε, then absolute count rate C and observed counting rate A may be related as

$$C = \frac{A}{\theta.K.G_\varepsilon} \tag{3.7}$$

where, A is the total number of counts of the induced activity recorded during the accumulation time t_3 and θ is the branching ratio of the characteristic γ-ray identified by the detecting spectrometer. λ in Eq. (3.6) is the decay constant of the activity. A fraction of the gamma rays emitted by the radioactive isotope produced in the irradiated sample is likely to be absorbed within the sample itself. The correction factor K for this self-absorption of gamma rays in the thickness of the target is given as $K = [\{1 - \exp(-\mu d)\}/\mu d]$, where d (gm/cm^2) is the sample thickness and μ (cm^2/gm) is the absorption coefficient of the gamma ray in the material of the sample. In Eq. (3.7), G_ε is the geometry dependent efficiency of the detector.

Thus, $\sigma_{X(a,b)Y}$ can be written as,

$$\sigma_{X(a,b)Y}(E) = \frac{A.\lambda e^{(\lambda t_2)}}{N_0.\Phi.\theta.K.G_\varepsilon.\left[1 - e^{(-\lambda t_1)}\right]\left[1 - e^{(-\lambda t_3)}\right]} \quad (3.8)$$

Moreover, the count rate when irradiation stops, $C_{t=0}$ may be given as

$$C_{t=0} = \frac{A.\lambda e^{(\lambda t_2)}}{\left[1 - e^{(-\lambda t_3)}\right]} \quad (3.9)$$

The reaction cross-section $\sigma_{X(a,b)}$ may be written with the help of the aforementioned equations as

$$\sigma_{X(a,b)}(E) = \frac{C_{t=0}}{N_0.\Phi.\theta.K.G_\varepsilon.\left[1 - \lambda e^{(-\lambda t_1)}\right]} \quad (3.10)$$

This expression has been used for the determination of reaction cross-sections in off-beam experiments.

In general, several gamma rays of different energies are emitted by a radioactive nuclide; the observed intensity of each emitted gamma may be independently used to calculate the cross-section for the production of the same radionuclide. In our measurements, cross-section for the production of a specific radioactive residue has been calculated from the measured intensities of different gamma rays associated with the specific residue; finally, the weighted average of the measured cross-section values has been taken. The following formulation has been used for the determination of the weighted average.

If $Y_1 \pm \Delta Y_1$, $Y_2 \pm \Delta Y_2$, $Y_3 \pm \Delta Y_3$, ... are the different measured values of the same quantity Y, then the weighted average[2] is given as

$$\overline{Y} = \frac{\sum W_i Y_i}{\sum W_i} \quad (3.11)$$

where,
$$W_i = \frac{1}{(\Delta Y_i)^2} \quad (3.12)$$

and the internal error $= \left[\sum W_i\right]^{-1/2} \quad (3.13)$

while the external error $= \left[\frac{\sum W_i(\overline{Y} - Y_i)^2}{n(n-1)\sum W_i}\right]^{-1/2} \quad (3.14)$

Equation (3.13) depends entirely on the errors of individual observations, whereas Eq. (3.14) also depends upon the differences between observations and the mean value. External error is, therefore, a function of what might be called the external consistency of observation whereas internal error depends only upon the internal consistency. The experimentally measured values of cross-sections reported in this monograph are the weighted average of cross-sections obtained from intensities of various gamma rays associated with the given residue and the error is either internal or external, whichever is larger.

In order to carry out fast calculations, one may develop a computer program based on the aforementioned formulations and determine the cross-section values for various reactions. In the present work, a FORTRAN program EXP-SIGMA[3] based on these equations has been used to determine the reaction cross-sections of the residues populating various projectile target interactions at the respective irradiated energies.

3.3   Experimental Details

Broadly speaking, the experiments reported in this monograph may be divided into two categories, namely, off-beam experiments and in-beam experiments. In-beam experiments are performed to study reactions where the residues of interest are either extremely short-lived or so stable that they may be identified only through their prompt gamma rays. In-beam experiments are more involved, since they require a lot of fast electronics to select the residue of interest. Further, the detectors used in in-beam experiments are subjected to a heavy flux of undesirable neutrons etc., produced in the various side reactions. Another important factor is the cost of the beam and its availability. It is, therefore, desired to keep in-beam experiments to a minimum so as to reduce the use of beam time. On the other hand, one may use the radioactive isotopes of measurable half-lives produced in an irradiation and follow their decay in an off-beam experiment to identify the reaction product. Off-beam experiments are relatively simple, may be continued for a few days after the irradiation stops and yield information of considerable value about the reaction dynamics.

3.3.1 Off-beam experiments

This technique is also referred to as the activation technique. The activation technique is a method of measuring nuclear reaction cross-sections by observing the activities induced in the sample by beam irradiations. The unique decay mode of each residual radioisotope provides a means to identify and measure it. Generally, several activities, due to various residual radio-isotopes of different reactions are produced in the irradiated sample. The cross-section for more than one reaction can, therefore, be measured in the same experiment. In the activation method, however, the analysis sometimes becomes complicated due to the interfering reactions. Still, the activation analysis is quite simple and accurate. It has been widely used in various fields of research work. This is mainly due to the substantial improvement in the detection techniques and nuclear electronics. The advantages of the technique include the fact that it is relatively simple, it is non-destructive and requires less expensive setups. However, the technique is limited only for reaction products having measurable half-lives. The stacked foil activation technique, which is an extension of the activation technique, is generally used for the measurement of cross-sections for specific reactions at different energies, i.e., for the measurement of excitation functions. The spinning wheel technique has also been used for the same purpose. In the stacked foil technique, a stack of sample foils with energy degraders in between them is irradiated in a fixed geometry. In this way, successive samples of the same stack are irradiated at decreasing incident energies. On the other hand, in the spinning wheel technique, individual samples with energy degraders are fixed separately round the face of a wheel at the same radius. The rotation of the wheel exposes, in turn, each sample to a beam for a fixed period of time. Since in the spinning wheel technique, only one sample is irradiated at a time, changes in the beam profile and fluctuations in the beam current etc., may introduce uncertainties in calculations for different samples. On the other hand, in the stacked foil technique, all the samples in a stack are irradiated during a single run and by the same beam; hence, beam parameters remain the same for all the samples in the stack. A typical sample stack contains three to four samples of the material of interest with degraders sandwiched between the samples. The stack assembly is fixed in a sample holder, which is kept normal to the beam direction. The beam with full energy E_0 irradiates the front sample; then, the energy of the beam is reduced by ΔE_1 when it passes through the sample as well as the first energy degrader; therefore, the second sample is irradiated by a beam of energy $(E_0 - \Delta E_1)$. Thus, successive samples in the stack are irradiated by beams of energies $(E_0 - \Delta E_1 - \Delta E_2)$, $(E_0 - \Delta E_1 - \Delta E_2 - \Delta E_3)$, etc., where ΔE_1, ΔE_2, ΔE_3, ... are the energy losses in respective degraders and/or foils. The degraders are either a single foil of some material like carbon or aluminium or a bunch of thin foils that reduce the beam energy but do not interfere with the reactions of interest in the sample. The energy loss ΔE_i may be calculated from the known stopping power of the degrading material and its thickness. In the present measurements for the excitation functions of various systems, the stacked foil technique has been employed. As mentioned earlier, in the stacked foil technique, the target foil and the degraders are successively arranged for the measurement of excitation functions, as shown in Figure 3.1.

The experiments are carried out using both alpha cluster as well as non-alpha cluster beams. The ion beams are produced from the Pelletron accelerator[4,5]. A brief description of the Pelletron accelerator setup at the Inter University Accelerator Centre (IUAC), New Delhi, India, used in the present work, is given here for the sake of completeness.

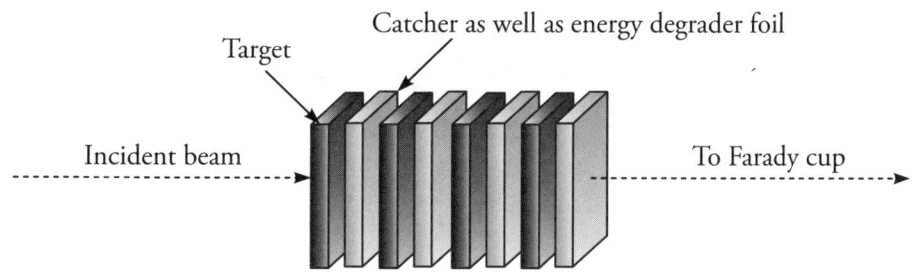

Figure 3.1 A typical stack arrangement for the measurement of excitation functions

3.3.2 Pelletron accelerator at the IUAC, New Delhi

A particle accelerator is a versatile instrument that can be used to study the nature of matter and energy. The IUAC, New Delhi, Pelletron[4,5] is a 15UD, tandem electrostatic accelerator and is capable of accelerating any ion (independent of its mass and charge state) from protons to uranium ions in the energy region from a few tens of MeV to a few hundred MeV, depending on the ion species. The accelerator is installed in a vertical configuration in a huge stainless steel tank 26.5 m in height and 5.5 m in diameter. The tank is filled with a high dielectric constant gas SF_6 at $\approx$ 7–8 atmospheric pressure to insulate the high voltage terminal from the tank wall and prevent the breakdown of high voltage. In the middle of the tank, there is a high voltage terminal, which can hold a potential up to 16 million volts (MV). Since 16 MV is quite a high potential, a special technique of charging the terminal is adopted using the Pelletron charging chain.

The basic principle of acceleration of charged particles with this accelerator is similar to that of the Van de Graaff generator, except for a novel feature in which it uses the accelerating voltage twice; hence, it is called a tandem accelerator. Once the terminal is charged to a high voltage, it may be used for accelerating any ion beam. A typical layout of the Pelletron is shown in Figure 3.2. An inside view of the tank of the Pelletron accelerator is shown in Figure 3.3.

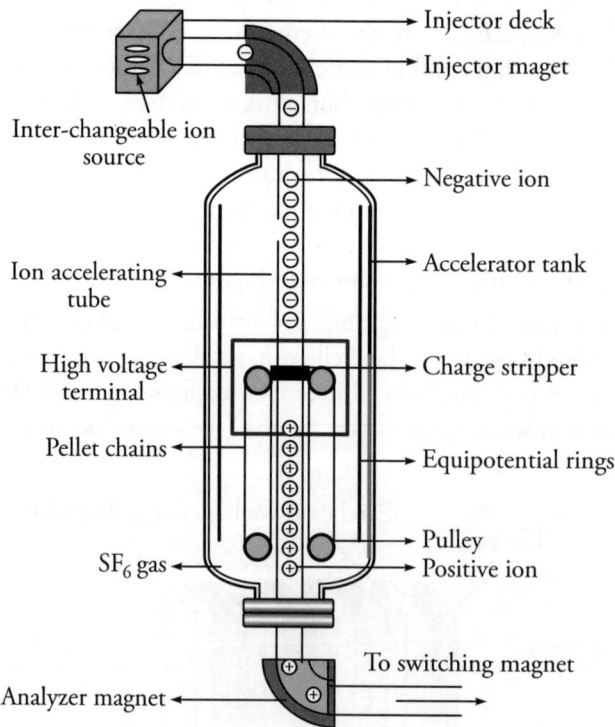

Figure 3.2 A schematic diagram of the Pelletron accelerator at the IUAC

Figure 3.3 Inside view of the tank of the Pelletron accelerator

By attaching an extra electron to the neutral atoms, negatively charged ions are produced in the ion source. The negatively charged ions are injected at the ground potential to the accelerator with the help of an injector magnet and the beam is accelerated towards the terminal at high positive potential, increasing its energy to eV_t (where, V_t is the terminal potential in million volts). At the terminal, these ions pass, either through a thin carbon foil or some gas used as a stripper, which strips off at least a few electrons from each negative ion, thereby, converting it to a positive ion. Since the terminal is at high positive potential, the positive ions formed after stripping are now repelled and accelerated below the terminal to ground potential. Thus, as a result of stripping, a positive ion beam with a distribution of charged states emerges from the stripping cell. Depending on the type of the ion, the type of the stripper, the pressure of the gas, if a gas stripper is used, etc, one or two charge states of the positive ions have maximum intensity, called the most probable charged state(s). A quadrupole magnet is used to select the ions of desired charged state for further acceleration. If the charged state of a positive ion after passing through the stripper at the terminal is q, then the energy gained in acceleration below the terminal to the ground potential is qV_t. Therefore, after passing through two stages of acceleration, the final energy of the ion beam is given by,

$$E_{final} = E_0 + (q+1)V_t \text{ MeV} \tag{3.15}$$

where, E_0 is the energy of the ion before acceleration by terminal voltage V_t and q is the charged state of the ion after passing through the stripper. Since, $E_0 << E_{final}$, it may be neglected. As such, Eq. (3.15) may be written as

$$E_{final} = (q+1)V_t \text{ MeV} \tag{3.16}$$

As an example, if the charged state of ^{16}O is 6+ and the terminal potential is 12 MV, then the final energy (E_{final}) of the ions obtained from Eq. (3.16) will be 84 MeV.

These high-energy ions are then passed through the analyzing magnet and energy slits which select particular ions of desired energy. With the help of switching magnets, the beam of ions is then directed towards the desired experimental area[5]. At the IUAC accelerator facility, dedicated experimental facilities are located in six beam lines for research in focused areas of nuclear physics, material science, biological science, atomic physics, etc.

In the present work, various complimentary experiments have been performed. In the first set of experiments, the excitation functions for several projectile–target combinations have been measured. The analysis of these excitation functions provided information of considerable value about the reaction dynamics. In order to confirm the incomplete fusion processes in HI (heavy ion) interactions, the recoil range distribution of heavy residues populated via CF and/or ICF have been measured. In another set of experiments, the angular distribution of heavy residues populated in CF and/or ICF processes have been measured. Further, in order to get information regarding the role of angular momentum in ICF reactions, in-beam experiments involving particle–gamma coincidence have been carried out. Details of the experimental setup required for each of these experiments are provided in the following sections of the chapter.

3.3.3 Experimental details for the measurement of excitation functions

The cross-section of a particular reaction depends on the incident beam energy. The variation of the cross-section for a particular reaction as a function of incident beam energy is generally referred to as the excitation function (EF). Several experiments have been carried out to measure excitation functions (EFs). Systems for which the excitation functions have been measured in the present set of experiments are listed in Table 3.1. As can be seen from this table, a comprehensive range of projectile-target combinations and energies have been covered in these measurements.

Table 3.1   List of reactions and relevant parameters for the study of excitation functions

S. No.	Projectile	Energy range (MeV)	Target	Abundance of target	Composite system	Reactions
1	$^{12}_{6}$C	$\approx$ 42–82	$^{128}_{52}$Te	87%	$^{140}_{58}$Ce	^{128}Te(^{12}C, 3n) ^{137m}Ce
						^{128}Te(^{12}C, 5n) ^{135}Ce
						^{128}Te(^{12}C, p4n) ^{135}La
						^{128}Te(^{12}C, α3n) ^{133m}Ba
						^{128}Te(^{12}C, α5n) ^{131}Ba
						^{128}Te(^{12}C, α4pn) ^{131m}Te
2.	$^{12}_{6}$C	$\approx$ 55–80	$^{165}_{67}$Ho	99.9%	$^{177}_{73}$Ta	^{165}Ho(^{12}C, 3n) ^{174}Ta
						^{165}Ho(^{12}C, 4n) ^{173}Ta

						^{165}Ho(^{12}C, 5n) ^{172}Ta
						^{165}Ho(^{12}C, p3n) ^{173}Hf
						^{165}Ho(^{12}C, p5n) ^{171}Hf
						^{165}Ho(^{12}C, α2n) ^{171}Lu
						^{165}Ho(^{12}C, α4n) ^{169}Lu
						^{165}Ho(^{12}C, α6n) ^{167}Lu
						^{165}Ho(^{12}C, 2α2n)^{167}Tm
						^{165}Ho(^{12}C, 2α4n)^{165}Tm
3.	$^{12}_{6}$C	≈ 59–88	$^{159}_{65}$Tb	100%	$^{171}_{71}$Lu	^{159}Tb(^{12}C, 3n) ^{168}Lu^{g+m}
						^{159}Tb(^{12}C, 4n) ^{167}Lu
						^{159}Tb(^{12}C, 6n) ^{165}Lu
						^{159}Tb(^{12}C, p3n) ^{167}Yb
						^{159}Tb(^{12}C, α2n) ^{165}Tm
						^{159}Tb(^{12}C, α4n) ^{163}Tm
						^{159}Tb(^{12}C, 2α2n) ^{161}Ho
						^{159}Tb(^{12}C, 2α3n) ^{160}Ho
4.	$^{12}_{6}$C	≈ 56–89	$^{169}_{69}$Tm	100%	$^{181}_{75}$Re	^{169}Tm(^{12}C, 5n) ^{176}Re
						^{169}Tm(^{12}C, p4n) ^{176}W
						^{169}Tm(^{12}C, α3n) ^{174}Ta
						^{169}Tm(^{12}C, 2α2n) ^{171}Lu
5.	$^{12}_{6}$C	≈ 55–85	$^{175}_{71}$Lu	97.41%	$^{187}_{77}$Ir	^{175}Lu(^{12}C, 4n) ^{183}Ir
						^{175}Lu(^{12}C, 5n) ^{182}Ir
						^{175}Lu(^{12}C, p4n) ^{182}Os
						^{175}Lu(^{12}C, α2n) ^{181}Re
						^{175}Lu(^{12}C, α4n) ^{179}Re
						^{175}Lu(^{12}C, 2α3n)^{176}Ta
6	$^{13}_{6}$C	≈ 65–86	$^{169}_{69}$Tm	100%	$^{182}_{75}$Re	^{169}Tm(^{13}C, 3n) ^{179}Re
						^{169}Tm(^{13}C, 4n) ^{178}Re
						^{169}Tm(^{13}C, 5n) ^{177}Re
						^{169}Tm(^{13}C, 6n) ^{176}Re
						^{169}Tm(^{13}C, p4n) ^{177}W
						^{169}Tm(^{13}C, α3n) ^{175}Ta
						^{169}Tm(^{13}C, α4n) ^{174}Ta
						^{169}Tm(^{13}C, α5n) ^{173}Ta
						^{169}Tm(^{13}C, 2α2n) ^{172}Lu
						^{169}Tm(^{13}C, 2α3n) ^{171}Lu

7.	$^{14}_{7}N$	≈ 64–90	$^{128}_{52}Te$ 87%	$^{142}_{59}Pr$	$^{128}Te(^{14}N, 4n)$ ^{138m}Pr
					$^{128}Te(^{14}N, 5n)$ ^{137}Pr
					$^{128}Te(^{14}N, p4n)$ ^{137g}Ce
					$^{128}Te(^{14}N, \alpha5n)$ ^{133}La
					$^{128}Te(^{14}N, \alpha6n)$ ^{132g}La
					$^{128}Te(^{14}N, \alpha2pn)$ ^{135m}Cs
					$^{128}Te(^{14}N, 2\alpha2pn)$ ^{131}I
					$^{128}Te(^{14}N, 3\alpha)$ ^{130g}I
8.	$^{16}_{8}O$	≈ 58–94	$^{27}_{13}Al$ 99.9%	$^{43}_{21}Sc$	$^{27}Al(^{16}O, 2\alpha n)$ ^{34}Cl
					$^{27}Al(^{16}O, 3\alpha3p)$ ^{28}Mg
					$^{27}Al(^{16}O, 3\alpha3pn)$ ^{27}Mg
					$^{27}Al(^{16}O, 4\alpha2pn)$ ^{24}Na
					$^{27}Al(^{16}O, 4\alpha3p)$ ^{24}Ne
9.	$^{16}_{8}O$	≈ 70–95	$^{159}_{65}Tb$ 100%	$^{175}_{73}Ta$	$^{159}Tb(^{16}O, 3n)$ ^{172}Ta
					$^{159}Tb(^{16}O, 4n)$ ^{171}Ta
					$^{159}Tb(^{16}O, 5n)$ ^{170}Ta
					$^{159}Tb(^{16}O, p3n)$ ^{171}Hf
					$^{159}Tb(^{16}O, p4n)$ ^{170}Hf
					$^{159}Tb(^{16}O, 2p2n)$ ^{171}Lu
					$^{159}Tb(^{16}O, \alpha n)$ ^{170}Lu
					$^{159}Tb(^{16}O, \alpha2n)$ ^{169}Lu
					$^{159}Tb(^{16}O, 2\alpha2n)$ ^{165}Tm
10.	$^{16}_{8}O$	≈ 71–95	$^{169}_{69}Tm$ 100%	$^{185}_{77}Ir$	$^{169}Tm(^{16}O, 3n)$ ^{182}Ir
					$^{169}Tm(^{16}O, 4n)$ ^{181}Ir
					$^{169}Tm(^{16}O, p2n)$ ^{182}Os
					$^{169}Tm(^{16}O, p3n)$ ^{181g}Os
					$^{159}Tb(^{16}O, 2p2n)$ ^{181}Re
					$^{169}Tm(^{16}O, \alpha3n)$ ^{178}Re
					$^{169}Tm(^{16}O, 2\alpha pn)$ ^{175}Hf
					$^{169}Tm(^{16}O, 3\alpha n)$ ^{172}Lu
11.	$^{16}_{8}O$	≈ 85–100	$^{181}_{73}Ta$ 100%	$^{197}_{81}Tl$	$^{181}Ta(^{16}O, 3n)$ $^{194}Tl^{g+m}$
					$^{181}Ta(^{16}O, 4n)$ $^{193}Tl^{g+m}$
					$^{181}Ta(^{16}O, 5n)$ $^{192}Tl^{g+m}$
					$^{181}Ta(^{16}O, p3n)$ $^{193}Hg^{g+m}$
					$^{181}Ta(^{16}O, p4n)$ ^{193}Hg
					$^{181}Ta(^{16}O, p5n)$ $^{191}Hg^{g+m}$
					$^{181}Ta(^{16}O, \alpha n)$ ^{192}Au

						^{181}Ta(^{16}O, α2n) ^{191}Au
						^{181}Ta(^{16}O, α3n) ^{190}Au
12.	$^{16}_{8}$O	≈ 50–85	$^{103}_{45}$Rh	100%	$^{119}_{53}$I	^{103}Rh(^{16}O, pn) ^{117}Te
						^{103}Rh(^{16}O, p2n) ^{116}Te
						^{103}Rh(^{16}O, p3n) ^{115}Te^{g+m}
						^{103}Rh(^{16}O, p4n) ^{114}Te
						^{103}Rh(^{16}O, 2p) ^{117}Te
						^{103}Rh(^{16}O, 2p2n) ^{115}Sb
						^{103}Rh(^{16}O, 2α) ^{111g}In
						^{103}Rh(^{16}O, 2αn) ^{110}In^{g+m}
						^{103}Rh(^{16}O, 2α2n) ^{109g}In
						^{103}Rh(^{16}O, 2α3n) ^{108}In^{g+m}
						^{103}Rh(^{16}O, 3αn) ^{106m}Ag
						^{103}Rh(^{16}O, 3α3n) ^{104g}Ag
						^{103}Rh(^{16}O, 3α4n) ^{103g}Ag
13.	$^{18}_{8}$O	≈ 70–90	$^{159}_{65}$Tb	100%	$^{177}_{73}$Ta	^{159}Tb(^{18}O, 3n) ^{174}Ta
						^{159}Tb(^{18}O, 4n) ^{173}Ta
						^{159}Tb(^{18}O, 5n) ^{172}Ta
						^{159}Tb(^{18}O, 6n) ^{171}Ta
						^{159}Tb(^{18}O, α2n) ^{171}Lu
						^{159}Tb(^{18}O, α5n) ^{168m}Lu
						^{159}Tb(^{18}O, α6n) ^{167}Lu
14.	$^{19}_{9}$F	≈ 80–110	$^{159}_{65}$Tb	100%	$^{178}_{74}$W	^{159}Tb(^{19}F, 4n) ^{174}W
						^{159}Tb(^{19}F, 5n) ^{173}W
						^{159}Tb(^{19}F, 6n) ^{172}W
						^{159}Tb(^{19}F, p4n) ^{173}Ta
						^{159}Tb(^{19}F, α3n) ^{171}Hf
						^{159}Tb(^{19}F, α4n) ^{170}Hf
						^{159}Tb(^{19}F, αp3n) ^{170}Lu
						^{159}Tb(^{19}F, 2α3n) ^{167}Yb
						^{159}Tb(^{19}F, 2αp2n) ^{167}Tm
						^{159}Tb(^{19}F, 2αp4n) ^{165}Tm
15.	$^{19}_{9}$F	≈ 80–110	$^{169}_{69}$Tm	100%	$^{188}_{78}$Pt	^{169}Tm(^{19}F, 3n) ^{185}Pt
						^{169}Tm(^{19}F, 4n) ^{184}Pt
						^{169}Tm(^{19}F, 5n) ^{183}Pt
						^{169}Tm(^{19}F, p3n) ^{184}Ir
						^{169}Tm(^{19}F, p4n) ^{183}Ir

					^{169}Tm(^{19}F, αn) ^{183}Os
					^{169}Tm(^{19}F, α2n) ^{182}Os
					^{169}Tm(^{19}F, α3n) ^{181}Os
					^{169}Tm(^{19}F, α5n) ^{179}Os
					^{169}Tm(^{19}F, 2α3n) ^{177}W
					^{169}Tm(^{19}F, 2α5n) ^{175}W
					^{169}Tm(^{19}F, 2α6n) ^{174}W
					^{169}Tm(^{19}F, 2αp3n) ^{176}Ta
					^{169}Tm(^{19}F, 2αp4n) ^{175}Ta

In most of the experiments reported in literature, alpha cluster beams have been used. However, in the present work, both types of ion beams, that is, alpha cluster type beams (like ^{12}C and ^{16}O) as well as non-alpha cluster beams (^{14}N and ^{19}F) have been used for initiating heavy ion reactions. Attempt has been made to look at the effect of non-alpha cluster beams on the reaction dynamics as compared to alpha cluster beams.

In nuclear physics experiments using accelerated ion beams, the target and, in particular, its thickness affects the radiations emitted by the reaction residues. Therefore, such targets need to be prepared with utmost care.

3.4   Target Preparation

Two types of targets are generally used for bombardment by ion beams – they may either be self-supporting or prepared as a deposition of a thin layer on a thin film of some backing material. Generally, targets made of isotopically pure materials are best suited for nuclear physics experiments as this reduces the number of interfering reactions. Self-supporting samples of isotopically pure materials of thickness $\approx$ 1.0 mg/cm^2 are prepared by the rolling method. A typical photograph of the machine at IUAC, New Delhi used for rolling samples is shown in Figure 3.4. By successive rolling of the sample and increasing the pressure each time, the samples are made thin enough to be used in experiments.

On the other hand, for those materials that are available in the powder form or could not be rolled to the required thickness, the vacuum evaporation technique is used for sample preparation. Here, the sample (in powder form) is evaporated by heating at high temperature inside a vacuum chamber on to a thin metallic or other suitable foil so that it gets deposited on this backing material. The vacuum evaporation technique is also used for preparing very thin targets and the catcher foils used in recoil range distribution measurements. The target thickness required in such measurements is a few tens of μg/cm^2. Since it is very difficult to prepare self-supporting target foils without any backing, for our experiments, thin aluminium foils of suitable thickness, were used as backing material. These backing aluminium foils were prepared by the vacuum evaporation method. For this purpose, cleaned glass slides were used as the substrate and Teepol as the parting agent. Aluminium was deposited on the slides

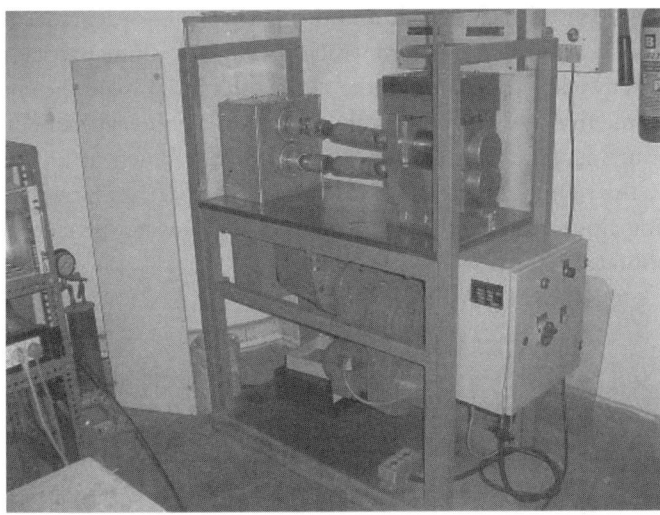

Figure 3.4 The rolling machine used for the preparation of thin samples at the target laboratory of the IUAC

using the electron gun (2 kW) bombardment technique, without disturbing the vacuum inside the chamber. Specially designed tantalum boats were used to keep the sample material (aluminium) inside the chamber for evaporation. A pictorial arrangement of the high vacuum chamber is shown in Figure 3.5. About 100 to 170 mA current was used during evaporation. The thickness of the deposited film was monitored on-line using the quartz crystal monitor setup. After the deposition of required aluminium film, the chamber was allowed to cool for about 4–5 hours; then, the slides were carefully taken out. The material deposited on the glass slides is then floated on water and taken to the sample holders to be used in experiments.

Figure 3.5 The high vacuum chamber used for making samples with the evaporation technique

Further, the thicknesses of each sample and the catcher/energy degrader foil was measured separately by the α-transmission method in which 5.486 MeV alpha particles obtained from a ^{241}Am source were allowed to pass through the sample foil to estimate the energy loss of alpha particles in the sample thickness. A typical block diagram of the experimental setup used for the thickness measurements of samples and catcher foils is shown in Figure 3.6. As may be seen in the figure, the setup consists of an alpha particle spectrometer. The spectrometer is first calibrated using alpha groups of known energies; in this way, the channel number of the spectrometer is calibrated in terms of the alpha energy.

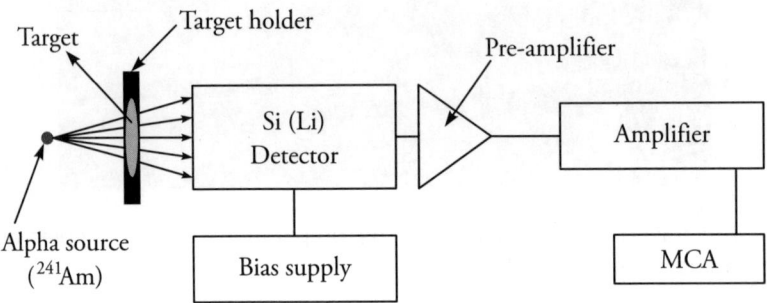

Figure 3.6 Block diagram of the alpha spectrometer used for target thickness measurements

As such, the peak due to 5.486 MeV alpha particles emitted from the ^{241}Am source show up at a particular channel number. Now when a sample whose thickness is to be measured is placed between the alpha source and the detector, the 5.486 MeV alphas lose a part of their energy, ΔE, in the sample thickness and therefore, the alpha particles reaching the detector have energy less than 5.486 MeV. The energy loss ΔE is related to the thickness of the sample. As a result of the energy loss, the position of the alpha peak shifts to a lower channel number. A typical alpha particle spectrum from ^{241}Am recorded using a surface barrier detector without any absorbers is shown in Figure 3.7.

The position of the energy shifted alpha peak is shown in Figure 3.8. From the magnitude of the shift of the peak position, channel number one obtains the energy shift (ΔE) in the peak position. This shift in energy, when divided by the stopping power (dE/dX) of the alpha particles in the material of the sample gives the thickness of the sample as $\Delta E/(dE/dX) = dX$. It may however, be pointed out that the source, the sample whose thickness is to be determined and the Si(Li) detector are kept in a vacuum chamber to avoid energy loss of alpha particles in the air. In general, both in the case of a sample prepared by the rolling method or by vacuum evaporation, the thickness of the sample is not uniform everywhere in the sample. Therefore, it is required to measure the sample thicknesses at various sections of the sample by shifting the sample position with respect to the alpha beam; a mean value of all measured thicknesses is then taken as the sample thickness.

Heavy ion beams carry large linear momentum as compared to the light ion beams; as a result, the reaction residues formed in the projectile target interaction recoil back with

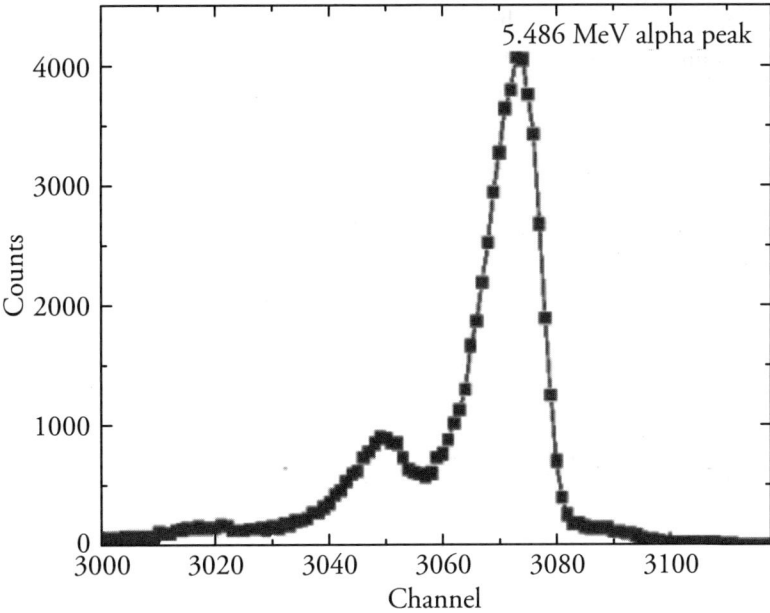

Figure 3.7 5.486 MeV alpha peak of a ^{241}Am source (without absorber)

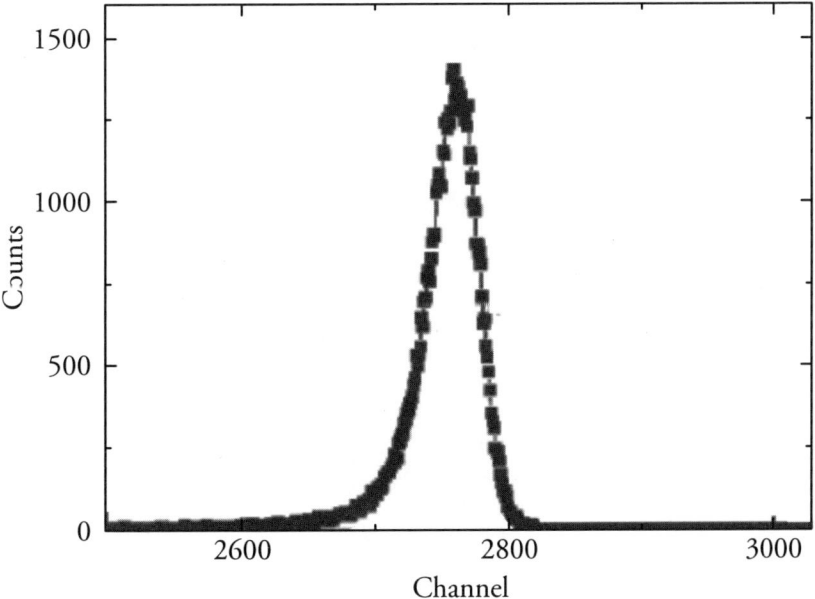

Figure 3.8 Shift in the position of the 5.486 MeV alpha peak of the ^{241}Am source after absorption in the sample

considerable momentum in case of heavy ion interactions. If the half thickness of the sample is large, the recoiling residue may be stopped within the sample. However, in the case of samples made by evaporation, the foil of the backing material on which the sample material was deposited serves both as a catcher for the recoiling residues as well as an energy degrader for reducing the energy of the incident beam. Since in our measurements, samples were made by depositing the sample material on thin aluminium foils, the Al-backing of targets served both as energy degraders as well as catchers for the residues recoiling out of the target foil during the irradiations. All samples were cut into 1.2×1.2 cm^2 squares and pasted on Al-holders having a concentric hole of diameter 1.0 cm. The Al-holders are generally used for rapid dissipation of heat produced during the irradiation. These holders were then screwed to the 'sample holding ladder' of the scattering chamber where irradiations were carried out.

3.5   Sample Irradiation by HI Beam

In the present set of experiments at the IUAC, New Delhi Pelletron facility[4], the irradiations were performed in the general purpose scattering chamber[6] (GPSC), which is installed at 45° beam line in Beam Hall I of the IUAC. The stainless steel scattering chamber has a diameter of 1.5 m. Rotating arms for holding detectors, etc, are provided in the chamber, which also has an in-vacuum target transfer system. Using this in-vacuum transfer facility[7] (ITF), it is possible to quickly remove irradiated samples from the GPSC without disturbing the vacuum inside the scattering chamber. Thus, the time lapse between the end of irradiation and the counting of the samples by the recording spectrometer may be reduced considerably and the induced activities of relatively short half-lives may be recorded. Typical photographs of the GPSC and the ITF are given in Figures 3.9 (a) and (b), respectively.

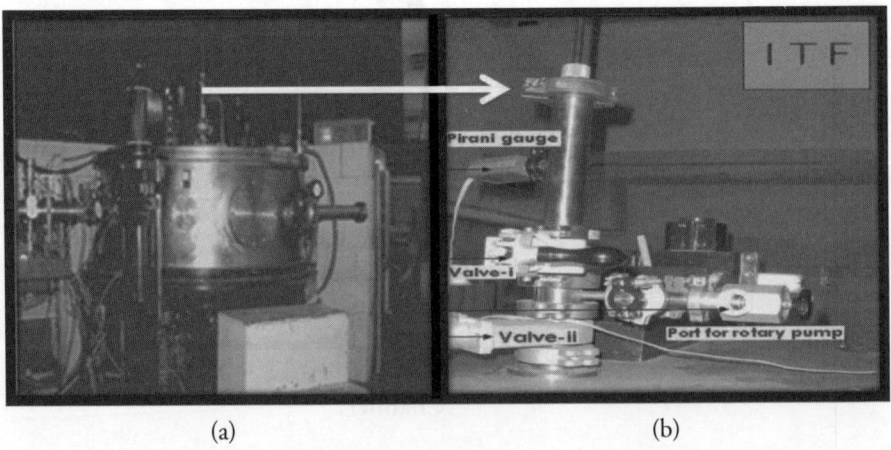

(a)   (b)

Figure 3.9   (a) The general purpose scattering chamber (GPSC); (b) typical arrangement of an in-vacuum transfer facility (ITF), used for quick transfer of irradiated samples without disturbing the vacuum inside the scattering chamber

The incident heavy ion beam enters the GPSC through a port in the chamber and after traversing the chamber along one of its diameter, emerges out into a Faraday cup. A sample holding vertical ladder with holes at different heights may be inserted in the path of the incident beam at the centre of the scattering chamber. It is possible to move the ladder up or down, so as to bring the desired hole in front of the incident beam, and also rotate it on its own axis, using remote control. The stack of the samples or a single sample, as the case may be, is fixed in one of the holes of the ladder and can be brought in a position so that the incident beam hits the sample stack normally, using the remote controls of the ladder.

The flux of the incident ion beam has been determined from the total charge collected in the Faraday cup at the back, behind the sample stack. An ORTEC current integrator has been used for digitalizing and recording of the total charge. In order to test the accuracy of the flux measurement by recording the total charge, an auxiliary experiment was carried out in which two silicon surface barrier detectors D_1 and D_2 (Rutherford monitors) were kept at 30° with respect to the incident beam at forward angles, to record the elastically scattered incident ions. The flux was calculated from the recorded count rates of Rutherford monitors using the well-known Rutherford formula. The magnitudes of the flux of the incident ion beam determined from the counts of Rutherford monitors and from the integrated counts of the Faraday cup were found to agree with each other within 5%.

Irradiation of the sample stack for various projectile–target combinations has been done using the ion beam from the Pelletron accelerator. A brief description of ^{12}C+^{175}Lu and ^{13}C+^{169}Tm systems[8] is given here. Four stacks each containing three samples of ^{175}Lu along with appropriate catcher foils were irradiated with ^{12}C beam at energies ≈ 67, 70, 82 and 84 MeV, whereas the stacks of samples of ^{169}Tm along with appropriate catcher foils were irradiated with ^{13}C beam at energies ≈ 65, 68, 82, 84 and 86 MeV. As the incident beam passes through the stack, it loses its energy both in the target material and in the Al-catcher. As such, successive targets (a combination of three target–catcher foils) of the stack get irradiated at different energies. The energies of the incident ion on successive targets have been calculated using stopping power values obtained from code SRIM[9] based on the range–energy formulations. A typical arrangement of the stack in the target ladder is shown in Figure 3.10. As mentioned earlier, the target ladder, which is attached to the bottom plate of the scattering chamber but electrically insulated from it, may be moved vertically and can be rotated by 360° about its axis. The ladder has several holes at different heights where targets can be mounted for irradiation. A quartz lens is also mounted in one of the ladder holes for viewing the beam. Since the target ladder is insulated from the body of the scattering chamber, it is possible to record the beam current. A co-axial cable attached to the target ladder carries the current information to the data room. The vertical and rotational motions of the target ladder can be controlled from the data room using remote control systems. Several targets may be mounted on the target ladder at different holes before irradiation and any one of them may be moved in front of the beam for irradiation from the data room using remote control devices. The scattering chamber has two rotatable detector arms with remote-controlled precise angle measuring devices. Detectors may be mounted on these arms at a separation of 6°. The angular position of the detector arms may be controlled from the data room with an accuracy of ± 0.05°. Apart from the detector arms, it is possible to mount two detectors at ±10° with respect to the beam at chamber walls

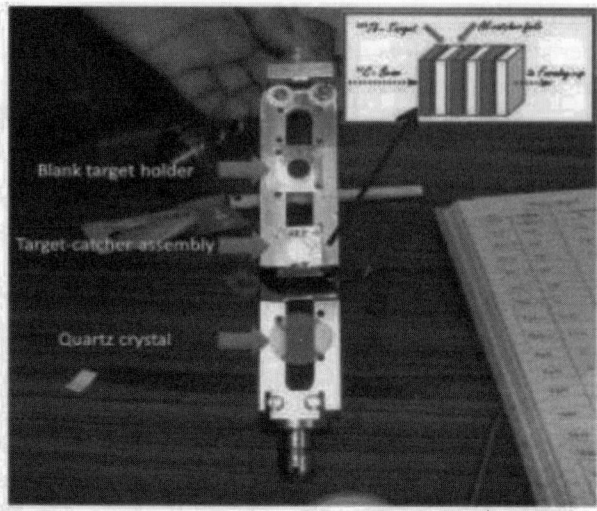

Figure 3.10 The target ladder with a blank holder, target–aluminium catcher assembly and a quartz crystal for tuning the beam. In the inset, an artistic view of the stack is shown

and another two detectors at $\mp 15°$ by hanging them from the scattering chamber top lid. There is provision in the scattering chamber to mount a beam collimator at the beam entrance port of the chamber. A beam dump with an electron-suppressed Faraday cup is attached to the beam exit port of the scattering chamber for beam current measurements.

Accelerated heavy ions carry considerable linear momentum and therefore, when they interact with the target nucleus, large amounts of linear momentum are transferred to the reaction residues. In case of thin targets, it is possible that recoiling residues may fly out of the target. For irradiation, therefore, the samples followed by *Al*-catchers were held normal to the beam direction to ensure that the recoiling residues, if any, may be trapped in catcher foils and are not lost. While counting the irradiated samples for induced activities due to radioactive reaction residues, after the end of irradiation, both the sample and the catcher foils together were placed in front of the detector so that residues both in the sample and in the catcher are counted. This ensures that all reaction residues are counted and no activity is lost. The typical target–catcher foil arrangement for the measurement of EFs is shown in Figure 3.10. As can be seen from this figure, catcher foils of sufficient thickness are placed just after each target to stop the recoils produced in the reaction. Further, as mentioned earlier, each target–catcher assembly was counted together by the detector. Successive targets are bombarded at decreasing energies because of the degradation of the beam energy through each sample and catcher foil; hence, with a single irradiation, one gets samples that are irradiated at different energies. As such, cross-sections at several different energies may be computed. The duration of irradiation in a particular case is decided by the half-lives of the induced activities of interest. As the duration of irradiation is increased, more residues are produced but being radioactive, they also decay in large numbers during the irradiation time. The thumb rule is that maximum activity is induced when the sample is irradiated for a time equal to about 10 times the half-life of interest. It is, however, not always possible to irradiate for such a long time, particularly if the half-life of interest is of a few hours or more. In the present measurements, each sample was irradiated for 8–10 hours with a beam current of ≈ 2–3 pnA, which was kept constant throughout the experiment.

3.6 Post-irradiation Analysis

Irradiation of samples with HI beams induced radioactivity in the samples. Irradiated sample stacks were quickly removed from the scattering chamber (GPSC) using the in-vacuum transfer facility and placed in front of the HPGe detector setup at such a distance that the dead time of counting remained less than 10%. Further details of the detector setup and post-irradiation analysis are provided in the following subsections.

3.6.1 Calibration of HPGe detector and efficiency measurement

After the irradiation of the stack, the target–catcher assembly was taken out from the general purpose scattering chamber (GPSC) employing the in-vacuum transfer facility (ITF) assembly. In order to identify the characteristic γ-rays of residues in the complex γ-ray spectrum, a detector of good resolution with proper calibration is required. In off-beam experiments, the post-irradiation analysis has been carried out using a pre-calibrated γ-ray spectrometer. The γ-ray spectrometer has an HPGe detector of 100 c.c. active volume coupled to a PC through CAMAC-based CANDLE software[10.] The HPGe detector was calibrated both for energy as well as for efficiency by using several standard γ-sources, i.e., ^{22}Na, ^{60}Co, ^{133}Ba, ^{137}Cs and ^{152}Eu of known strengths. A list of the prominent γ-rays of the standard ^{152}Eu source used for energy and efficiency calibration of the γ-ray spectrometer in the present measurements are given in Table 3.2.

Table 3.2   A list of energies and intensities of some of the prominent γ-rays from standard radioactive ^{152}Eu source

γ-ray energy (keV)	Absolute intensity (%)
121.78	28.58
244.69	7.58
344.27	26.54
443.96	2.82
778.90	12.94
867.37	4.24
964.07	14.60

1089.73	1.72
1112.07	13.64
1212.94	1.42
1299.14	1.62
1408.00	21.00

A typical gamma-ray spectrum of a standard gamma source ^{152}Eu, recorded using the HPGe spectrometer is shown in Figure 3.11. The geometry dependent efficiency (G_ε) of the HPGe detector at a given energy has been determined using the following expression;

$$G_\varepsilon = \frac{N_0}{N_m \exp(-\lambda t) I_\gamma} \tag{3.17}$$

where, N_0 is the disintegration rate of the standard γ-source at the time of measurement, N_m is the disintegration rate at the time of manufacture of the source, λ is the decay constant, t is the time elapsed between the manufacture of the source and the start of the counting and I_γ is the branching ratio of the characteristic γ-ray. The resolution of the spectrometer was ≈ 2 keV for a 1.33 MeV γ-ray of ^{60}Co source.

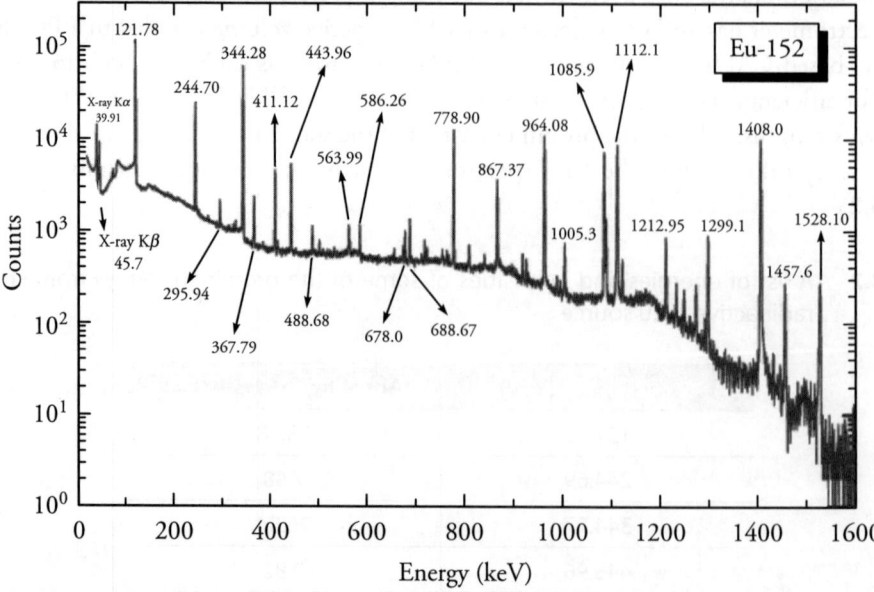

Figure 3.11 Gamma-ray spectrum of a ^{152}Eu source recorded using the HPGe gamma-ray spectrometer. The gamma peak energies are indicated in keV

A typical geometry dependent efficiency plot of the spectrometer, as a function of γ-ray energy for a 1 cm source–detector separation is shown in Figure 3.12. It may be noted that the geometry dependent efficiency of the detector depends on the relative distance between the source and the detector. Curves similar to the one shown in Figure 3.12 but for several different source–detector separations were obtained. This was necessary because the irradiated samples were counted by keeping them at different distances from the detector, depending on the strength of the activities induced in the sample as a result of irradiation. If the strength of the induced activity in the sample was large, the irradiated sample was kept at a larger distance from the detector to reduce the dead time of counting. The guiding criteria was to keep the counting dead time less than 10% for all samples. The experimental data of geometry dependent efficiency was found to be best fitted with a 5th order polynomial function of the type

$$G_\varepsilon = a_0 + a_1 E + a_2 E^2 + a_3 E^3 + a_4 E^4 + a_5 E^5 \qquad (3.18)$$

where, E is the energy of the γ-ray and a_0, a_1, a_2, a_3, a_4 and a_5 are the coefficients having different values for each source–detector separation. In the present work, the standard γ sources used to determine efficiency and the irradiated target–catcher foil assemblies were counted in the same geometry with respect to the detector in order to avoid errors due to solid angle effect. As mentioned earlier, care was taken to keep the dead time of the detector ≤10% by suitably adjusting the source–detector separation for each irradiated sample.

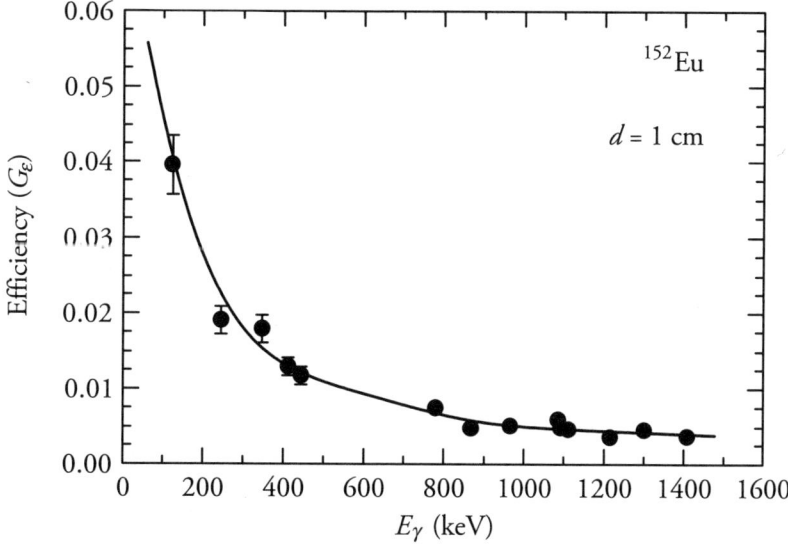

Figure 3.12 A typical geometry dependent efficiency curve as a function of γ-ray energy at a source–detector separation of 1 cm. The solid line represents the best polynomial fit for the data points

3.6.2 Identification of reaction residues

The residues populated via complete and/or incomplete fusion reactions of the target and the incident heavy ions, if radioactive, decay by emitting their characteristic radiations and with specific half-lives. They were identified first by identifying their characteristic gamma rays in the gamma-ray spectrum of the irradiated sample recorded by the calibrated gamma-ray spectrometer. In order to further confirm the identity of the residue, the intensity of the photo peak of the identified characteristic gamma ray was plotted as a function of time to draw the decay curve of the activity. The analysis of the decay curve gave the half-life of the activity. The identity of the residue is finally confirmed if both the measured energy of the characteristic gamma ray and the measured half-live of the recorded activity match literature data. Identification of residues both by its characteristic radiation and its half-live is a fool-proof method of identification. Identifying a residue only by identifying its characteristic gamma ray is not enough as in the activation technique, several residues are produced in the irradiated sample from many different reactions and some of them may have characteristic gamma rays of nearly the same energy. However, when both the characteristic gamma and the half-life of the residue are measured, the identification of the residue becomes unambiguous.

Typical γ-ray spectra of (a) ^{175}Lu bombarded by $\approx$ 78 MeV ^{12}C^{5+} beam[8], and (b) ^{169}Tm bombarded by $\approx$ 83 MeV ^{13}C^{6+} beam[8] are shown in Figures 3.13 and 3.14, respectively.

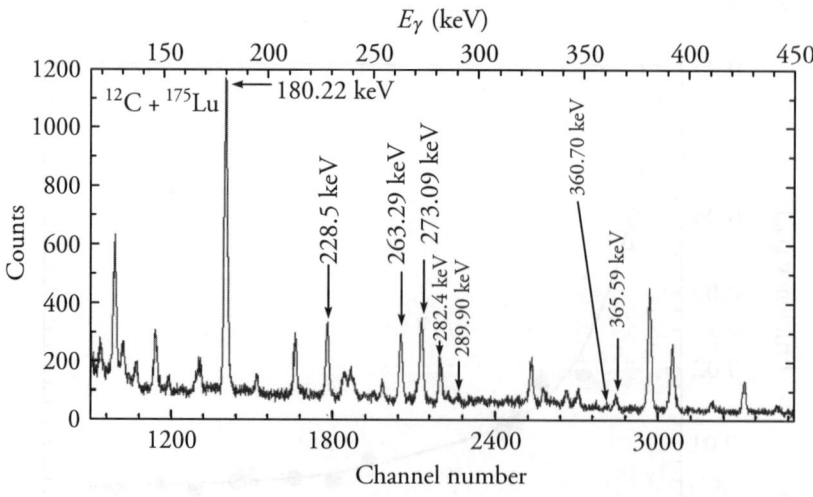

Figure 3.13 Gamma-ray spectrum of residues populated in the reaction ^{12}C+^{175}Lu at $\approx$ 78 MeV beam energy

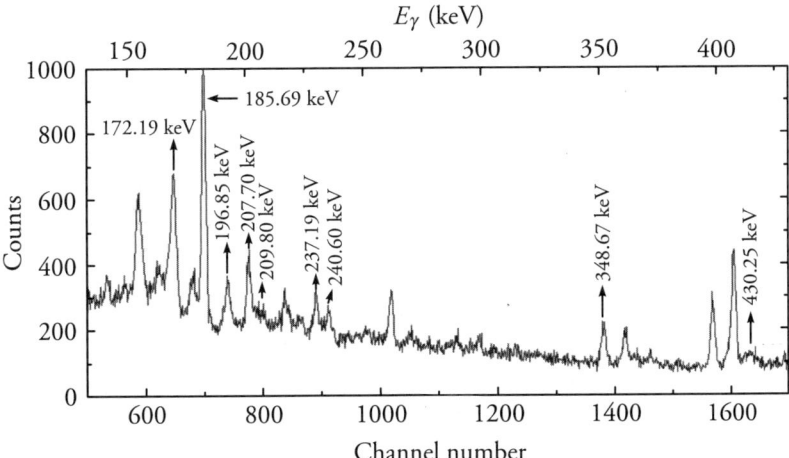

Figure 3.14 Gamma-ray spectrum of residues populated in the reaction ^{13}C+^{169}Tm at $\approx$ 83 MeV beam energy

The gamma-ray spectrum of (a) ^{159}Tb bombarded by $\approx$ 87 MeV ^{12}C beam and (b) ^{159}Tb bombarded by $\approx$ 88 MeV ^{13}C beam are shown in Figure 3.15 and Figure 3.16, respectively. The various peaks in the observed γ-ray spectra have been assigned to different radio nuclides. As a typical example, the observed decay curve for the rhenium isotope (^{178}Re) at $\approx$ 83 MeV

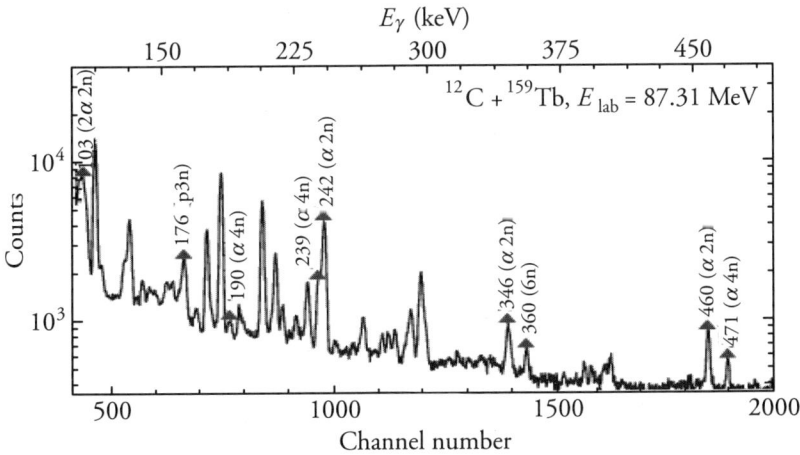

Figure 3.15 Gamma-ray spectrum of residues populated in the reaction ^{12}C+^{159}Tb at $\approx$ 87 MeV beam energy

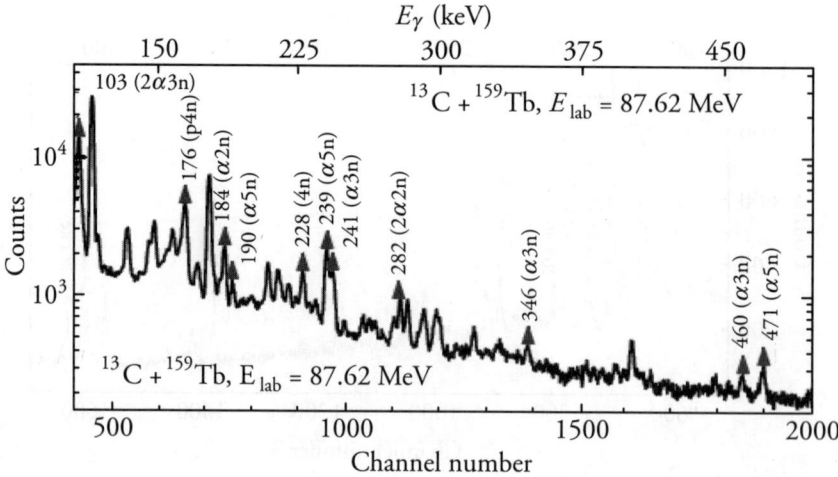

Figure 3.16 Gamma-ray spectrum of residues populated in the reaction ^{13}C+^{159}Tb at ≈ 88 MeV beam energy

beam energy populated in the ^{13}C+^{169}Tm reaction is shown in Figure 3.17. Similarly, the decay curve for ^{168}Lu isotope populated via the 4n channel and produced in the interaction of ^{13}C+^{159}Tb is shown in Figure 3.18. As may be observed from these decay curves, the measured half-lives of these residues match well with the corresponding literature values. Nuclear data, like half-lives, γ-ray energies, branching ratios etc., have been taken from the Table of Isotopes[12,13] and Nuclear Wallet Card[14]. The identified residues for various systems are listed in Table 3.1. As already mentioned, the half-lives of the identified residues determined in the present work were found to be in good agreement with literature values. Data analysis has been performed using CANDLE software[10]. The cross-sections have been determined employing the formulations discussed earlier.

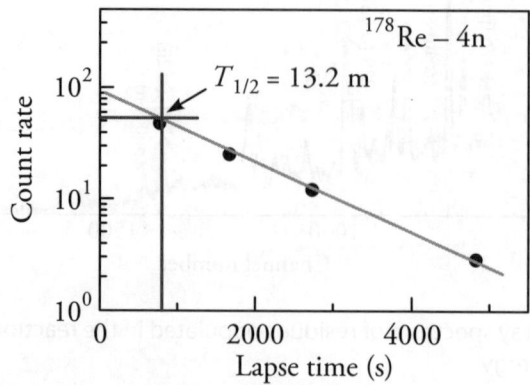

Figure 3.17 Decay curve of ^{178}Re residue populated via 4n channel in the interaction(^{13}C+^{169}Tm) at ≈ 83 MeV beam energy

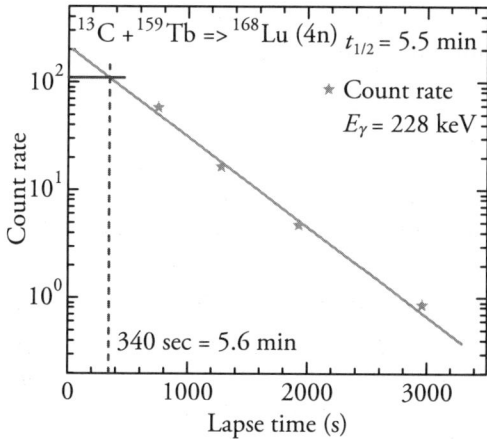

Figure 3.18 Observed decay curve of ^{168}Lu residues populated via 4n channel in the ^{13}C+^{159}Tb reaction

3.7 Measurement of Recoil Range Distribution (RRD) of Heavy Residues

As we will see later, there has been a significant enhancement in the production cross-section of the alpha-emitting channel as compared to the theoretical predictions based on the statistical compound reaction model. This enhancement has been attributed to the additional contribution coming from the incomplete fusion process. In order to understand the reaction dynamics for the fusion processes, it is reasonable to assume that the residues produced in the HI induced reactions may originate via two different processes viz., (a) from the compound nucleus formed via full linear momentum transfer events and (b) from the incompletely fused composite nucleus formed due to partial linear momentum transfer events. In case of energies little above the Coulomb barrier, complete momentum transfer events are expected to be dominant. However, at relatively higher projectile energies, both complete and partial momentum transfer events may contribute significantly to the production cross-section for the alpha-emitting channels. One simple way of measuring the amount of linear momentum transfer for a particular reaction residue, is by recording the linear recoil range of the residues in some absorbing material. Obviously, a larger recoil range will mean a larger value of the linear momentum transfer. As such, different momentum transfer events may be differentiated on the basis of an experiment where the measurement of recoil ranges is done. In the following, we will describe how one may separate CF and ICF events on the basis of linear momentum transfer from the projectile to the target nucleus by measuring the forward recoil ranges of the residues produced in different projectile–target combinations. The forward recoil ranges of CF and ICF reaction products depend on the recoil velocity of the composite nuclear system associated with the degree of linear momentum transfer from the projectile to the target nucleus. In the case of CF reactions, the projectile completely fuses with the target nucleus and transfers its full linear momentum. On the other hand, during ICF reactions, a part of the projectile fuses with the target nucleus and thus, partial linear momentum is transferred. The degree of linear momentum transfer may be given as

$$\rho_{LMT} = P_f/P_P \qquad (3.19)$$

Here P_P is the total linear momentum of the projectile at a given energy and P_f is the linear momentum of the fused part (or the fused fraction) of the projectile. Since ρ_{LMT} is proportional to the mass of the fused part of the projectile, transfer of a larger linear momentum will produce residues with larger recoil velocity and hence, with larger kinetic energy. If reaction residues are made to pass through some absorbing medium, those with larger kinetic energy will have larger range in the medium. Thus, measurement of linear recoil ranges of reaction residues is a promising way to separate full and partial momentum transfer events corresponding to CF and ICF processes respectively. Maximum linear momentum is transferred from the projectile to the target nucleus in CF events and therefore, for the given entrance channel, the compound nucleus formed in CF events has a predetermined mass, energy and momentum. On the other hand, in case of ICF reactions, only partial linear momentum transfer takes place and the compound nucleus formed in ICF events may not have unique values of mass, energy and momentum. This may be because in different ICF events, different masses may be transferred from the projectile to the target and also due to different possible interaction trajectories. The velocity distribution of a given type of reaction products is expected to be symmetric about V_o, having a width that depends on the evaporation process and in particular, on the particles evaporated from the CN. The mean velocity V_o, may be given as

$$V_o = V_{cn} = \sqrt{(2M_P.E)}/M_{(P+T)} \qquad (3.20)$$

where M_P is the projectile mass, $M_{(P+T)}$ is the total mass of the composite system, i.e., projectile + target nucleus, and E is the projectile energy in the laboratory frame.

As explained earlier, the experimentally measured forward recoil ranges of the final reaction products, in a stopping medium, may give information about the degree of linear momentum transfer involved in each case. The measurement of forward recoil range distribution (FRRDs) is one of the most direct methods to distinguish the different ICF components, where the same kind of residues may be formed by more than one fusion channel. Though the differences in the velocities or ranges of CF and ICF reaction products are not very large, by using very thin catcher foils ($\approx$ few tens of $\mu g/cm^2$), it is possible to separate the CF and ICF events. For nuclei that are formed via partial linear momentum transfer of the projectile, the recoil velocity is less than that for complete momentum transfer events. Therefore, the reaction products populated via ICF will show relatively smaller ranges in the stopping medium as compared to CF products. For different degrees of linear momentum transfer events, the residues may have different recoil ranges in the stopping medium. The FRRD (or RRD) measurements may thus be used to separate the relative contributions of various partial fusion components in the formation of particular reaction residues. In the present experiments, the recoil–catcher technique followed by off-line gamma-ray spectroscopy has been used to determine the recoil range distributions.

3.7.1 Target and catcher foil preparation for RRD measurements

In RRD measurements, the thickness of the target is kept small so that the recoiling residues are *not stopped* in the thickness of the target. As such, thin targets prepared by using vacuum evaporation method with Al-backing have been used in such measurements. In order to trap the recoiling residues, thin Al-catchers prepared by ultra-high vacuum evaporation techniques have been used. This is a common method of thin-film deposition. The source material is evaporated in a vacuum chamber. The vacuum allows vapour material particles to travel directly to the glass plate substrate, where they condense back to a solid state. A rough estimate of the deposition thickness is made with the help of a quartz monitor. The glass slides with Al-deposition on them are taken out of the vacuum chamber. Then, as mentioned earlier, the thin deposition of Al material is floated on water and attached to the holders very carefully. Great care is required to handle these thin foils. The thicknesses of these samples were measured using the already discussed alpha-transmission method.

As shown in Figure 3.19, for the RRD experiment, the target is followed by a stack of thin Al-catcher foils (sufficient to stop the compound nucleus formed via full momentum transfer) to trap the recoiling heavy residues formed as a result of CF and ICF interactions and travelling with different velocities (energies). A typical target–catcher foil setup used for the measurement[11] of forward recoil ranges in the ^{12}C+^{159}Tb system is shown in Figure 3.19. As can be seen in the figure, CF residues will be trapped at higher cumulative catcher thickness while ICF residues populated due to partial momentum transfer will be trapped at shorter ranges.

In experiments that aim to measure recoil range distributions, thin catcher foils should be used to enhance the power of resolving the recoil ranges. As an example, the thicknesses of the successive catcher foils in the experiment for measuring the recoil range distribution for the system ^{16}O+^{181}Ta at three distinct energies are given in Table 3.3.

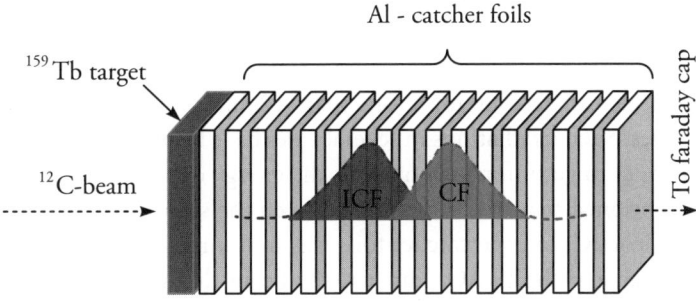

Figure 3.19 Schematic representation of the absorber stack used for trapping recoiling residues used in the measurement of forward recoil range distributions

Table 3.3   List of catcher foil thicknesses used in RRD experiments for the system $^{16}O+^{181}Ta$ at different incident energies

S. No.	Thickness ($\mu g/cm^2$) for the Irradiation at ≈ 85 MeV	Thickness ($\mu g/cm^2$) for the Irradiation at ≈ 94 MeV	Thickness ($\mu g/cm^2$) for the Irradiation at ≈ 100 MeV
1	9.66	23.33	10.01
2	15.19	23.52	16.17
3	30.86	23.78	18.17
4	30.94	23.89	18.32
5	31.05	24.35	19.00
6	31.15	25.06	19.29
7	32.05	25.35	19.90
8	32.05	26.56	21.59
9	32.09	29.08	23.33
10	32.27	29.08	40.16
11	32.31	29.40	40.17
12	33.21	35.45	40.56
13	33.50	36.07	50.70
14	33.53	36.35	50.73
15	33.53	56.04	50.89
16	33.55	54.74	51.68
17	33.59	54.94	56.77
18	41.47	55.02	64.40

The target–catcher foil assembly, made up of a thin target followed by a stack of thin Al-catcher foils to trap the recoiling reaction products, is irradiated at a given beam energy for sufficiently long time to achieve the desired statistics. In the present experiments, the irradiations have been carried out for more than 12 hours with beam currents of about 20–30 nano Amperes. Such irradiations in several cases have been carried out generally at three incident beam energies to get information about the energy dependence of the ICF processes. It may again be mentioned that the delay time between the stop of irradiation and the start of the counting of radioactivity in the catcher foils was minimized by employing the in-vacuum transfer facility available at the IUAC. Further, the beam flux used for determination of the cross-sections has been calculated from the total charge collected in the Faraday cup placed behind the target–catcher assembly arrangement in the experiment. After the irradiation, the target– catcher assembly was disassembled to record the activity collected in each Al-catcher foil. The residues populated via CF and/or ICF processes are expected to be trapped at different catcher foil thickness, depending on the recoil velocity of the residues or on the degree of linear momentum transfer associated with the mode of reaction channel. The activities produced in different catcher foils of the catcher stack were counted separately using a pre-calibrated (HPGe) spectrometer coupled to CAMAC-based CANDLE software[10]. The characteristic gamma-ray radiations originating from different residues have been used to identify them and to determine their production probability. The gamma-ray spectrum of each foil has been recorded at increasing times and the decay curve analysis has been done to measure the half-life of the residues. A typical gamma-ray spectrum of the induced activity in one of the Al-catcher foil for a ^{12}C+^{159}Tb system[11] at 87 MeV beam energy is shown in Figure 3.20. The measured intensity of the characteristic gamma lines in each Al-catcher foil has been used to determine the population cross-section in that foil.

Some of the systems for which the recoil range distributions have been measured by our group are listed in Table 3.4. The recoil range distributions (RRDs) have been measured at several beam energies for different systems using ^{12}C and ^{16}O beams.

In order to get the normalized yield of various reaction residues as a function of cumulative thickness of the Al-catcher foils, the cross-section of the populated reaction products in each catcher foil was divided by respective catcher foil thickness. The normalized yields of the reaction products thus obtained have been plotted against the cumulative catcher foil thickness to obtain the differential recoil range distributions. In the present work, the recoil range distributions of residues populated via CF and/or ICF process have been obtained for several projectile–target combinations. The details of some of these are given in the next chapter, where analysis of the data is presented.

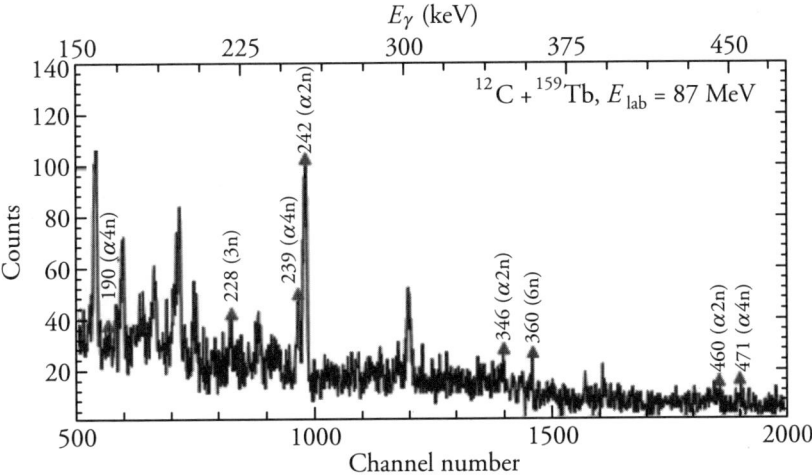

Figure 3.20 Typical gamma-ray spectrum at one of the aluminium catcher foil in the interaction of the ^{12}C+^{159}Tb system at 87 MeV beam energy

Table 3.4   List of reactions and other relevant parameters used for the recoil range distributions studies

S. No	Projectile	Incident energy (MeV)	Target	Purity of target	Composite system	Reactions
1.	$^{12}_{6}C$	71 and 80	$^{159}_{65}Tb$	99.9%	$^{177}_{73}Ta$	$^{159}Tb(^{12}C, 3n)\ ^{168}Lu$
						$^{159}Tb(^{12}C, 4n)\ ^{167}Lu$
						$^{159}Tb(^{12}C, 6n)\ ^{165}Lu$
						$^{159}Tb(^{12}C, p3n)\ ^{167}Yb$
						$^{159}Tb(^{12}C, \alpha2n)\ ^{165}Tm$
						$^{159}Tb(^{12}C, \alpha4n)\ ^{163}Tm$
						$^{159}Tb(^{12}C, 2\alpha2n)\ ^{161}Ho$
						$^{159}Tb(^{12}C, 2\alpha3n)\ ^{160}Ho$
2.	$^{12}_{6}C$	71 and 80	$^{165}_{67}Ho$	99.9%	$^{177}_{73}Ta$	$^{165}Ho(^{12}C, 3n)\ ^{174}Ta$
						$^{165}Ho(^{12}C, 4n)\ ^{173}Ta$
						$^{165}Ho(^{12}C, 5n)\ ^{172}Ta$
						$^{165}Ho(^{12}C, p3n)\ ^{173}Hf$
						$^{165}Ho(^{12}C, p5n)\ ^{171}Hf$
						$^{165}Ho(^{12}C, \alpha2n)\ ^{171}Lu$
						$^{165}Ho(^{12}C, \alpha4n)\ ^{169}Lu$
						$^{165}Ho(^{12}C, \alpha6n)\ ^{167}Lu$
						$^{165}Ho(^{12}C, 2\alpha2n)^{167}Tm$
						$^{165}Ho(^{12}C, 2\alpha4n)^{165}Tm$
3.	$^{16}_{8}O$	87	$^{169}_{69}Tm$	100%	$^{185}_{77}Ir$	$^{169}Tm(^{16}O, 3n)\ ^{182}Ir$
						$^{169}Tm(^{16}O, p2n)\ ^{182}Os$
						$^{169}Tm(^{16}O, p3n)\ ^{181}Os$
						$^{159}Tb(^{16}O, 2p2n)\ ^{181}Re$
						$^{169}Tm(^{16}O, \alpha2n)\ ^{179}Re$
						$^{169}Tm(^{16}O, \alpha3n)\ ^{178}Re$
						$^{159}Tb(^{16}O, \alpha4n)\ ^{177}Re$
						$^{159}Tb(^{16}O, \alpha p3n)\ ^{177}W$
						$^{169}Tm(^{16}O, \alpha2pn)\ ^{178}Ta$
						$^{169}Tm(^{16}O, \alpha3pn)\ ^{177}Hf$
						$^{169}Tm(^{16}O, 2\alpha pn)\ ^{175}Hf$
						$^{169}Tm(^{16}O, 3\alpha n)\ ^{172}Lu$
						$^{169}Tm(^{16}O, 3\alpha2n)\ ^{171}Lu$

Experimental Details and Formulations **119**

4.	$^{16}_{8}O$	90	$^{159}_{65}Tb$	100%	$^{175}_{73}Ta$	$^{159}Tb(^{16}O, 3n)\,^{172}Ta$
						$^{159}Tb(^{16}O, 4n)\,^{171}Ta$
						$^{159}Tb(^{16}O, p3n)\,^{171}Hf$
						$^{159}Tb(^{16}O, p4n)\,^{170}Hf$
						$^{159}Tb(^{16}O, p3n)\,^{171}Hf$
						$^{159}Tb(^{16}O, 2p2n)\,^{171}Lu$
						$^{159}Tb(^{16}O, \alpha n)\,^{170}Lu$
						$^{159}Tb(^{16}O, \alpha 3n)\,^{168m}Lu$
						$^{159}Tb(^{16}O, \alpha 4n)\,^{167}Lu$
						$^{159}Tb(^{16}O, \alpha p3n)\,^{167}Yb$
						$^{159}Tb(^{16}O, \alpha 2n)\,^{169}Lu$
						$^{159}Tb(^{16}O, 2\alpha)\,^{167}Tm$
						$^{159}Tb(^{16}O, 2\alpha n)\,^{166}Tm$
						$^{159}Tb(^{16}O, 2\alpha 2n)\,^{165}Tm$
						$^{159}Tb(^{16}O, 3\alpha n)\,^{162}Ho$
5.	$^{16}_{8}O$	76 and 81	$^{169}_{69}Tm$	100%	$^{185}_{77}Ir$	$^{169}Tm(^{16}O, 3n)\,^{182}Ir$
						$^{169}Tm(^{16}O, p2n)\,^{182}Os$
						$^{169}Tm(^{16}O, p3n)\,^{181}Os$
						$^{169}Tm(^{16}O, \alpha)\,^{181}Re$
						$^{169}Tm(^{16}O, 2\alpha pn)\,^{175}Hf$
						$^{169}Tm(^{16}O, 2\alpha p5n)\,^{171}Hf^{g}$
						$^{169}Tm(^{16}O, 3\alpha n)\,^{172}Lu$
						$^{169}Tm(^{16}O, 3\alpha 2n)\,^{171}Lu$
6.	$^{16}_{8}O$	81, 90 and 96	$^{181}_{73}Ta$	100%	$^{197}_{81}Tl$	$^{181}Ta(^{16}O, 3n)\,^{194}Tl$
						$^{181}Ta(^{16}O, 4n)\,^{193g}Tl$
						$^{181}Ta(^{16}O, 5n)\,^{192}Tl$
						$^{181}Ta(^{16}O, p3n)\,^{194}Tl^{m+g}$
						$^{181}Ta(^{16}O, p4n)\,^{192}Hg$
						$^{181}Ta(^{16}O, p5n)\,^{191m}Hg$
						$^{181}Ta(^{16}O, \alpha n)\,^{192g}Au$
						$^{181}Ta(^{16}O, \alpha 2n)\,^{191g}Au$
						$^{181}Ta(^{16}O, \alpha 3n)\,^{190g}Au$
						$^{181}Ta(^{16}O, 2\alpha 3n)\,^{186g}Ir$

3.8 Measurement of Angular Distribution of Residues

In order to get the complimentary information regarding CF and/ICF reactions, the angular distributions for various radioactive residues produced in the interaction of ^{16}O beam with ^{169}Tm target nucleus have been measured at 81 MeV beam energy. Figure 3.21 shows the schematic arrangement used in angular distribution measurements[15]. The ^{169}Tm sample for irradiation by the ^{16}O beam was made by vacuum evaporation (of ≈ 47.25 $\mu g/cm^2$) of ^{169}Tm of purity > 99.9% on Al foil. In the GPSC, the sample was mounted normal to the direction of the incident beam such that the Al surface (on which the target material was deposited) faces the beam. The beam energy incident on the Al surface was ≈ 85 MeV. The incident beam of ≈ 85 MeV first passes through the Al backing of the sample and loses ≈ 4 MeV of energy; thus, the incident beam energy gets reduced to ≈ 81 MeV on the Tm material. In order to trap residues recoiling at different angles, a stack of annular aluminium catcher foils each of thickness 0.3 mm with concentric holes respectively of diameter 0.81, 1.29, 1.95, 2.64, 3.27, 5.46, and 6.4 cm were used. This arrangement of annular catchers was placed at a distance of 1.8 cm behind the target. The angular ranges covered by the catcher foils of the annular stack were: 0–13° , 13–21°, 21–30°, 30–39°, 39–45°, 45–60° and 60–64°.

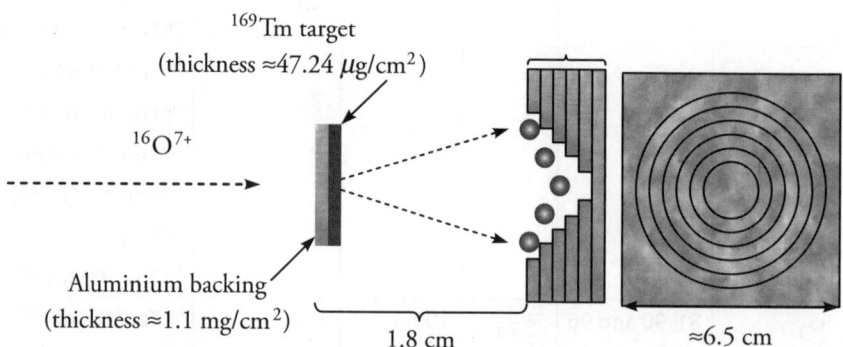

Figure 3.21 Schematic representation of the experimental setup used for the measurement of angular distribution of heavy residues

The target–annular catcher stack assembly was irradiated for 11 hours with a ^{16}O^{7+} beam of ≈ 7 pnA. The pre-calibrated high resolution (2 keV for 1.33 MeV γ-ray of ^{60}Co) HPGe detector of 100 c.c. active volume of CANBERRA at the IUAC was used to record the activities induced in each foil of the catcher stack. Induced activities were followed for about 15 days. The observed gamma-ray spectra of Al-catcher rings forming the angular range 0–13°, 13–21° and 21–30° are shown in Figure 3.22. The spectra covering the angular range 30–39°, 39–45°, 45–60° and 60–64° are shown in Figure 3.23. Various peaks in these spectra correspond to different reaction channels. The gamma-ray peaks corresponding to alpha emission channels are shown in the inset in Figure 3.24 for a clearer view.

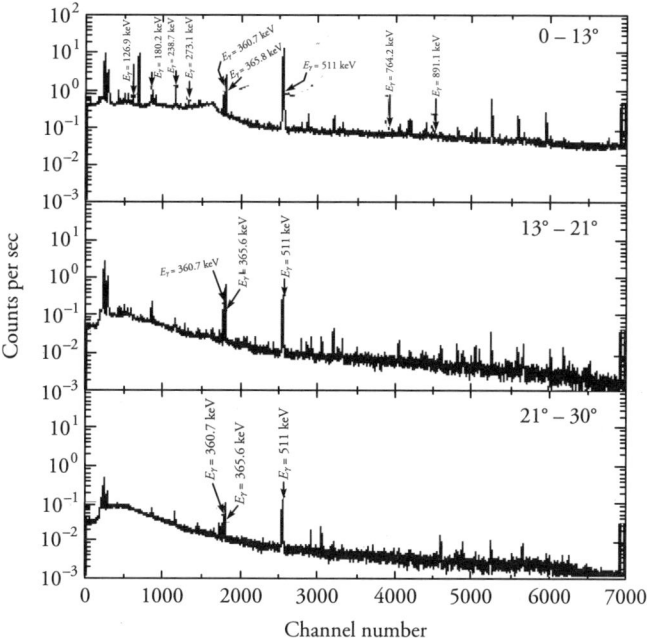

Figure 3.22 Typical gamma-ray spectra of Al-catcher rings covering angular ranges 0–13°, 13–21° and 21–30°

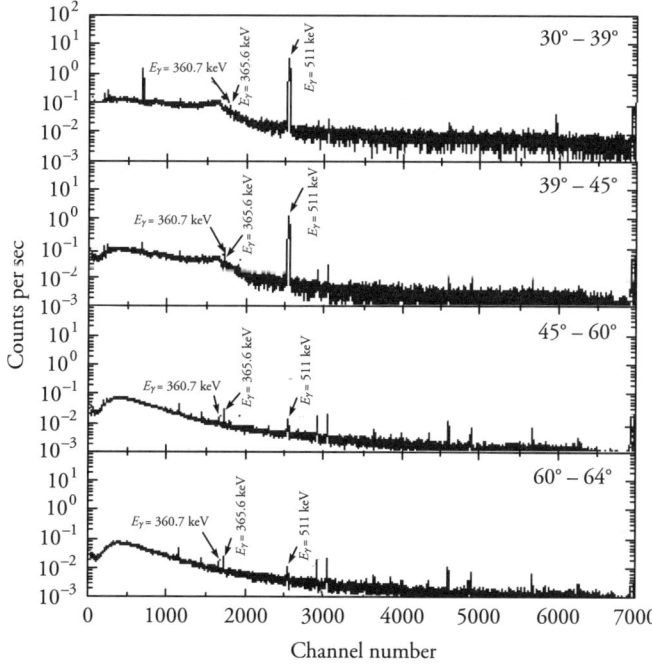

Figure 3.23 Gamma-ray spectra of Al-catcher rings covering angular ranges 30–39°, 39–45°, 45–60° and 60–64°

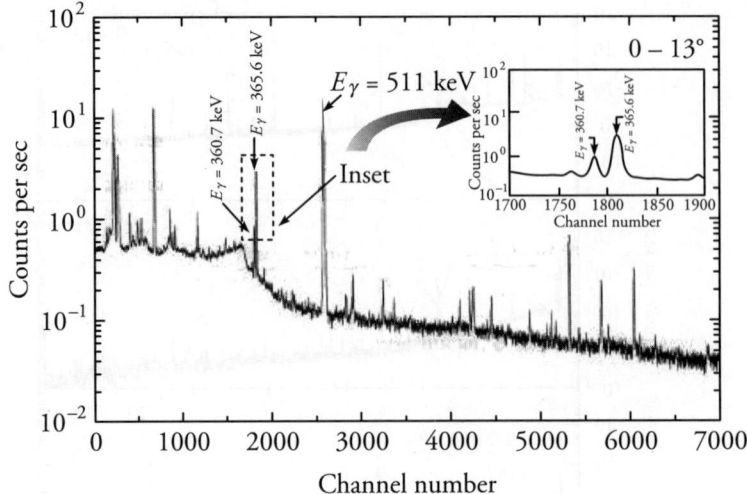

Figure 3.24 Gamma-ray spectra of Al-catcher ring covering the angle range 0–13°. Inset shows 360.7 keV and 365.6 keV gamma rays corresponding to the residue ^{181}Re produced in the interaction of ^{16}O+^{169}Tm at 81 MeV beam energy

The measured cross-sections (σ) for a particular reaction product in different catcher foils were obtained using Eq. (3.10). Measured angular distributions for various residues populated via CF and ICF processes are presented in Chapter 4.

3.9   In-beam Experiments

With a view to obtain information about reaction dynamics and the role of angular momenta involved in complete and incomplete fusion reactions, particle–gamma ray coincidence experiments have been performed for several systems at different energies at the Inter-University Accelerator Centre (IUAC), New Delhi using gamma detector array (GDA) along with charged particle detector array (CPDA) setups[16,17]. The coincidences have been recorded between prompt γ-rays using HPGe detectors and charged particles ($Z = 1,2$) detected by the charged particle detector (CPD) system. These experiments have been performed to explore the reaction dynamics and investigate the effect of beam energy on the entry state spin populations. A brief description of the target preparation, experimental setup and the details of the electronics used in these in-beam experiments are given in the following subsections.

3.9.1 Target preparation

In in-beam experiments, it is customary to use spectroscopic pure target foils of uniform thickness. In the present work, thin self-supporting targets have been prepared employing the rolling technique. Spectroscopic pure and self-supporting targets of thickness ≈ 1.00 mg/cm^2 were prepared. The thickness of each target was measured by the α-transmission technique, details of which were given earlier in section 3.4. Further, the projectile–target combinations selected for these studies were such that their reaction residues were well-known rotational nuclei and data on their prompt γ-transitions is available in literature.

3.9.2 Experimental setup used

Considerable improvement in in-beam particle–gamma coincidence technique has been achieved in recent years due to the availability of advanced multi-detector systems, different energy filter detectors and anti-Compton spectrometers. This advancement has become possible as a result of large-scale scientific efforts made to improve upon the spectroscopic limitations of single detector systems, where the total counting rate is not, in general, allowed to exceed 10,000 counts per second due to the excessive pile up. However, it may be reduced to 3,000 counts per second with the use of Compton scattering filters. The counting rate in a single HPGe detector may be estimated as

$$N_S = N_r (\Omega / 4\pi) M G_\varepsilon \qquad (3.21)$$

where N_r is the number of residual nuclei produced per second. The term $\Omega / 4\pi$ is the fraction of solid angle covered by each detector, M, is the number of photons emitted per event and G_ε is the gamma ray detection efficiency of the detector after Compton correction. In case of in-beam gamma-ray spectroscopy, it is required to collect the coincidence data to improve the resolving power. This technique is one of the most reliable criteria for assigning γ-transitions and/or cleaning up the spectrum. The main disadvantage of coincidence experiments is that the count rate goes down drastically with the degree of coincidence or with the number of fold(s). For example, the data rate may be very less for an α–γ coincidence as compared to a p–γ coincidence in a nuclear reaction. As such, with any gating condition, the count rate in coincidence with one detector may be too small for analysis and/or to draw any significant conclusion. The count rate in an experiment may be increased by increasing the number of residual nuclei produced per second (N_r). This may be done by increasing the beam current and/or the target thickness. However, it may be pointed out that the aforementioned methods of increasing the count rate may have constrains due to the pile-up of events. Further, the coincidence count rate cannot be increased greatly by increasing the solid angle of individual detectors, since with this increase, the Doppler broadening of γ-rays also increases. Therefore, the coincidence count rate can only be increased by using multi-detector systems without affecting the quality of data. As such, in the present work, the gamma detector array (GDA) consisting of 12 HPGe detectors in specified geometry has been used to record characteristic γ-rays. However, for particle identification and to generate the particle–γ-coincidences, a charged particle detector array (CPDA) setup has been used along with a gamma detector array. A detailed description of the GDA and CPDA setups[16,17] used in the present work and their specifications are given in the following sections.

3.9.3 The gamma detector array (GDA) setup

The gamma detector array (GDA), at the IUAC, New Delhi, India, contains an array of 12 Compton-suppressed HPGe γ-detectors of high counting efficiency. It is a dedicated array for γ-spectroscopy and reaction dynamics studies at the Pelletron accelerator facility. The schematic view of the GDA–CPDA setup at the IUAC is shown in Figure 3.25, while a photograph of the side view of GDA set up in the beam line is given in Figure 3.26. In this array, 12 n-type, intrinsic, HPGe detectors are housed. The n-type Ge semiconductor is used almost universally as a detector material for γ-spectroscopy, because of its good timing and high resistance to neutron damage. Further, in the mechanism of γ-ray detection, the photo-electric effect is important, as it results in full energy deposition. Generally, only 15–20% of photons result in full energy peak in a typical (100 c.c.) HPGe detector.

In the particle–γ-coincidence experiments, about 4% events are expected to take place with full energy deposition. As such, it is customary and important to improve the signal-to-background ratio for coincidence experiments. Here, each HPGe detector of the array is surrounded by an anti-Compton shield (ACS). The signal-to-background ratio can be increased by detecting the Compton scattered γ-rays coming out of the HPGe detector, using

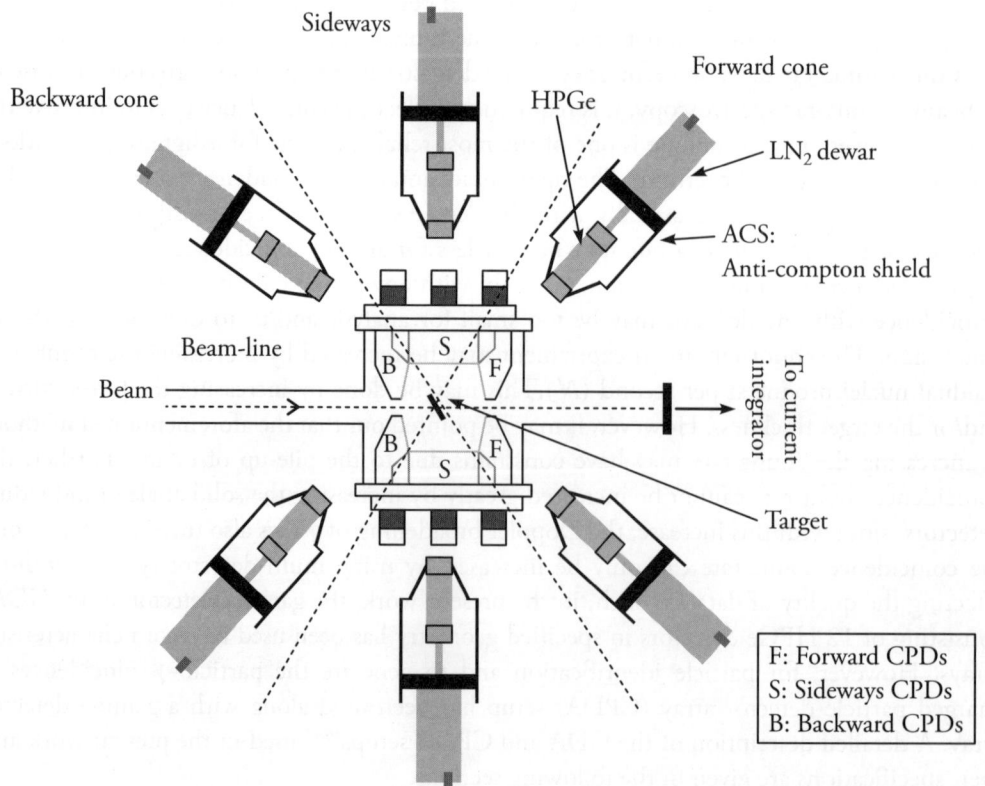

Figure 3.25 Schematic representation of the GDA–CPDA setup at the IUAC, New Delhi

Figure 3.26 Side view of the gamma detector array at the IUAC, New Delhi, India

another scintillation detector and vetoing them out electronically. Both NaI and BGO crystals (which have absorption thickness about 2.2 times that of NaI) have been used to construct an ACS. A typical schematic diagram of an ACS is shown in Figure 3.27. As can be seen from this figure, the ACS is of symmetric design, and has mainly two parts, i.e., (i) the main cylindrical body, which consists of eight optically separated 15 cm long BGO crystals; there is a separate photo-multiplier tube (PMT) attached at an angle of 20° to each of these BGOs, and (ii) the front panel of the ACS, which is made up of a piece of NaI crystal that is optically coupled to BGO crystals.

Since, the energy of the back-scattered γ-rays from the HPGe detector is ≤300 keV, very high density material like BGO is not needed to detect these. Therefore, NaI crystals

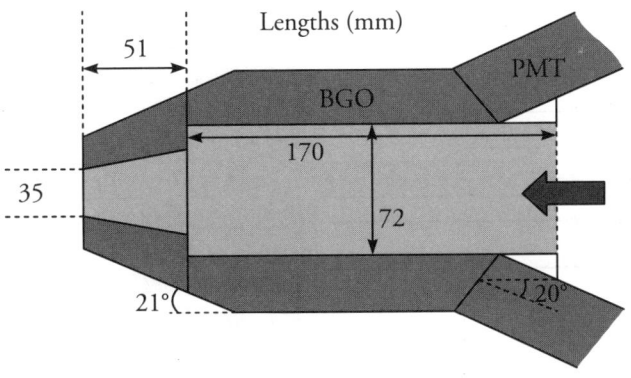

Figure 3.27 The symmetric anti-Compton shield (ACS) as used in the GDA setup

are used for the front part of the anti-Compton shields (ACS), which has a higher light output than BGO, particularly at lower γ-ray energies. As shown in Figure 3.27, the HPGe detectors are inserted axially through the insert port inside the ACS and mounted on two rigid mechanical structures on either side of the 30° beam line. Each ACS is shielded from in-beam γ-rays from the target by a 3 cm thick tantalum collimator. The structures are movable employing the rails fixed on the floor vertical to the beam line. Each of the structure consists of six HPGe detectors in two rows, 25° above and below the horizontal plane with three detectors in each row at 45°, 99° and 153° in plane angles. This makes approximately 50°, 98° and 144° actual angles, respectively, with respect to the incident beam direction; there are four detectors at each of these angles. Needless to mention that the HPGe detectors are required to be cooled to liquid nitrogen temperature (i.e., 77 K) to reduce leakage current.

3.9.4 Charged particle detector array (CPDA) setup

In the study of CF and ICF processes, the detection of charged particles (Z = 1, 2) in coincidence with prompt γ-rays is quite useful and important. One of the main reasons for this is the fact that in heavy ion fusion, the excited compound nucleus formed as a result of projectile–target interaction de-excites predominantly via light particle (n, p, t, α, ...) emission. Coincidences are essential for selecting the desired reaction channels and/or for removing the background. In view of this, it is desired to have a detector system for charged particle detection that may cover a 4π solid angle.

The CPDA at the IUAC, New Delhi, consists of 14 charged particle detectors (CPDs) arranged in two truncated hexagonal pyramid shapes[3]. The base of all the pyramids is in a horizontal plane with each having a trapezoidal shape. The remaining spaces at the top and bottom are covered by hexagonal detectors, which together with the trapezoids cover nearly 90% of the total solid angle. The corners of the trapezoids are cut in a V-shape for beam entrance and exit. These V-cuts are also useful for providing support to the target mounting. The full CPDA is held in place from the top and bottom flanges of the scattering chamber, and is shown in Figure 3.28. There are four charged particle detectors at the forward angles

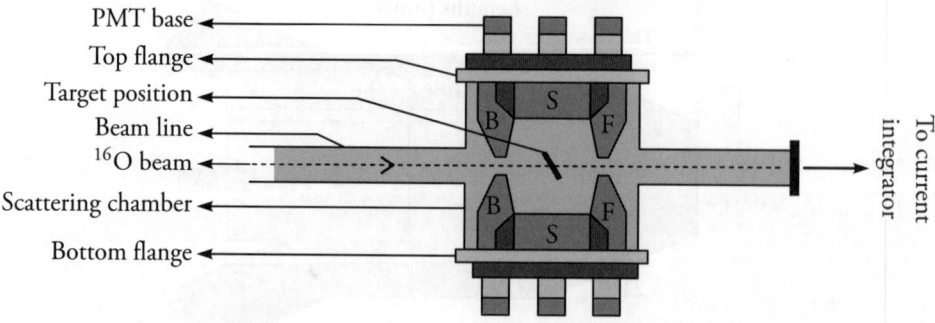

Figure 3.28 The schematic representation of the charged particle detector array inside a small scattering chamber

(F) covering 10–60°, four in the backward (B) covering 120–180° and six in the sideway (S) at an angle around 90°. In Table 3.5, some of the features of the CPDA are given. Various components of the CPD are shown in Figure 3.28. A brief description of these components is given in the following subsections.

Table 3.5 Important parameters of a charged particle detector array

1	Geometry of CPDA	Two truncated hexagonal pyramids
2	Detector material	Plastic scintillator
3	Solid angle covered	90% of 4π
4	Number of detectors	14
5	Count rate for each element	50,000 counts/s
6	ΔE detector	BC 400, thickness $\approx$ 100 μm
7	E detector	BC 444, thickness $\approx$ 5 mm
8	Optical guide	BC 800 UVT lucite material
9	PMT	1" diameter, R1924- HAMAMATSU
10	Glue for PMT and optical guide	Optical cement (BC600)

3.9.4.1 Phoswich detector

The Phoswich detector is a combination of two detectors in optical contact. The front detector consists of a thin (BC 400 ~100 μm), ΔE scintillation detector that has a fast (~5 ns) decay component to its light output, which is particle dependent. Typically, the heavier the detected particle, the larger and faster the signal. The second detector (thick, BC 444 ~5 mm), placed behind the thin ΔE detector, is used to stop the particle and usually gives out light output with a much longer decay time (~1 μs), which is independent of particle type. The total signal from

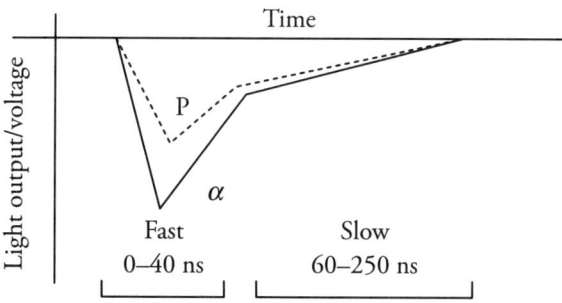

Figure 3.29 Schematic diagram of the operation of a Phoswich detector for discrimination between protons and alpha particles

the two detectors is taken through a light pipe, into a photomultiplier tube (PMT). The anode signal from the PMT is then integrated into two different time regions by setting two separate time discriminator gates on short (ΔE) and long (E) components of the signal (see Figure 3.29). By making a 2-dimensional array of ΔE versus E, different detected particles can be identified. In order to stop the signals arising from the scattered beam that enters the Phoswich detector, thin absorber foils of Al, are placed on the front face of the detector. These foils are thick enough to stop the (higher Z) beam particles, but thin enough to allow evaporated protons and α-particles pass through.

3.9.4.2 Optical guide

The Phoswich detector and the PMT are coupled through an optical guide. The PMT converts the light signal into an electrical signal for further processing. The light, emitted by a plastic scintillator lies in the ultra-violet (UV) region. As such, the UV transmitting (UVT) lucite material (BC800 UVT acrylic) has been used for better transmission of light. The length of the optical guide has to be kept at a minimum in order to have maximum light collection efficiency at the PMT. To join the optical guide to the Phoswich detector and also with the PMT, typical optical cement BC 600 has been used, which is a clear optically transparent epoxy. The optical guide along with the Phoswich detector is covered with the aluminized mylar foil of about 6 µg/cm² for good transmission of light. The vacuum seal is used with the PMT, as it avoids the leakage of light and vacuum problems to a large extent.

3.9.4.3 Photo-multiplier tube (PMT)

Keeping in view the space availability in the scattering chamber, the rise time compatibility between the scintillator, the photo-multiplier tube (PMT), stability and low noise, the HAMAMATSU PMT R1924 is used as the photomultiplier tube for the charged particle detector. The proper working of the PMT is ensured by keeping the anodes and plates successively at a positive stable voltage with respect to the photo-cathode. Since the gain of the PMT is extremely sensitive to changes in voltage, it is very important that sources of high voltage must be well regulated and free of ripples.

3.9.5 Irradiations for spin distribution measurement

In the present experiments, natural targets were irradiated separately using $^{12}C^{5+}$ and $^{16}O^{7+}$ ion beams from the 15 UD Pelletron accelerator at the IUAC, New Delhi. Irradiations were carried out in an aluminium scattering chamber of 16 cm diameter housing the CPDA. The target is placed at an angle 45° with beam direction, in order to increase the effective number of target nuclei. Position of the beam has been monitored with the help of GDA and CPDA collimator(s). The GDA collimator is housed in the beam line, just before the CPDA scattering chamber and the CPDA collimator is placed just before the target. Special care has been taken so that the centres of the GDA and that of CPDA coincide. The chamber has two rotatable flanges for alignment purposes. As the CPDA is held and aligned by the top and bottom flanges of the target chamber, a specially designed set of two flanges hold each set of seven charged particle detectors (CPDs) in a proper geometry to reliably reproduce the centre (see Figure 3.25).

The present experiments are based on the particle–γ coincidence technique using GDA and CPDA setups. For coincidence purposes, the array is divided into three angular zones, i.e., (a) forward angles (F): 10–60°; (b) sideways (S): 60–120° and (c) backward angles (B): 120–170°. Moreover, in order to remove scattered beam particles during the experiment, the CPDs have been covered by Al-absorbers of appropriate thickness. The Al-absorbers also reduce the number of random coincidences due to elastically scattered beams and protect the detector array from direct radiation damage. In these particle–gamma coincidence experiments, several reactions have been studied at different energies. A list of reactions for which spin distribution measurements have been done is given in Table 3.6.

Table 3.6 List of reactions studied for spin distribution measurements at different energies

S. No.	Projectile	Incident energy (MeV)	Target	Purity of target	Composite system	Reactions
1.	$^{12}_{6}$C	≈ 67 and 78	$^{169}_{69}$Tm	100%	$^{181}_{75}$Re	^{169}Tm(^{12}C, 3n) ^{177}Re
						^{169}Tm(^{12}C, 4n) ^{176}Re
						^{169}Tm(^{12}C, 5n) ^{176}Re
						^{169}Tm(^{12}C, 6n) ^{175}Re
						^{169}Tm(^{12}C, p3n) ^{177}W
						^{169}Tm(^{12}C, p4n)^{176}W
						^{169}Tm(^{12}C, α2n) ^{175}Ta
						^{169}Tm(^{12}C, α3n)^{174}Ta
						^{169}Tm(^{12}C, α4n)^{173}Ta
						^{169}Tm(^{12}C, αp4n)^{172}Hf
						^{169}Tm(^{12}C, 2α2n)^{171}Lu
2.	$^{16}_{8}$O	≈ 89	$^{169}_{69}$Tm	100%	$^{185}_{77}$Ir	^{169}Tm(^{16}O, 3n) ^{182}Ir
						^{169}Tm(^{16}O, 4n) ^{181}Ir
						^{169}Tm(^{16}O, 5n) ^{180}Ir
						^{169}Tm(^{16}O, α2n) ^{179}Re
						^{169}Tm(^{16}O, α3n) ^{178}Re
						^{169}Tm(^{16}O, α4n) ^{177}Re
						^{169}Tm(^{16}O, αp3n) ^{177}W
						^{169}Tm(^{16}O, αp4n) ^{177}W
						^{169}Tm(^{16}O, αp6n) ^{178}W
						^{169}Tm(^{16}O, 2α3n)^{174}Ta
						^{169}Tm(^{16}O, 2αp4n) ^{172}Hf
						^{169}Tm(^{16}O, α3pn) ^{177}Hf

3.	$^{16}_{8}O$	≈ 99	$^{159}_{65}Tb$	100%	$^{175}_{73}Ta$	$^{159}Tb(^{16}O, 4n)\,^{171}Ta$
						$^{159}Tb(^{16}O, p4n)\,^{170}Hf$
						$^{159}Tb(^{16}O, \alpha n)\,^{170}Lu$
						$^{159}Tb(^{16}O, \alpha 3n)\,^{168}Lu$
						$^{159}Tb(^{16}O, \alpha 4n)\,^{167}Lu$
						$^{159}Tb(^{16}O, \alpha p4n)\,^{166}Yb$

All HPGe γ-detectors of the GDA setup were calibrated using various standard γ-sources of known strength. The efficiencies of these high resolution HPGe γ-detectors have been measured by putting a ^{152}Eu standard γ-source at the target position. Further, to fix the CPDA gain matching, a ^{241}Am α-source has been used. Further details of data acquisition and analysis to deduce information about spin distributions are provided in Chapter 4 and 5 of the monograph.

Measurements

4.1 Measurement of Excitation Functions and their Analysis

In the introductory part of this monograph, it has already been mentioned that various interesting phenomena are associated with heavy ion interactions; they have attracted the attention of many researchers during the last couple of decades. In heavy ion reactions, when the projectile energy is more than the Coulomb barrier, the fusion of incident ion and target nucleus is the most likely process. The composite nucleus so formed is excited and is likely to decay initially via particle emission; when the excitation energy decreases, it decays by emitting gamma radiations. Such reactions in which the projectile completely fuses with the target nucleus are referred to as the complete fusion (CF) reactions, as already mentioned in earlier chapters. These complete fusion reactions are dominant at energies slightly above the Coulomb barrier. On the other hand, at considerably higher energies, the interaction between the incident and the target heavy ions proceeds in a different way; only a part of the incident ion fuses with the target nucleus while the remaining unfused part moves on without any interaction. This is referred to as incomplete fusion (ICF), which is likely to dominate at considerably higher incident energies. However, in recent years, it has been observed that incident ions such as ^{12}C and ^{16}O that have an alpha cluster structure exhibit a significant contribution of incomplete fusion (ICF) even at low energies where the CF is expected to dominate.[1–9] Further, in some recent experiments where non-alpha cluster beams like ^{19}F were used, significant contributions by ICF were observed[10–11]. With the objective to study the dynamics of complete and incomplete fusion reactions in heavy ion interactions in a variety of projectile–target combinations, several experiments have been carried out using both alpha cluster as well as non-alpha cluster projectiles. Since a direct evidence of incomplete fusion may be obtained from the measurement of the excitation function of a specific reaction channel, excitation functions for a large number of reaction channels have been measured using the stacked foil activation technique. Table 4.1 lists the systems for which excitation functions have been measured, along with the energy range of study and the height of the Coulomb barrier for each system. The specified energy range covers from near the Coulomb barrier to well above it for each system.

Table 4.1 List of systems for which excitation functions have been measured; the energy range and the Coulomb barrier energy of the systems are also listed

S.No.	System	Energy range (MeV)	Coulomb barrier (MeV)
1.	$^{12}C + ^{128}Te$	≈42–82	40.06
2.	$^{12}C + ^{165}Ho$	≈55–80	49.72
3.	$^{12}C + ^{159}Tb$	≈59–88	48.53
4.	$^{12}C + ^{169}Tm$	≈56–89	51.04
5.	$^{12}C + ^{175}Lu$	≈55–85	52.22
6.	$^{13}C + ^{169}Tm$	≈65–86	50.49
7.	$^{14}N + ^{128}Te$	≈64–90	46.32
8.	$^{16}O + ^{27}Al$	≈58–94	15.95
9.	$^{16}O + ^{159}Tb$	≈70–95	63.69
10.	$^{16}O + ^{169}Tm$	≈71–95	66.99
11.	$^{16}O + ^{181}Ta$	≈85–100	70.09
12.	$^{16}O + ^{103}Rh$	≈50–85	47.09
13.	$^{18}O + ^{159}Tb$	≈70–99	62.66
14.	$^{19}F + ^{159}Tb$	≈80–110	70.64
15.	$^{19}F + ^{169}Tm$	≈80–110	74.32

The excitation functions for various reactions have been determined using the stacked foil activation method, while the cross-sections for various reactions have been measured employing the offline gamma-ray spectroscopy technique. The relevant formulations for the determination of the cross-section are already given in Chapter 3. The samples of a stack after bombardment by the heavy ion beam were taken out of the scattering chamber and the gamma-ray spectrum of each sample was recorded. The intensity of the characteristic gamma lines have been used to determine the production cross-sections of the residues populated via CF and/or ICF processes. Some typical examples of recorded gamma ray spectra are already given in Chapter 3. The decay data (residues, half life, spin, parity, gamma ray energy and corresponding intensity) used for the determination of the cross-section for various reactions in different systems are given in Table.4.2. These decay data have been taken from the Table of Isotopes[12].

Table 4.2   Decay data of residues identified in the reactions

System: $^{12}C + ^{128}Te$					
Reactions	**Residue**	**Half-life**	J^π	$E\gamma$ **(keV)**	**Intensity (%)**
$^{128}Te(^{12}C, 3n)$	^{137m}Ce	1.433 d	$11/2^-$	254.2	11
$^{128}Te(^{12}C, 5n)$	^{135}Ce	17.8 h	$1/2^+$	265.5 and 300	42 and 22.7
$^{128}Te(^{12}C, p4n)$	^{135}La	19.8 h	$5/2^+$	480.5	11
$^{128}Te(^{12}C, \alpha3n)$	^{133m}Ba	1.62 d	$11/2^-$	276.1	17.5
$^{128}Te(^{12}C, \alpha5n)$	^{131}Ba	11.8 d	$1/2^+$	216	20
$^{128}Te(^{12}C, \alpha4pn)$	^{131m}Te	30.0 h	$3/2^+$	240.9	7.6
System: $^{12}C + ^{165}Ho$					
$^{165}Ho(^{12}C, 3n)$	^{174}Ta	1.18 h	3^+	206.38	57.70
$^{165}Ho(^{12}C, 4n)$	^{173}Ta	3.65 h	$5/2^-$	172.19 and 180.58	17 and 2.10
$^{165}Ho(^{12}C, 5n)$	^{172}Ta	3.68 m	3^+	213.96 and 318.75	52 and 4.96
$^{165}Ho(^{12}C, p3n)$	^{173}Hf	23.6 h	$1/2^-$	139.63	12.30
$^{165}Ho(^{12}C, p5n)$	^{171}Hf	12.1 h	$7/2^+$	147.11	16.20
$^{165}Ho(^{12}C, \alpha2n)$	^{171}Lu	8.24 d	$7/2^+$	739.83	48.10
$^{165}Ho(^{12}C, \alpha4n)$	^{169}Lu	1.42 d	$7/2^+$	191.21	20.70
$^{165}Ho(^{12}C, \alpha6n)$	^{167}Lu	51.5 m	$7/2^+$	239.14	8.20
$^{165}Ho(^{12}C, 2\alpha2n)$	^{167}Tm	9.24 d	$1/2^+$	207.79	41
$^{165}Ho(^{12}C, 2\alpha4n)$	^{165}Tm	1.25 d	$1/2^+$	218.79	2.40
System: $^{12}C + ^{159}Tb$					
$^{159}Tb(^{12}C, 3n)$	$^{168}Lu^g$	5.5 m	3^+	198.86 and 228.58	180* and 70*
$^{159}Tb(^{12}C, 3n)$	$^{168}Lu^m$	6.7 m	6^-	198.86 and 228.58	180* and 70*
$^{159}Tb(^{12}C, 4n)$	^{167}Lu	51.5 m	$7/2^+$	213.21	3.5
$^{159}Tb(^{12}C, 6n)$	^{165}Lu	10.74 m	$1/2^+$	120.58 and 360.51	25 and 8.2
$^{159}Tb(^{12}C, p3n)$	^{167}Yb	17.5 m	$5/2^-$	176.2 and 177.26	20.4 and 2.4
$^{159}Tb(^{12}C, \alpha2n)$	^{165}Tm	30.06 h	$1/2^+$	242.85 and 346.44	35 and 3.9
$^{159}Tb(^{12}C, \alpha4n)$	^{163}Tm	1.81 h	$1/2^+$	190.07 and 239.67	1.28 and 4.1
$^{159}Tb(^{12}C, 2\alpha2n)$	^{161}Ho	2.48 h	$7/2^-$	103.03	3.6
$^{159}Tb(^{12}C, 2\alpha3n)$	$^{160}Ho^g$	25.6 m	5^+	645.25	16.20
$^{159}Tb(^{12}C, 2\alpha3n)$	$^{160}Ho^m$	5.02 h	2^-	728.18	30.8
System: $^{12}C + ^{169}Tm$					
$^{169}Tm(^{12}C, 4n)$	^{177}Re	14 m	$5/2^-$	196.85	100
$^{169}Tm(^{12}C, 5n)$	^{176}Re	5.2 m	3^+	240.6	100
$^{169}Tm(^{12}C, \alpha3n)$	^{174}Ta	1.05 h	3^+	206.50	58

^{169}Tm(^{12}C, α4n)	^{173}Ta	3.65 h	5/2$^-$	172.19	17
^{169}Tm(^{12}C, 2αn)	^{172}Lu	6.7 d	4$^-$	181.56	20
^{169}Tm(^{12}C, 2α2n)	^{171}Lu	8.24 d	7/2$^+$	739.8	48.1

System: ^{12}C + ^{175}Lu

^{169}Tm(^{12}C, 4n)	^{183}Ir	57 m	5/2$^-$	228.5	100*
^{169}Tm(^{12}C, 5n)	^{182}Ir	15 m	5$^+$	273.09	43.0
^{169}Tm(^{12}C, p4n)	^{182}Os	21.6 h	0$^+$	180.22	34.7
^{169}Tm(^{12}C, α2n)	^{181}Re	19.9 h	5/2$^+$	360.70	20.0
^{169}Tm(^{12}C, α4n)	^{179}Re	19.7 m	5/2$^+$	289.98	26.9
^{169}Tm(^{12}C, 2α3n)	^{176}Ta	8.08 h	1$^-$	1159.3	24.6

System: ^{13}C + ^{169}Tm

^{169}Tm(^{13}C, 3n)	^{179}Re	19.7 m	3$^+$	430.25	28
^{169}Tm(^{13}C, 4n)	^{178}Re	13.2m	7/2$^+$	106.06	23.1
^{169}Tm(^{13}C, 5n)	^{177}Re	14 m	1/2$^+$	196.85	100*
^{169}Tm(^{13}C, 6n)	^{176}Re	5.2 m	5/2$^-$	108.9	100*
^{169}Tm(^{13}C, p4n)	^{177}W	132 m	1/2$^-$	115.05	59
^{169}Tm(^{13}C, α3n)	^{175}Ta	10.5 h	7/2$^+$	207.70	13.3
^{169}Tm(^{13}C, α4n)	^{174}Ta	1.18 h	3$^+$	206.38	57.7
^{169}Tm(^{13}C, α5n)	^{173}Ta	3.65 d	5/2$^-$	172.19	17
^{169}Tm(^{13}C, 2α2n)	^{172}Lu	6.7 d	4$^-$	900.70	28.8
^{169}Tm(^{13}C, 2α3n)	^{171}Lu	8.24 d	7/2$^+$	739.82	48.1

System: ^{14}N + ^{128}Te

^{128}Te(^{14}N, 4n)	^{138m}Pr	2.12 h	7$^-$	302.7	80
^{128}Te(^{14}N, 5n)	^{137}Pr	1.28 h	5/2$^+$	434.3	1.3
^{128}Te(^{14}N, p4n)	^{137g}Ce	9.0 h	3/2$^+$	447.2	2.2
^{128}Te(^{14}N, α5n)	^{133}La	3.9 h	5/2$^+$	302.4	1.2
^{128}Te(^{14}N, α6n)	^{132g}La	4.8 h	2$^-$	540.4	7.8
^{128}Te(^{14}N, α2pn)	^{135m}Cs	53 m	19/2$^-$	786.9	99.7
^{128}Te(^{14}N, 2α2pn)	^{131}I	8.04 d	7/2$^+$	284.3	6.1
^{128}Te(^{14}N, 3α)	^{130g}I	12.36 h	5$^+$	536.1	99

System: ^{16}O + ^{27}Al

^{27}Al(^{16}O, 2αn)	^{34m}Cl	32.2 m	3$^+$	146.5	40.5
^{27}Al(^{16}O, 3α3p)	^{28}Mg	20.9 h	0$^+$	400.5	36
^{27}Al (^{16}O, 3α3pn)	^{27}Mg	9.4 m	1/2$^+$	843.7	73
^{27}Al (^{16}O, 4α2pn)	^{24}Na	14.6 m	4$^+$	1368	100
^{27}Al (^{16}O, 4α3p)	^{24}Ne	3.8 m	0$^+$	472.2	100

System: $^{16}O + ^{159}Tb$

$^{159}Tb(^{16}O, 3n)$	^{172}Ta	36.8 m	3^+	213.9, 318.7	52, 49
$^{159}Tb(^{16}O, 4n)$	^{171}Ta	23.3 m	$5/2^-$	152.2, 166.1	5.8, 19.8
$^{159}Tb(^{16}O, 5n)$	^{170}Ta	6.76 m	3^+	860.4, 986.9	7.3, 3.3
$^{159}Tb(^{16}O, p3n)$	^{171}Hf	12.1 h	$7/2^+$	122, 137.6	11.5, 12.7
$^{159}Tb(^{16}O, p4n)$	^{170g}Hf	16.01 h	0^+	120.1, 164.6	19, 33
$^{159}Tb(^{16}O, 2p2n)$	^{171}Lu	8.24 d	$7/2^+$	667, 739.8	11, 48.1
$^{159}Tb(^{16}O, \alpha n)$	^{170}Lu	2.0 d	0^+	193.1	2.07
$^{159}Tb(^{16}O, \alpha 2n)$	^{169}Lu	1.41 d	$7/2^+$	191.2	20.7
$^{159}Tb(^{16}O, 2\alpha 2n)$	^{165}Tm	1.25 d	$1/2^+$	242.8, 296.0	35, 23

System: $^{16}O + ^{169}Tm$

$^{169}Tm(^{16}O, 3n)$	^{182}Ir	15 m	5^+	126.9, 273.1	34.4, 43
$^{169}Tm(^{16}O, 4n)$	^{181}Ir	4.90 m	$5/2^-$	107.6, 123.5	15.2, 8.7
$^{169}Tm(^{16}O, p2n)$	^{182}Os	22.10 h	0^+	180.22, 263.29	34.7, 6.6
$^{169}Tm(^{16}O, p3n)$	^{181g}Os	105 m	$1/2^-$	238.68, 826.74	44, 20.2
$^{159}Tb(^{16}O, 2p2n)$	^{181}Re	19.9 h	$5/2^+$	360.7, 365.59	20, 57
$^{169}Tm(^{16}O, \alpha 3n)$	^{178}Re	13.2 m	3^+	237.19	45
$^{169}Tm(^{16}O, 2\alpha pn)$	^{175}Hf	70 d	$5/2^-$	343.4	87
$^{169}Tm(^{16}O, 3\alpha n)$	^{172}Lu	6.70 d	4^-	1093.6	63.5

System: $^{16}O + ^{181}Ta$

$^{181}Ta(^{16}O, 3n)$	$^{194}Tl^g$	32.8 m	7^+	636.1	99
$^{181}Ta(^{16}O, 3n)$	$^{194}Tl^m$	33 m	2^-	636.1	15.3
$^{181}Ta(^{16}O, 4n)$	$^{193}Tl^g$	2.1 m	$9/2^-$	365	90.1
$^{181}Ta(^{16}O, 4n)$	$^{193}Tl^m$	21.6 m	$1/2^+$	324.4, 1044.7	15.2, 8.99
$^{181}Ta(^{16}O, 5n)$	$^{192}Tl^g$	10.6 m	7^+	422.9	31.1
$^{181}Ta(^{16}O, 5n)$	$^{192}Tl^m$	9.6 m	2^-	422.9	31.1
$^{181}Ta(^{16}O, p3n)$	$^{193}Hg^g$	3.8 h	$3/2^-$	381.6, 539	11.0, 1.2
$^{181}Ta(^{16}O, p3n)$	$^{193}Hg^m$	11.8 h	$3/2^+$	258.1	60
$^{181}Ta(^{16}O, p4n)$	^{193}Hg	4.85 h	0^+	274.8	50
$^{181}Ta(^{16}O, p5n)$	$^{191}Hg^g$	49 m	$3/2^-$	224.6, 241.2	17.4, 8.9
$^{181}Ta(^{16}O, p5n)$	$^{191}Hg^m$	50.85 m	$13/2^+$	420.3, 578.7	17.9, 17
$^{181}Ta(^{16}O, \alpha n)$	^{192}Au	4.94 h	1^-	295.5, 316.5	22.7, 58
$^{181}Ta(^{16}O, \alpha 2n)$	^{191}Au	3.18 h	$3/2^+$	283.9, 399.8	6.3, 4.5
$^{181}Ta(^{16}O, \alpha 3n)$	^{190}Au	42.8 m	1^-	295.9, 301.9	71, 25.1

System: $^{16}O + ^{103}Rh$

$^{103}Rh(^{16}O, pn)$	^{117g}Te	62 m	$1/2^+$	719.7, 1090.7	64.7, 6.9
$^{103}Rh(^{16}O, p2n)$	^{116}Te	2.49 h	0^+	628.7	1.0
$^{103}Rh(^{16}O, p3n)$	^{115g}Te	5.8 m	$7/2^+$	1326.8, 1380.5	22.7, 23
$^{103}Rh(^{16}O, p3n)$	^{115m}Te	6.7 m	$1/2^+$	770.3	34.2
$^{103}Rh(^{16}O, p4n)$	^{114}Te	15.2 m	0^+	244.6, 726.5	33, 43
$^{103}Rh(^{16}O, 2p)$	^{117}Sb	2.8 h	$5/2^+$	158.6	85.9
$^{103}Rh(^{16}O, 2p2n)$	^{115}Sb	32.1 m	$5/2^+$	497.4, 489.1	98, 1.3
$^{103}Rh(^{16}O, 2\alpha)$	^{111g}In	2.8 d	$9/2^+$	171.3, 245.4	90.2, 94
$^{103}Rh(^{16}O, 2\alpha n)$	^{110g}In	4.9 h	7^+	641.6, 884.6	25.9, 92.9
$^{103}Rh(^{16}O, 2\alpha n)$	^{110m}In	1.152 h	2^+	657.7, 1235.6	98, 0.26
$^{103}Rh(^{16}O, 2\alpha 2n)$	^{109g}In	4.2 h	$9/2^+$	203.2, 623.6	73.5, 6
$^{103}Rh(^{16}O, 2\alpha 3n)$	^{108g}In	58 m	7^+	242.7	38
$^{103}Rh(^{16}O, 2\alpha 3n)$	^{108m}In	39.6 m	2^+	311.9, 968	1.01, 4.38
$^{103}Rh(^{16}O, 3\alpha n)$	^{106m}Ag	8.28 d	6^+	451, 717.4	27.6, 29
$^{103}Rh(^{16}O, 3\alpha 3n)$	^{104g}Ag	69.2 m	5^+	767.8, 555.8	65.9, 92.8
$^{103}Rh(^{16}O, 3\alpha 4n)$	^{103g}Ag	65.7 m	$7/2^+$	146	28.3

System: $^{18}O + ^{159}Tb$

$^{159}Tb(^{18}O, 3n)$	^{174}Ta	1.05 h	3^+	206.38, 206.38	57.7, 1.26
$^{159}Tb(^{18}O, 4n)$	^{173}Ta	3.14 h	$5/2^-$	160.4, 172.19	4.8, 17
$^{159}Tb(^{18}O, 5n)$	^{172}Ta	36.8 m	3^+	213.9, 318.7	52, 4.96
$^{159}Tb(^{18}O, 6n)$	^{171}Ta	23.3 m	$5/2^-$	116.9, 166.3	5.5, 19.2
$^{159}Tb(^{18}O, \alpha 2n)$	^{171}Lu	8.24 d	$7/2^+$	667, 689.2	11, 2.37
$^{159}Tb(^{18}O, \alpha 5n)$	^{168m}Lu	6.7 m	3^+	198.82, 298.6	28, 2.6
$^{159}Tb(^{18}O, \alpha 6n)$	^{167}Lu	57.5 m	$7/2^+$	239.1, 317.4	8.2, 1.5
$^{159}Tb(^{18}O, 3p7n)$	$\alpha p5n$	17.5 m	$5/2^-$	113.34, 106.1	55.4, 22.6

System: $^{19}F + ^{159}Tb$

$^{159}Tb(^{19}F, 4n)$	^{174}W	35.2 m	0^+	323.68	100*
$^{159}Tb(^{19}F, 5n)$	^{173}W	29 m	$5/2^-$	457.6	100*
$^{159}Tb(^{19}F, 6n)$	^{172}W	7.6 m	0^+	623.48	102*
$^{159}Tb(^{19}F, p4n)$	^{173}Ta	3.14 h	$5/2^-$	172.2	17.5
$^{159}Tb(^{19}F, \alpha 3n)$	^{171}Hf	12.1 h	$7/2^+$	662.2	100*
$^{159}Tb(^{19}F, \alpha 4n)$	^{170}Hf	16.01 h	0^+	164.6	26
$^{159}Tb(^{19}F, \alpha p3n)$	^{170}Lu	2 d	0^+	193.1	2.14
$^{159}Tb(^{19}F, 2\alpha 3n)$	^{167}Yb	17.5 m	$5/2^-$	176.2	20.4
$^{159}Tb(^{19}F, 2\alpha p2n)$	^{167}Tm	9.25 d	$1/2^+$	207.7	42
$^{159}Tb(^{19}F, 2\alpha p4n)$	^{165}Tm	30.06 h	$1/2^+$	242.85	35

System: ¹⁹F + ¹⁶⁹Tm						
¹⁶⁹Tm(¹⁹F, 3n)	¹⁸⁴Pt	1.18 h	1/2⁻	135.3	80*
¹⁶⁹Tm(¹⁹F, 4n)	¹⁸⁵Pt	17.3 m	0⁺	154.4	31
¹⁶⁹Tm(¹⁹F, 5n)	¹⁸³Pt	6.5 m	1/2⁻	254.6	30
¹⁶⁹Tm(¹⁹F, p3n)	¹⁸⁴Ir	3.09 h	5⁻	254.6	67.5*
¹⁶⁹Tm(¹⁹F, p4n)	¹⁸³Ir	58 m	5/2⁻	228.1	6.9
¹⁶⁹Tm(¹⁹F, αn)	¹⁸³Os	13 h	9/2⁺	381.7	77
¹⁶⁹Tm(¹⁹F, α2n)	¹⁸²Os	22.10 h	0⁺	274.5	1.81
¹⁶⁹Tm(¹⁹F, α3n)	¹⁸¹Os	105 m	1/2⁻	118.0	12.9
¹⁶⁹Tm(¹⁹F, α5n)	¹⁷⁹Os	6.5 m	1/2⁻	165.6	34*
¹⁶⁹Tm(¹⁹F, 2α3n)	¹⁷⁷W	2.2 h	1/2⁻	115.65	59
¹⁶⁹Tm(¹⁹F, 2α5n)	¹⁷⁵W	35.2 m	1/2⁻	270.2	12.6
¹⁶⁹Tm(¹⁹F, 2α6n)	¹⁷⁴W	29 m	0⁺	125.1	81
¹⁶⁹Tm(¹⁹F, 2αp3n)	¹⁷⁶Ta	8.08 h	1⁻	201.8	6
¹⁶⁹Tm(¹⁹F, 2αp4n)	¹⁷⁵Ta	10.5 h	7/2⁺	179.2	1.22

* Relative intensities

As already mentioned, the cross-sections for the population of the aforementioned reaction residues have been determined employing the formulations presented in Chapter 3. The measured cross-sections for different reactions in various systems are also given in tabular from in this chapter. In these tables, the errors shown in the cross-section values refer to the cumulative sum due to various factors. Critical evaluation of uncertainties in various quantities that may introduce errors in the measured cross-sections reflects the quality of measurements. Some of the factors that may introduce errors in the measurement of cross-sections in the present work are described in the following.

1. Non-uniformity of the sample foil may lead to uncertainty in the determination of the number of target nuclei in the sample. Though it is quite difficult to determine the uncertainty in the target thickness in the rolled foil or those samples prepared by vacuum evaporation, the thicknesses of the samples were measured at different positions of the same sample by the α-transmission technique to check the uniformity of the sample foil. It has been estimated that the error in the target thickness is ≤1%.

2. During irradiations, the beam current may fluctuate, which may result in the variation of the incident flux during the bombardment of the sample foil. Many tests were performed to check the time-integrated beam fluctuations and it was estimated that beam fluctuations may introduce errors of not more than 4–5% in the measured cross-sections. It may further be pointed out that the Pelletron crew provided a very stable beam current during the run.

3. Uncertainty in the determination of the geometry-dependent efficiency of the γ-ray spectrometer may introduce additional error in the measured cross-sections. The efficiency of the HPGe detector was measured before and very frequently during the

run, using calibrated sources of ^{60}Co and ^{152}Eu. The efficiency curves were corrected for the acquisition system dead time. Moreover, proper care was taken to keep the dead time of the detector < 10% by suitably adjusting the source–detector separation. Statistical errors in counting of the standard sources may also introduce errors in the efficiency, which can be minimized by accumulating a large number of counts for comparatively longer times ($\approx$ 5000 s). Experimental data on geometry-dependent efficiencies with γ-ray energy at a fixed source–detector separation was fitted onto a power law curve. The uncertainty due to the fitting of the efficiency curve is estimated to be < 3%. The uncertainty in determining the efficiency may also appear due to the solid-angle effect because the irradiated samples were not point sources like the standard source but had a diameter of $\approx$ 4 mm. It is estimated that the error in the efficiency due to the solid-angle effect is less than 5%.

4. The statistical errors in the evaluation of the γ-ray intensity and the background subtraction were different for different observed residues and were separately evaluated by employing the peak fitting software.

5. During the irradiation, the beam traverses the thickness of the material; thus, the initial beam intensity may get reduced. It is estimated that the error due to the decrease in beam intensity is less than 2%.

6. Dead time of counting is likely to introduce errors in determining the count rates. As already mentioned, in all the measurements, the dead time of the spectrometer was kept <10% by suitably adjusting the sample–detector distance; the corrections for it were applied to the counting rate.

7. During bombardment, the residues populated will recoil in the forward direction. If the recoiling velocity is large enough, they may not be trapped in the sample itself. This is especially important for those residues that are populated towards the other extreme of the sample foil. The loss of the product nuclei recoiling out of the sample may introduce large errors in the measured cross-sections. In the present work, the thickness of the catcher foil placed just behind the target was kept sufficient to stop even the most energetic residues populated via the CF process. Moreover, in the present measurements, both the sample and the next catcher foil were counted together and hence, the losses due to the recoiling of nuclei are totally avoided.

Further, the uncertainties of the nuclear data like branching ratio, decay constant etc., which have been taken from the Table of Isotopes,[12] have not been taken into consideration. Considering all the possible sources of errors described here, the uncertainty in the absolute values of the cross-sections is estimated to be less than 15%, including those of the statistical errors.

In the past few decades, energetic heavy ion (HI) beams have been used to study the reaction mechanism in complex nuclei. In medium energy HI reactions, one expects interplay between the compound nucleus and direct processes, including pre-equilibrium (PE) emission. The interaction between two heavy ions may be understood in terms of the interaction potential between the centres of mass of the two colliding nuclei—the potential is due to an infinite range Coulomb repulsion and short range attractive nuclear and repulsive centrifugal potentials. Such reactions have provided a unique way of populating and studying a system of nucleons under extreme degrees of freedom such as angular momentum and excitation energy. At energies around the Coulomb barrier to well above it, complete fusion (CF) and incomplete fusion (ICF) reactions along with pre-equilibrium (PE) emission have been found to have significant contributions in HI reactions. In order to study the dynamics of heavy ion reactions, a programme of precise measurement and analysis of excitation functions in a large number of systems have been undertaken. The results of these measurements are presented in this chapter.

4.1.1 Reactions initiated by ^{12}C beam

4.1.1.1 ^{12}C+^{128}Te system

Since most of the experiments on incomplete fusion in the early stages were carried out using beams like ^{12}C, ^{16}O and ^{20}Ne, which have alpha cluster structure, it was assumed that ICF occurs only in such beams. However, in recent years, it has been observed that reactions initiated by non-alpha cluster beams (e.g., ^{14}N, ^{18}O and ^{19}F) also show significant contribution from incomplete fusion processes. As such, in order to make a comparison, a programme of measurement and analysis of excitation functions for reactions employing both types of beams, i.e., beams with alpha cluster structure as well as non-alpha cluster beams have been used to study incomplete fusion reactions. In the present work, five different systems, i.e., ^{12}C+^{128}Te, ^{12}C+^{165}Ho, ^{12}C+^{159}Tb, ^{12}C+^{169}Tm and ^{12}C+^{175}Lu have been studied employing ^{12}C beams. The experimentally measured cross-section for these five systems in the energy range from near the Coulomb barrier to well above it are given in Tables 4.3–4.7.

Table 4.3 Experimentally measured production cross-sections for residues populated via CF and /or ICF processes in various systems

System: ^{12}C + ^{128}Te						
Energy (MeV)	^{137m}Ce σ (mb)	^{135}Ce σ (mb)	^{135}La σ (mb)	^{133m}Ba σ (mb)	^{131}Ba σ (mb)	^{131m}Te σ (mb)
42.2 ± 1.4	14.4 ± 1.7	1.7 ± 0.7	–	0.08 ± 0.01	–	4.4 ± 0.4
57.7 ± 1.2	114.3 ± 13.0	2.2 ± 0.7	1.2 ± 0.1	7.8 ± 0.8	–	4.9 ± 0.5
70.8 ± 1.0	18.2 ± 2.3	208.2 ± 22.0	56.1 ± 6.2	43.6 ± 4.8	0.73 ± 0.17	22.2 ± 2.5
82.0 ± 0.9	8.9 ± 1.3	292.7 ± 32.0	82.4 ± 9.1	37.6 ± 4.2	7.1 ± 0.8	30.9 ± 3.5

In the ^{12}C+^{128}Te system, excitation functions (EFs) for the reactions ^{128}Te(^{12}C,3n)^{137m}Ce, ^{128}Te(^{12}C,5n)^{135}Ce, ^{128}Te(^{12}C, p4n)^{135}La, ^{128}Te(^{12}C,α3n)^{133m}Ba, ^{128}Te(^{12}C,α5n)^{131}Ba and ^{128}Te (^{12}C,α4pn)^{131m}Te have been measured in the energy range ≈ 42–82 MeV, using the stacked foil activation technique[7]. The analysis of the measured excitation functions has been carried out using the computer codes CASCADE[13] and ALICE-91[14]. The code CASCADE is purely statistical in nature, while the code ALICE-91 also includes PE emissions. However, ICF is not considered in the calculations done by both these codes. Though, the details of the codes

CASCADE and ALICE are already given in Chapter 2 of this monograph, for the sake of completeness, a brief description of these codes and particularly the parameters used in them are described here.

The code CASCADE is based on the Hauser–Feshbach theory[15] and does not consider ICF and/or PE emission. In this code, the level density parameter a of the compound system, the ratio of the actual moment of inertia of the excited system to the rigid body moment of inertia F_θ and level density parameter at the fission saddle point a_f are some of the important parameters. The transmission coefficients in these calculations are generated using the optical model potentials of Becchetti and Greenlees[16] for neutrons and protons; those of Satchler[17] are used for alpha particles. The Fermi gas model is used in this code to calculate the level density parameter a for the product nuclei, using the expression $a = A/K$, where A is the atomic number and K is the free parameter that may be adjusted to reproduce experimental data. The default value of K in the code is 10; however, in the present work, the calculations have been done by varying the value of K in the range 10–14. The effect of variation in the values of K on the calculated excitation functions is shown in Figure 4.1. As can be observed from these figures, in general, $K = 14$ reproduces the experimental data satisfactorily. In model-based calculations, the value of parameter F_θ is varied from 0.55 to its default value 0.85. The effect of variation in parameter F_θ on calculated excitation functions is shown in Figure 4.2, and is found to have negligible effect on the calculated excitation functions for the reactions under investigation. This code also takes into account the fission channel. The level density parameter a_f at the saddle point may be obtained from the relation $a_f = A/D_{AF}$, where A is the mass number of the nucleus and D_{AF} is a sort of free parameter whose default value in the code is 8. The parameter D_{AF} is found to influence the calculated EFs considerably. As such, the influence of variation of D_{AF} from 8 to 11 on calculated EFs has also been studied. The resulting excitation functions using these values of parameter $D_{AF} = 8$–11, $K = 14$ and $F_\theta = 0.55$ are shown in Figure 4.3. A value of $D_{AF} = 11$ gives a good agreement, in general, with the experimental data, even in the peak region of the excitation function. As can be seen from these figures, EFs for the reactions $^{128}Te(^{12}C,3n)^{137m}Ce$, $^{128}Te(^{12}C,5n)^{135}Ce$ and $^{128}Te(^{12}C, p4n)^{135}La$ are qualitatively in good agreement with the theoretical calculations done with code CASCADE in the peak region. The higher values of the experimental cross-sections in the tail portion of the EF for the reaction $^{128}Te(^{12}C,3n)^{137m}Ce$ [Figure 4.3(a)] as compared to the theoretical calculations may be attributed to PE emission, which contributes significantly to the reaction cross-section at higher energies and is not considered in the CASCADE calculations. In the $^{128}Te(^{12}C,5n)$ channel, the experimental data is satisfactorily reproduced by the CASCADE[13] calculations as is indicated in Figure 4.3(b). This shows that at 82 MeV energy, there is negligible contribution from PE emission in the (C,5n) channel. This is expected as PE emission is more likely in the first step of de-excitation that leaves the residual system in an excited state from where the emission of four neutrons may take place. It may, however, be pointed out that the PE emission in the (C,5n) channel may appear at still higher excitation energies. The calculations done for the reactions $^{128}Te(^{12}C,\alpha3n)^{133m}Ba$ and $^{128}Te(^{12}C,\alpha5n)^{131}Ba$ are not in good agreement with experimentally measured EFs; they reproduce only the qualitative trend of the measured EFs. One of the plausible reasons for this discrepancy may be the large contribution of ICF processes in these cases. In Figure 4.1(f), the calculations done using code CASCADE are not plotted as the theoretical calculations give cross-section values less than 0.1 mb at energies of interest for the reaction ^{128}Te(^{12}C,α4pn)^{131m}Te. This indicates that ICF is the dominant mode of reaction for this channel at the energies studied in the present experiment.

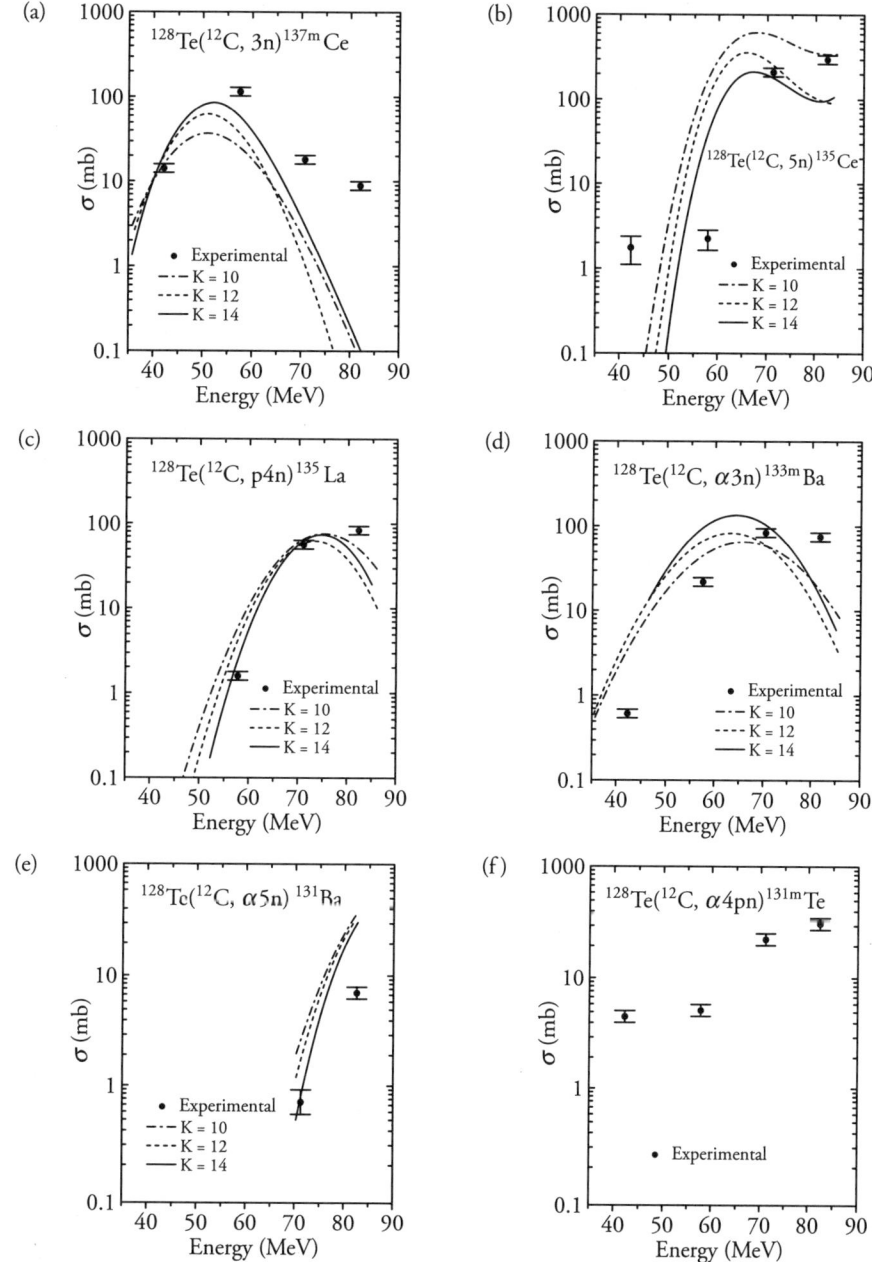

Figure 4.1 Experimentally measured and theoretically calculated excitation functions using code CASCADE: Effect of variation of level density free parameter (K) on calculated excitation functions with $F_\theta = 0.65$ and $D_{AF} = 8$

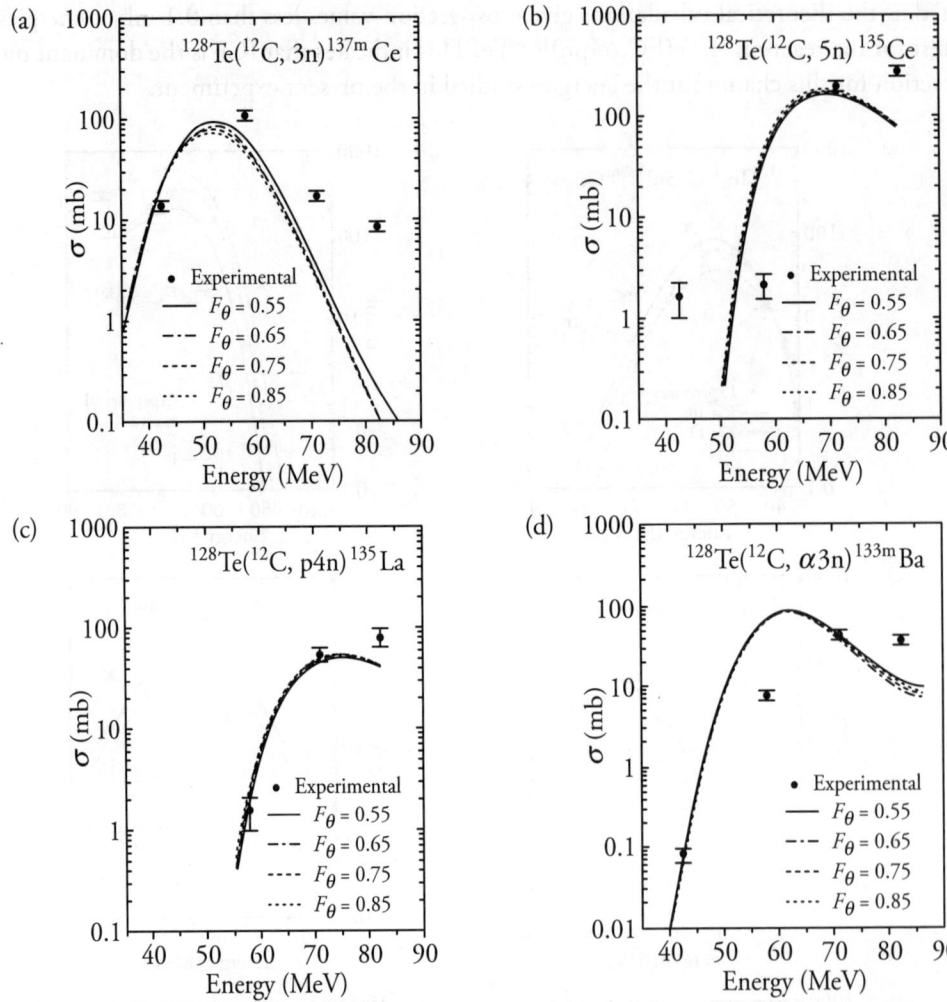

Figure 4.2 Experimentally measured and theoretically calculated excitation functions using code CASCADE: Effect of variation of parameter F_θ on calculated excitation functions with $K = 14$ and $D_{AF} = 8$

Code ALICE-91 developed by M. Blann has also been used to analyse the data for the excitation functions for this system, particularly to account for equilibrium (CN) as well as PE emissions in heavy ion reactions. The CN calculations in this code are performed using the Weisskopf–Ewing model[18]; while the PE component is simulated using the hybrid/geometry dependent hybrid model[19]. In this code, the level density parameter a, the initial exciton number n_0 and the mean free path multiplier COST are some of the important parameters. The level density parameter mainly affects the equilibrium component, while the initial exciton number n_0 and mean free path multiplier COST govern the PE component. The level density parameter a is calculated from the expression $a = A/K$, where K is a parameter that may be varied to match the experimental data. Calculations have been performed with different values of these parameters. The effect of variation of the parameter K on calculated EFs is presented in Figure 4.4. As can be seen from these figures, in the present calculations, a value of $K = 18$ and COST = 2, in general, satisfactorily reproduce the experimental data for all the reactions.

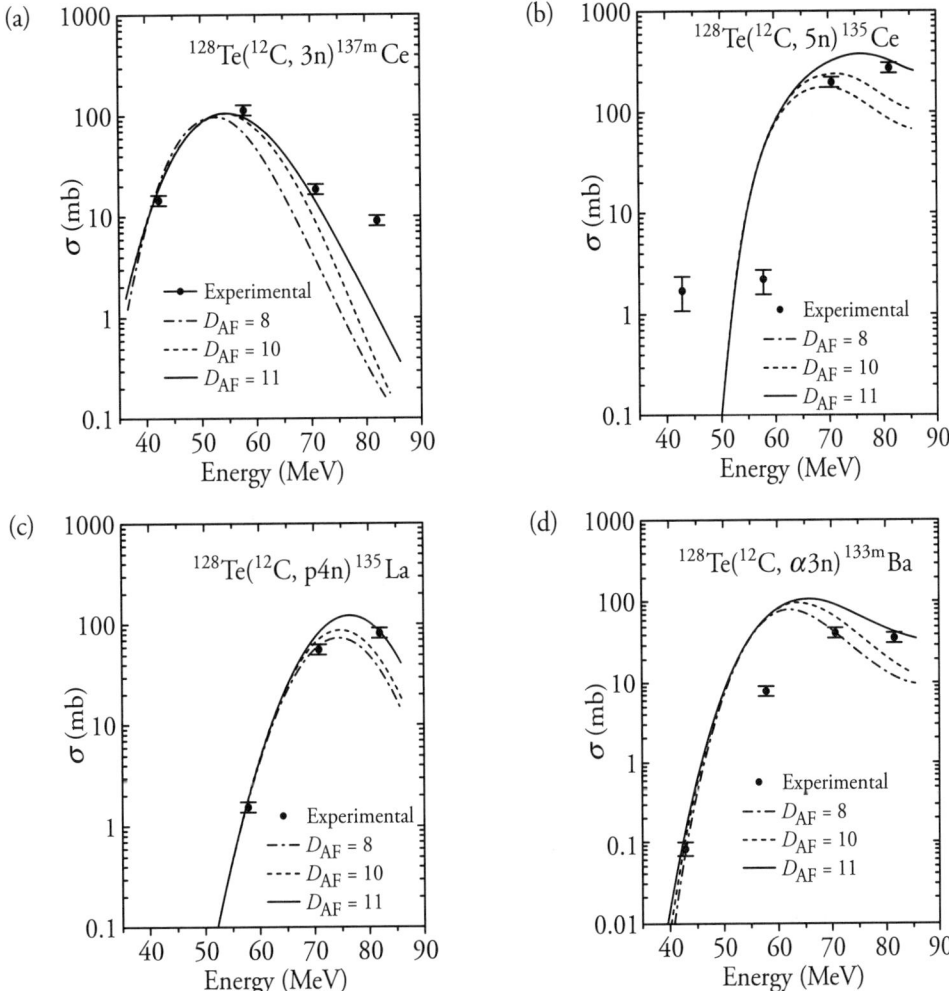

Figure 4.3 Experimentally measured and theoretically calculated excitation functions using code CASCADE: Effect of variation of parameter D_{AF} on calculated excitation functions with $K = 14$ and $F_\theta = 0.55$

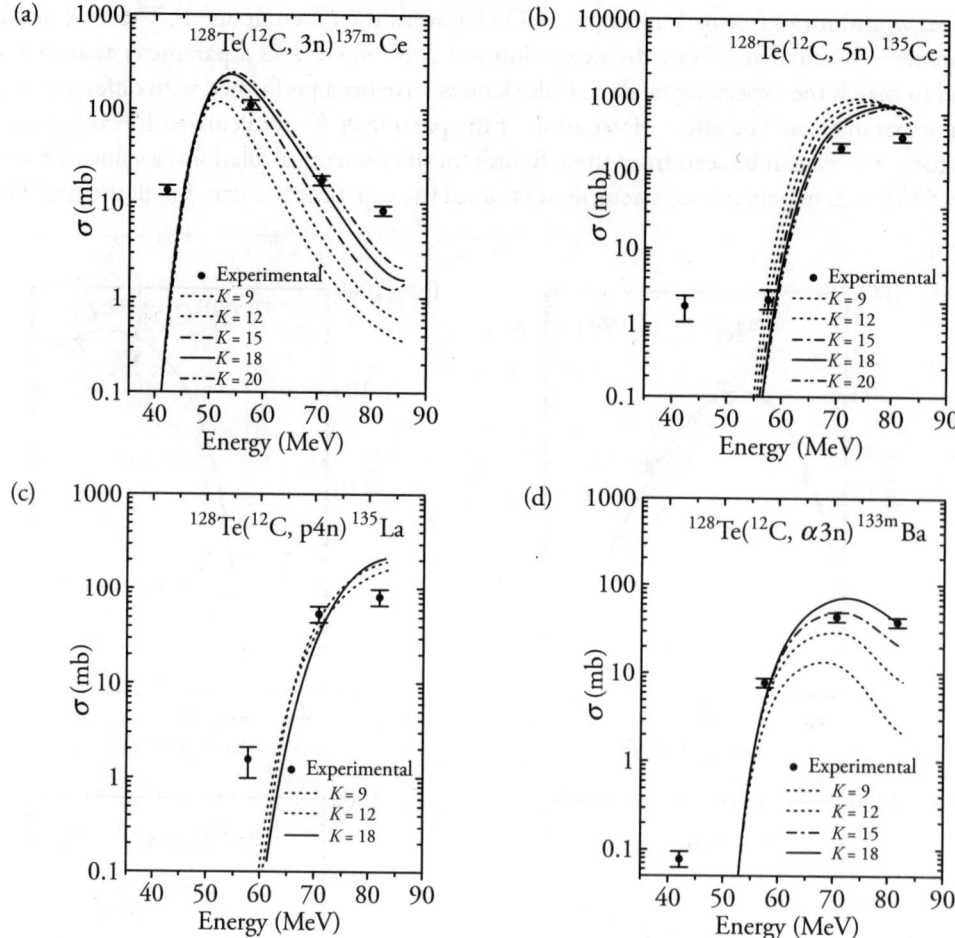

Figure 4.4   Experimentally measured and theoretically calculated excitation functions using code ALICE-91: Effect of variation of level density parameter (K) on calculated excitation functions with initial exciton number $n_0 = 12$ and mean free path multiplier COST = 2

In the hybrid model,[19] the intermediate states of the system are characterised by the excitation energy E, number n_p of excited particles and n_h of excited holes. Particles and holes are defined relative to the ground state of the nucleus and are called excitons. The initial configuration of the compound system defined by the exciton number $n_0 = n_p + n_h$ is an important parameter of PE formalism. It is of particular interest to determine the initial exciton number required to reproduce the data. In order to see the effect of variation in the values of the initial exciton number n_0 on calculated EFs, calculations for different initial exciton configurations were performed. As a representative case, the calculations for the reactions ^{128}Te(^{12}C,3n)^{137m}Ce and ^{128}Te(^{12}C, α3n)^{133m}Ba for $n_0 = 12(6p+6n+0h)$ and $n_0 = 14(6p+7n+1h)$ are shown in Figure 4.5, respectively.

(a)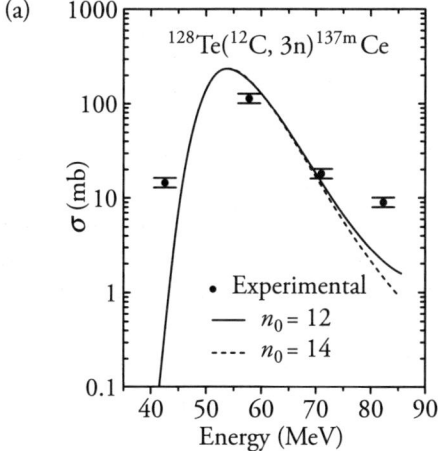

(b)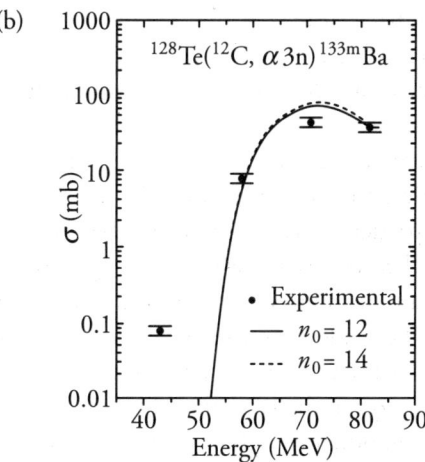

Figure 4.5 The experimentally measured and theoretically calculated excitation functions using code ALICE-91: Effect of variation of parameter n_0 on calculated excitation function with $K = 18$ and mean free path multiplier COST = 2

It may be seen from these figures that, as expected, lower values of initial exciton number, in general, give larger PE contributions. This is because lower value of n_0 means larger number of two-body interactions prior to the establishment of an equilibrium characteristic of CN resulting in larger PE contribution. Further, it has been found that the parameter COST does not influence the calculated EFs to any significant extent. As a representative case the effect of variation of parameter COST on the calculated EF for the reaction ^{128}Te(^{12}C,3n)^{137m}Ce is shown in Figure 4.6.

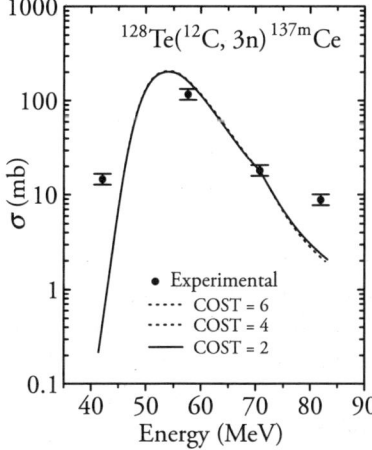

Figure 4.6 The experimentally measured and theoretically calculated excitation function using code ALICE-91: Effect of variation of parameter COST on calculated excitation function with $K = 18$ and initial exciton number $n_0 = 12$

Further, it may be pointed out that the maxima of the measured EFs were found to peak at energies higher than the corresponding calculated EFs. This is expected, since in ALICE-91 calculations, the angular momentum effects have not been taken into account. In HI induced reactions, the incident ion imparts relatively larger angular momentum to the composite system. If, in the last stages of nuclear de-excitation, higher angular momentum inhibits particle emission more than it does gamma emission, then, the peak of EFs corresponding to the particle emission mode will be shifted to higher energies[20]. The effect is more pronounced in HI reactions as compared to light ion reactions since the rotational energy is much greater in case of HI reactions. An estimate of the possible shift due to angular momentum effects may be made from the nuclear rotational energy E_{rot}. For a rigid body $E_{rot} \sim (m/M)E_{lab}$, where m/M is the ratio of the projectile and the target nucleus masses and E_{lab} is the incident energy of the ion in the laboratory frame[20]. In the present case, at incident energies 42.2 to 82 MeV, the rotational energies may vary from 3.9 to 7.7 MeV. Since the angular momentum effects have not been considered in the Weisskopf–Ewing calculations of the present version of code ALICE-91, it is desirable to shift the calculated EFs by an amount approximately equal to E_{rot} as calculated earlier. It has been observed that the ALICE-91 calculations satisfactorily reproduce the experimental data when the energy scales of the calculated excitation functions are shifted by their respective E_{rot} values. The enhancement of cross-sections in the measured EFs for the reactions ^{128}Te(^{12}C, α3n)^{133m}Ba, ^{128}Te(^{12}C,α5n)^{131}Ba and ^{128}Te(^{12}C,α4pn)^{131m}Te, in general, as compared to the theoretical predictions can be attributed to the fact that these channels may be populated not only by the CF of ^{12}C but may also have a significant contribution from the ICF of ^{8}Be or of ^{12}C with ^{128}Te. It may, however, be noted that for the ^{128}Te(^{12}C,α3n)^{133m}Ba reaction, the measured cross-section value at energy ~42.2 MeV is larger than what is predicted by the calculations.

The reaction ^{128}Te(^{12}C,3n) produces residual isotope ^{137}Ce, which has both a ground ($t_{1/2}$ = 9 h) as well as a metastable state ($t_{1/2}$ = 1.43 d). The metastable state ^{137m}Ce decays to the ground state by the emission of 254.2 keV (11%) γ-rays. Since the ^{137}Ce emits γ-rays of very low intensities, it could not be observed in the present experiment. As such, only the contribution of the metastable state ^{137m}Ce has been measured. In the present case, the residue ^{135}La may be produced independently via the reaction ^{128}Te(^{12}C,p4n); the same residue (^{135}La) may also be produced by the β^+ decay of its higher charge isobar precursor (^{135}Ce) produced via (^{12}C,5n) reaction. The independent yield of ^{135}La could not be measured in the present analysis because the half-lives of the residue ^{135}La and its precursor ^{135}Ce are not very different (19.8 h and 17.7 h respectively). However, in such cases, the formulations developed by Evans[21] may be followed, according to which the ratio of the activities of the parent (^{135}Ce) and the daughter (^{135}La) having nearly the same half-lives would increase linearly for some time. Using these formulations, the yield of ^{135}La via the precursor decay of ^{135}Ce has been found to be less than 1 mb at 82 MeV. The experimentally measured cross-sections for the reaction (^{12}C,p4n) also contain the contributions of the precursor decay. The reaction ^{128}Te(^{12}C,α4pn) produces both the ground state ^{131}Te($t_{1/2}$ = 25 min) as well as the isomeric state ^{131m}Te ($t_{1/2}$ = 1.2 d). The isomeric state decays to ground state. Since the counting of the irradiated samples was started after a considerable time lapse after irradiation on account of the high activity of the irradiated samples, the ^{131}Te ground state contribution could not be separated for the (^{12}C,α4pn) reaction.

4.1.1.2 ^{12}C+^{165}Ho system

Excitation functions for several reactions in the ^{12}C+^{165}Ho system have also been measured in the energy range ≈ 55–80 MeV. The excitation functions for the reactions ^{165}Ho(^{12}C,3n)^{174}Ta, ^{165}Ho(^{12}C,4n)^{173}Ta, ^{165}Ho(^{12}C,5n)^{172}Ta, ^{165}Ho(^{12}C,p3n)^{173}Hf, ^{165}Ho(^{12}C,p5n)^{171}Hf, ^{165}Ho (^{12}C,α2n)^{171}Lu, ^{165}Ho(^{12}C, α4n)^{169}Lu, ^{165}Ho(^{12}C, α6n)^{167}Lu, ^{165}Ho(^{12}C, 2α2n)^{167}Tm and ^{165}Ho(^{12}C, 2α4n)^{165}Tm have been measured. The analysis of the excitation functions have been carried out within the framework of the statistical model code CASCADE[13]. The analysis of data with code ALICE[14] has also been done, where the pre-compound emission has also been taken into account. The measured cross-section values for the various residues populated via different reaction channels are presented in Table 4.4

Table 4.4   Measured values of cross-sections for various residues populated via different reaction channels in the system ^{12}C+^{165}Ho

System: ^{12}C + ^{165}Ho						
Energy (MeV)	**^{174}Ta** σ (mb)	**^{173}Ta** σ (mb)	**^{172}Ta** σ (mb)	**^{173}Hf** ^{ind} σ (mb)	**^{171}Hf** ^{cum} σ (mb)	**^{171}Lu** σ (mb)
55 ± 1.1	46 ± 6	13 ± 2	–	–	–	–
62 ± 0.9	62 ± 8	200 ± 27	–	11 ± 2	–	11 ± 2
71 ± 1.0	11 ± 2	336 ± 44	83 ± 11	446 ± 60	–	19 ± 3
80 ± 0.9	2 ± 0.3	122 ± 16	287 ± 38	404 ± 54	145 ± 22	140 ± 21
Energy (MeV)	**^{169}Lu** σ (mb)	**^{167}Lu** σ (mb)	**^{167}Tm** σ (mb)	**^{165}Tm** σ (mb)		
62 ± 0.9	14 ± 2	6 ± 1	5 ± 1	99 ± 14		
71 ± 1.0	19 ± 3	7 ± 1	34 ± 4	336 ± 47		
80 ± 0.9	21 ± 3	36 ± 5	109 ± 15	252 ± 37		

The experimental data for the xn and pxn channels in this system has been satisfactorily reproduced by the calculations done using the two computer codes, CASCADE and ALICE-91, however, enhancement of measured cross-sections as compared to the calculated ones have been observed for the alpha-emitting channels, viz., ^{165}Ho(^{12}C,α2n)^{171}Lu, ^{165}Ho(^{12}C,α4n)^{169}Lu, ^{165}Ho(^{12}C,α6n)^{167}Lu, ^{165}Ho(^{12}C, 2α2n)^{167}Tm and ^{165}Ho(^{12}C,2α4n)^{165}Tm. As a representative case, measured and calculated excitation functions for the reactions ^{165}Ho(^{12}C,α2n)^{171}Lu, and ^{165}Ho(^{12}C,α4n)^{169}Lu are shown in Figure 4.7. As may be seen from these figures, theoretical calculations underestimate the measured cross-sections. This enhancement of cross-sections with respect to the theoretical calculations (which does not take the ICF into consideration) may be attributed to incomplete fusion processes. Further details of this system are presented in the work by Gupta, et al.[22]

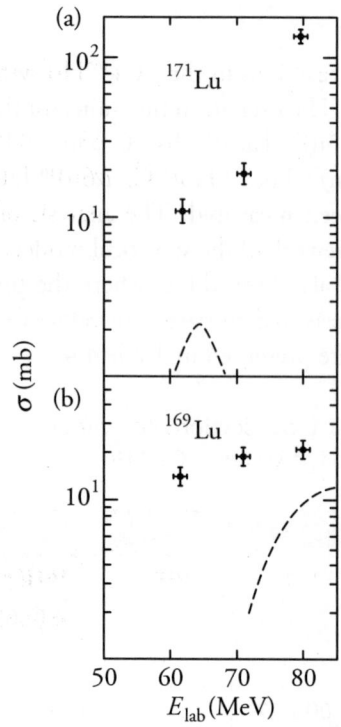

Figure 4.7 Experimentally measured and theoretically calculated excitation functions for reactions ^{165}Ho(^{12}C,α2n)^{171}Lu, and ^{165}Ho(^{12}C, α4n)^{169}Lu; the enhancement in the measured cross-sections as compared to the theoretical calculations is attributed to incomplete fusion process. The dashed lines represent theoretical calculations

4.1.1.3 ^{12}C+^{159}Tb *System*

Another experiment using ^{12}C projectile has been carried out with ^{159}Tb target nucleus at energies ~4–7 MeV/nucleon, using the Pelletron accelerator facility of IUAC, New Delhi, India. At these energies, the complete fusion process is likely to be the sole contributor to the total reaction cross-section; however, a significant ICF contribution has been observed at these low energies[8]. This start of ICF at energies slightly above the Coulomb barrier has regenerated interest in the study of ICF reactions. Collisions between the two heavy ions with relative angular momentum values of $\ell < \ell_{crit}$ may result in the formation of a completely fused system due to the fusion of the entire projectile with the target ion. This gives rise to the distribution of kinetic energy and angular momenta of the projectile among all the internal degrees of freedom, leading to the formation of a fully equilibrated compound nucleus. On the other hand, if $\ell > \ell_{crit}$, the pocket in the entrance channel potential vanishes, resulting in hindrance for complete fusion and giving rise to ICF processes. Here a part of the projectile is ejected as a spectator to release the excess angular momentum. Such partial fusion of the projectile gives rise to an incompletely fused composite system. Due to the alpha cluster structure of the projectile, the ICF reaction mechanism is likely to affect decay channels, which involve alpha particles. It may again be pointed out that due to non-availability of any reliable theoretical model for predicting ICF contributions at low energies, the experimental study of ICF is still an open field of investigation.

In view of the aforementioned, excitation functions for the reactions $^{159}Tb(^{12}C,3n)^{168}Lu^{g+m}$, $^{159}Tb(^{12}C,4n)^{167}Lu$, $^{159}Tb(^{12}C,6n)^{165}Lu$, $^{159}Tb(^{12}C,p3n)^{167}Yb$, $^{159}Tb(^{12}C,\alpha2n)^{165}Tm$, $^{159}Tb(^{12}C,\alpha4n)^{163}Tm$, $^{159}Tb(^{12}C,2\alpha2n)^{161}Ho$, and $^{159}Tb(^{12}C,2\alpha3n)^{160}Ho$, which are expected to proceed through CF and/or ICF of ^{12}C, have been measured and are presented in Table 4.5

Table 4.5 Experimental values of reaction cross-sections for various residues populated in the system $^{12}C+^{159}Tb$ at different energies

System: $^{12}C + ^{159}Tb$						
Energy (MeV)	$^{168g+m}Lu$ σ (mb)	^{167}Lu σ (mb)	^{165}Lu σ (mb)	^{167}Yb σ (mb)	^{165}Tm σ (mb)	^{163}Tm σ (mb)
54.8 ± 0.5	184.5 ± 29.6	12.2 ± 2	–	0.2 ± 0.0	–	–
58.5 ± 0.5	177.5 ± 24.2	155.2 ± 23.2	–	1 ± 0.1	13.6 ± 1.9	–
61.3 ± 0.6	118 ± 15.2	340.1 ± 51	–	10 ± 1.9	16.5 ± 2.5	–
62.6 ± 0.8	94.5 ± 12.8	545.2 ± 79	–	15 ± 2.1	19.6 ± 2.6	1 ± 0.0
65.4 ± 0.6	47.5 ± 6.9	635.2 ± 93.5	–	19.5 ± 2.4	17.4 ± 2.1	8 ± 1.0
67.2 ± 1.1	45 ± 6.2	649.3 ± 85	–	49 ± 6.5	12.5 ± 1.5	22 ± 2.6
69.1 ± 0.8	14 ± 3.0	699. 5 ± 102.9	–	55.3 ± 7.9	10.1 ± 1.3	58.4 ± 7.9
72.2 ± 0.7	7.5 ± 1.2	499.6 ± 70.6	0.03 ± 0.0	45.8 ± 6.6	8.7 ± 1.5	108.1 ± 15.6
74.9 ± 0.5	2.5 ± 0.2	298.4 ± 42.1	2 ± 0.3	19.2 ± 2.5	8.2 ± 0.9	129.6 ± 18.5
77.7 ± 0.6	1 ± 0.1	149.6 ± 19.5	36 ± 4.5	29 ± 4.2	8.9 ± 1.1	202.5 ± 32
79.6 ± 0.6	0.5 ± 0.08	150 ± 19.8	58 ± 8.5	11 ± 1.5	9.9 ± 1.3	189.9 ± 28.6
82.4 ± 0.8	0.15 ± 0.03	89.9 ± 12.6	301 ± 46	8 ± 1.0	12.0 ± 1.5	211.2 ± 26.6
84.4 ± 0.5	0.1 ± 0.02	35.3 ± 4.5	408 ± 58.9	4.5 ± 0.5	17.7 ± 2.6	204.1 ± 29.4
87.3 ± 0.6	0.01 ± 0.0	23.2 ± 3.9	745 ± 109.5	4 ± 0.4	18.9 ± 2.3	212.2 ± 25.6
Energy (MeV)	^{161}Ho σ (mb)	^{160g}Ho σ (mb)	^{160m}Ho σ (mb)			
54.8 ± 0.5	–	–	–			
58.5 ± 0.5	–	–	–			
61.3 ± 0.6	–	–	–			
62.6 ± 0.8	–	–	–			
65.4 ± 0.6	–	–	–			

67.2 ± 1.1	–	–	–		
69.1 ± 0.8	1.3 ± 0.1	–	–		
72.2 ± 0.7	4.4 ± 0.5	–	–		
74.9 ± 0.5	6.2 ± 1.1	2.3 ± 0.2	1.4 ± 0.2		
77.7 ± 0.6	7.5 ± 0.9	5.4 ± 0.7	1.5 ± 0.2		
79.6 ± 0.6	6.3 ± 0.8	5.7 ±0.8	1.8 ± 0.2		
82.4 ± 0.8	5.3 ± 0.7	6.4 ± 0.9	2.2 ± 0.3		
84.4 ± 0.5	4.5 ± 0.6	7.8 ± 1.2	2.5 ± 0.3		
87.3 ± 0.6	2.1 ± 0.3	6.0 ± 0.8	2.0 ± 0.3		

In order to determine to what extent the decay of the produced residues may be explained by the CN mechanism, the measured EFs have been analysed by employing the theoretical statistical model code PACE4[23], which is based on the Hauser–Feshbach theory of CN decay. It uses the statistical approach of CN decay by the Monte Carlo procedure. In the PACE4 code, the angular momentum projections are calculated at each stage of de-excitation, which enables the determination of the angular distribution of the emitted particles; the conservation of angular momentum is explicitly considered at each decay stage. The Bass model[24] is used to calculate the fusion cross-sections while the default optical model parameters for neutrons, protons and alpha particles are taken. The level density parameter ($a = A/K$ MeV^{-1}) is one of the important parameters of this code. The value of parameter K may be varied to match the experimental excitation functions. It may be mentioned that the enhancement in the measured EFs compared to PACE4 values may be regarded as if it is due to ICF processes. In Figure 4.8(a), the measured excitation functions for the reactions ^{159}Tb(^{12}C,3n)^{168}Lu^{g+m}, ^{159}Tb(^{12}C,4n)^{167}Lu, ^{159}Tb(^{12}C,6n)^{165}Lu, ^{159}Tb(^{12}C,p3n)^{167}Yb are shown. The solid lines through the data points are drawn to guide the eyes. The decay curve analysis indicated that the residues ^{167}Yb populated in the p3n reaction are strongly fed (through β+ emission) from their higher charge isobar precursor ^{167}Lu formed in the 4n reaction. The half-life of the precursor is larger than that of the daughter nuclei. As such, the independent cross-section of ^{167}Yb residues has been deduced using successive decay formulations[25]. In order to compare the experimental excitation functions of xn and pxn channels with PACE4 predictions and to choose the proper value of the level density parameter, different values of parameter K (8–12) have been tested. Figure 4.8(b) shows the measured EFs for all (xn + pxn) channels ($\Sigma\sigma_{CF}$) and the corresponding theoretical predictions of the PACE4 code. As can be seen from this figure, the sum of cross-sections for complete fusion is well reproduced by the corresponding PACE4 predictions ($K = 8$) indicating the production of these residues through the de-excitation of the fully equilibrated compound nucleus formed in the CF reaction.

Moreover, the experimentally measured EFs for reactions ^{159}Tb(^{12}C,α2n)^{165}Tm, ^{159}Tb (^{12}C,α4n)^{163}Tm, ^{159}Tb(^{12}C,2α2n)^{161}Ho and ^{159}Tb(^{12}C,2α3n)^{160}Ho are presented in Figure 4.9(a)–(d). Because alpha particles are involved in these reactions, they are expected to be populated via both the CF and/or ICF decay routes. In case of the residues populated via

(a)
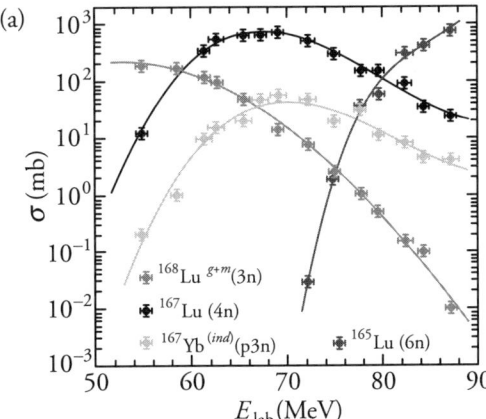

(b)
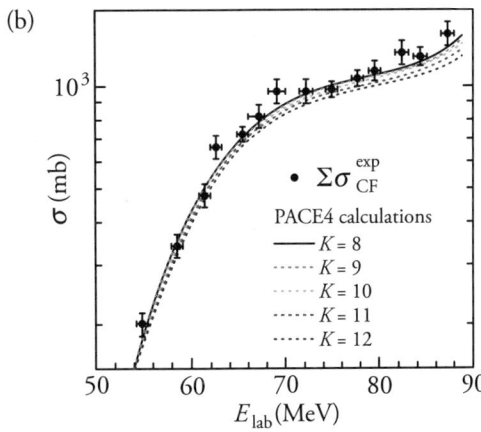

Figure 4.8 (a) Experimental EFs of ^{168}Lu^{g+m}(3n), ^{167}Lu(4n), ^{165}Lu(6n) and ^{167}Yb(p3n) residues populated in the ^{12}C+^{159}Tb system. The solid lines through the data points are drawn to guide the eyes. (b) Sum of the experimentally measured EFs of all (xn + pxn) channels ($\Sigma\sigma_{CF}$) are compared with that predicted by PACE4 for different values of the level density parameter ($a = A/K$ MeV^{-1}), where the effect of variation of the free parameter K from 8 to 12 is also shown

CF processes, the entire projectile, i.e., ^{12}C, fuses with the target nucleus ^{159}Tb forming a fully equilibrated CN, that may undergo decay via αxn channels. On the other hand, for the residues involving ICF process, only a fraction of the incident ion ^{12}C (i.e., ^{8}Be+α) fuses with the target nucleus ^{159}Tb forming an incompletely fused composite system, where the remnant ^{8}Be or ^{4}He (α particle) gets projected in the forward direction as a spectator. This fraction of the proceedings via the α-emitting channel may be accounted for by analysing the corresponding EFs within the framework of the statistical model code PACE4[23]. PACE4 does not take ICF into account; therefore, any enhancement in the measured EFs with respect to the PACE4 predications may be attributed to the contribution expected from ICF processes. As such, the experimental EFs of individual α-emitting channels likely to be produced via both CF and/or ICF are compared with the outcomes of the PACE4 code in Figures 4.9(a)–(d).

It may be pointed out here that the calculations using code PACE4 have been performed employing the same set of parameters that have been used to reproduce the xn+pxn channels populated through CF processes. The PACE4 predictions in these figures is shown by solid black lines using the level density parameter $a = A/8$ MeV^{-1}. It is quite clear from Figure 4.9 that, in general, PACE4 underestimates the measured EFs of these alpha emitting reaction channels. This observed enhancement in the measured cross-sections with respect to the PACE4 predictions may be attributed to the ICF processes, in such a way that its contributions are distributed over the full energy range of EFs, with a different behaviour depending on the residues. It may be noted that the trends in the measured EF for ^{165}Tm residues, which basically reflect the interplay between CF and ICF processes through different decay channels, are as follows.

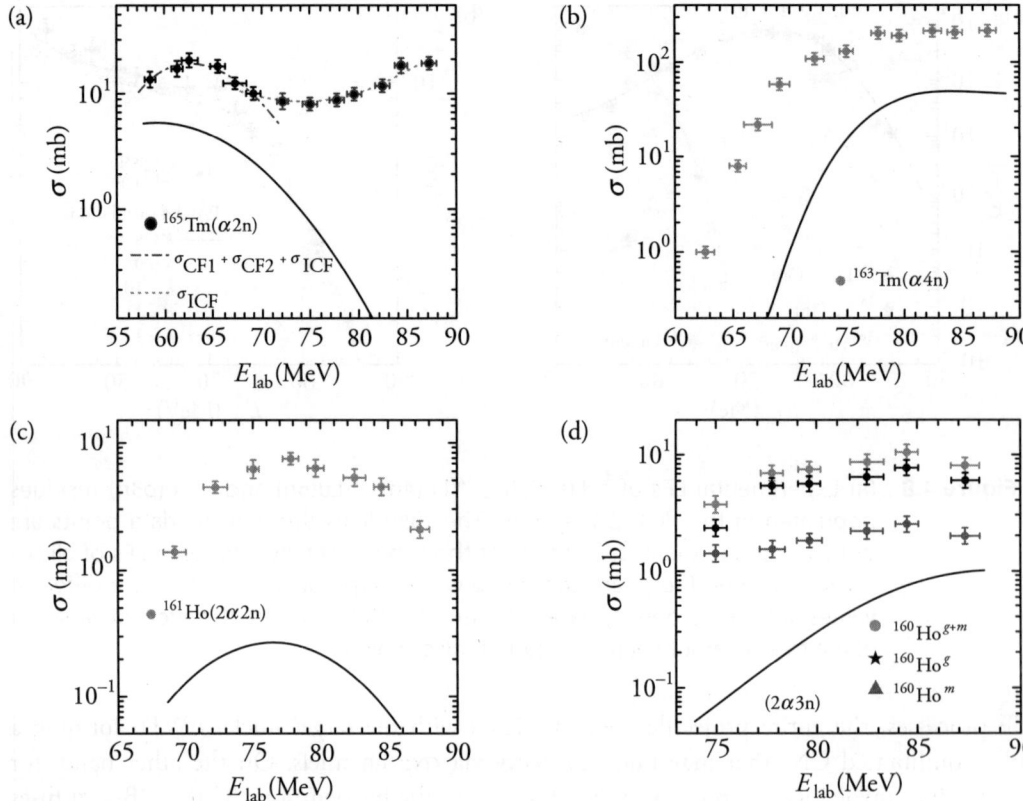

Figure 4.9   Experimentally measured EFs of evaporation residues ^{165}Tm(α2n), ^{163}Tm(α4n), ^{161}Ho(2α2n), and ^{160}Ho^{g+m}(2α3n) are compared with the PACE4 predictions. Solid black curves represent PACE4 predictions performed for $a = A/8$ MeV^{-1}. In Figure 4.9(a), the dash dotted and dotted lines through the data points are drawn to explain the trend of the excitation function

i.   CF 1: The CF of ^{12}C with ^{159}Tb leads to an excited nucleus ^{171}Lu*, which may decay via two protons and four neutrons (2p4n channel) as

$$^{12}\text{C} + {}^{159}\text{Tb} \Rightarrow {}^{171}\text{Lu}^* \Rightarrow {}^{165}\text{Tm} + 2\text{p}4\text{n}$$

Q value = -53.47 MeV

E_{thr^*} = 57.50 MeV

ii.   CF 2: The excited ^{171}Lu* nucleus formed in a CF reaction may decay through an α-cluster and two neutrons (α2n channel) as

$$^{12}\text{C} + {}^{159}\text{Tb} \Rightarrow {}^{171}\text{Lu}^* \Rightarrow {}^{165}\text{Tm} + \alpha2\text{n}$$

Q value = -25.17 MeV

E_{thr^*} = 27.07 MeV

iii. ICF: Only a part of the projectile ^{12}C (i.e., ^{8}Be) fuses with ^{159}Tb to form an incompletely fused composite system (^{167}Tm*) while an α-cluster flows in the forward direction as a spectator. The excited ^{167}Tm* may then decay via two neutrons (2n) as

^{12}C(^{8}Be + α) $\Rightarrow$ ^{8}Be +^{159}Tb $\Rightarrow$ ^{167}Tm* $\Rightarrow$ ^{165}Tm + α + 2n, (α particle as a spectator)

Q value = –17.80 MeV

E_{thr^*} = 18.70 MeV

The various contributions likely to be expected from different decay routes, as mentioned earlier are marked by dash dotted (CF 1 + CF 2 + ICF) and dotted lines (mainly ICF) drawn through the data points in Figure 4.9(a). As indicated in this figure, the contributions coming from CF 1 and CF 2 (i.e., the contributions of 2p4n and/or α2n) peak at ~63 MeV. Further, at energies above ~70 MeV, PACE4 gives very small values of cross-sections compared to the measured data points. In view of this, it may be safely inferred that the ICF contributes significantly to the population of residues ^{165}Tm via the α2n channel, where the alpha particle acts as a spectator. As such, it may be observed that in the ^{12}C interaction with ^{159}Tb, there is a significant contribution from the ICF process. For better indication of the ICF contribution in alpha-emitting channels in Figure 4.10, the sum of the cross-section of all identified α channels is compared with corresponding values predicted by the PACE4 code.

As can be seen from this figure, measured EFs for αxn+2αxn channels are much higher than the model code predictions for the level density parameter, which were used to reproduce the CF residues in the present work. As such, since the PACE4 does not take ICF into account, the observed enhancement in the measured EFs over the theoretical ones points towards the contribution of ICF processes in the population of these radioactive residues.

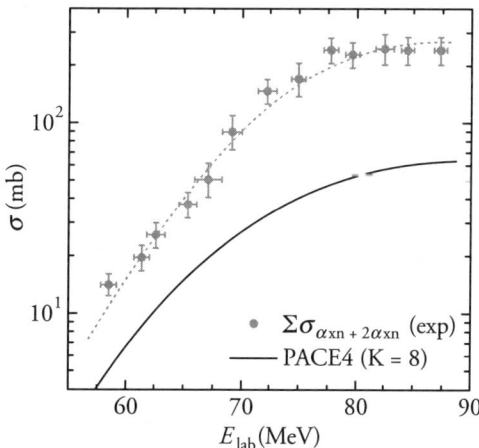

Figure 4.10 Comparison of experimentally measured and theoretically predicted EFs of all α-emitting channels; physically justified level density parameter $a = A/8$ MeV^{-1} is used in PACE4 calculations. The value of ($\Sigma\sigma_{exp}$ αxn+2αxn) is significantly higher than that predicted by PACE4, which may be attributed to the contribution of ICF. Lines through the data points are drawn to guide the eyes

4.1.1.4 ^{12}C+^{169}Tm System

To study the influence of incomplete fusion on complete fusion at energies from near the Coulomb barrier to well above it, the absolute cross-sections for different reaction channels in the system ^{12}C+^{169}Tm have also been measured, employing the general purpose scattering chamber (GPSC) at the Pelletron accelerator facility of IUAC, New Delhi, India. The details of the experimental conditions are already given in Chapter 3. As already mentioned, the residues have been identified using the decay curve analysis and characteristic gamma lines. The most intense gamma lines have been used for decay curve analysis and for the production cross-section measurements. In this system, ^{169}Tm(^{12}C,4n)^{177}Re and ^{169}Tm(^{12}C,5n)^{176}Re reactions have been identified and their cross-sections measured. The cross-sections for 4n and 5n channels have also been calculated using the complete fusion model code PACE4. As may be seen in Figure 4.11, the excitation functions for 4n and 5n channels agree well with the calculations, indicating that these channels are populated through CF mode. Moreover, some prominent alpha-emitting channels ^{169}Tm(^{12}C,α3n)^{174}Ta, ^{169}Tm(^{12}C,α4n)^{173}Ta, ^{169}Tm(^{12}C,2α2n)^{172}Lu, ^{169}Tm(^{12}C,2α3n)^{171}Lu have also been identified, where the contribution from incomplete fusion is expected. Here, the enhancements of the measured excitation functions as compared to their theoretical calculations have been attributed to contributions from ICF processes. The measured excitation functions for these reactions are given in Table 4.6 and are shown in Figure 4.12.

Table 4.6 Measured values of cross-sections for residues populated at different energies in the system ^{12}C+^{169}Tm

System: ^{12}C + ^{169}Tm						
Energy (MeV)	**^{177}Re** σ (mb)	**^{176}Re** σ (mb)	**^{174}Ta** σ (mb)	**^{173}Ta** σ (mb)	**^{172}Lu** σ (mb)	**^{171}Lu** σ (mb)
56.1 ± 0.9	7 ± 0.9	–		–	–	–
59.0 ± 0.9	95 ± 10.0	–	1.02 ± 0.1	–	11.41 ± 1.7	4.1 ± 0.6
59.9 ± 0.9	129.5 ± 19.4	–	2.09 ± 0.3	–		6.4 ± 1.0
62.0 ± 0.9	296 ± 27	–	6.6 ± 1.01	–	14.5 ± 1.9	5.96 ± 1.0
65.4 ± 0.8	515.6 ± 77.4	–	18.2 ± 2.7	7.2 ± 1.0	17.2 ± 2.8	9.94 ± 1.59
66.1 ± 0.9	572 ± 50.9	–	23.6 ± 2.8	1.51 ± 0.2	18.51 ± 2.1	11.51 ± 1.4
68.6 ± 0.8	487.1 ± 73	38.9 ± 5.8	26.6 ± 4	6.19 ± 0.9		–
71.4 ± 0.8	651 ± 68	169 ± 21.2	46.4 ± 5.8	28.1 ± 3.5	25.3 ± 3.9	23.6 ± 3.7
74.2 ± 0.9	588.6 ± 88.3	225.1 ± 33.7	62.3 ± 9.3	14.7 ± 2.5	–	–
77.1 ± 0.8	300 ± 24	600 ± 75	69.5 ± 8.3	70.8 ± 9.5	24.43 ± 2.1	–
79.0 ± 0.8	276.1 ± 41.4	470 ± 70.5	–	–	–	–
83.2 ± 0.8	90 ± 10	757 ± 94.4	64.3 ± 7.9	128.8 ± 18.4	–	–

83.6 ± 0.9	145 ± 21.7	739.3 ± 110.9	55.7 ± 8.3	68.4 ± 10.2	–	–
89.2 ± 0.7	22 ± 4.06	522 ± 72.6	45.7 ± 6.4	–	–	54.5 ± 5.2

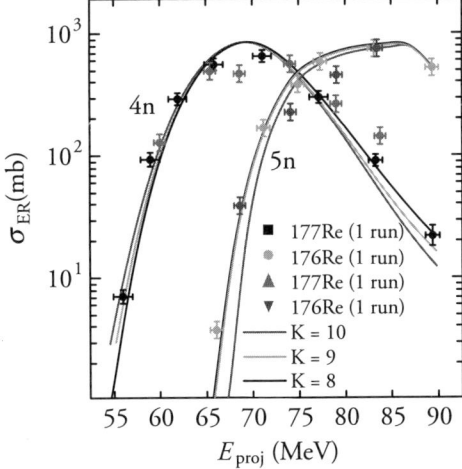

Figure 4.11 Experimentally measured and theoretically calculated excitation functions for the reactions ^{169}Tm(^{12}C,4n)^{177}Re and ^{169}Tm(^{12}C,5n)^{176}Re

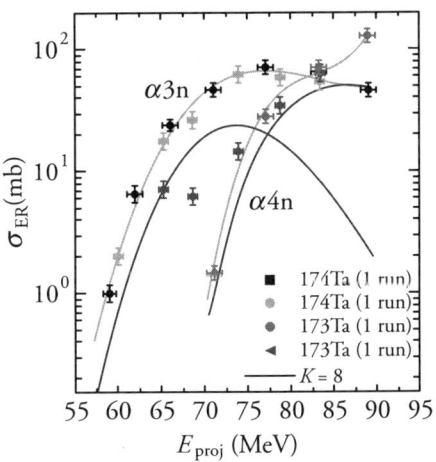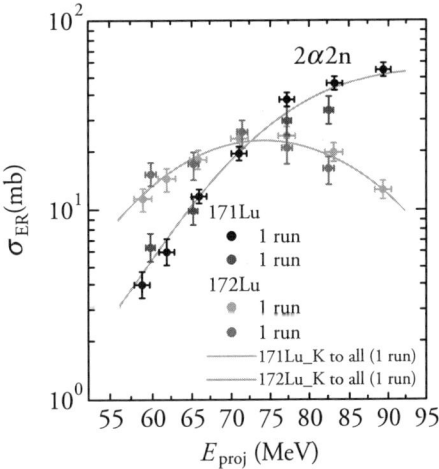

Figure 4.12 Experimentally measured and theoretically calculated excitation functions for ^{169}Tm(^{12}C,α3n)^{174}Ta, ^{169}Tm(^{12}C,α4n)^{173}Ta, ^{169}Tm(^{12}C,2α2n)^{172}Lu, ^{169}Tm(^{12}C,2α3n)^{171}Lu reactions

In Figure 4.13(a) and Figure 4.13(b), the $\Sigma\sigma_{CF}$ and $\Sigma\sigma_{ICF}$ cross-sections have been plotted along with the $\sigma_{TF} = \Sigma\sigma_{CF} + \Sigma\sigma_{ICF}$ on both the logarithmic as well as linear scales. As may be observed in these figures, as the projectile energy increases, the ICF contribution also increases indicating its strong energy dependence.

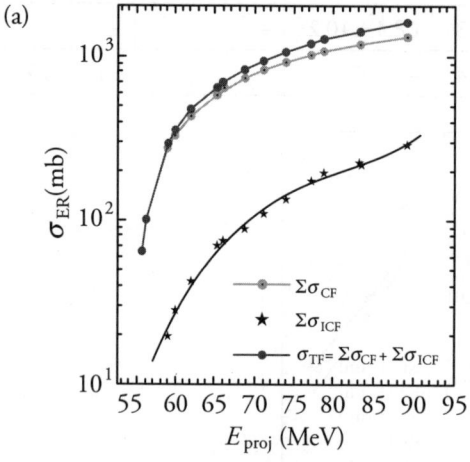
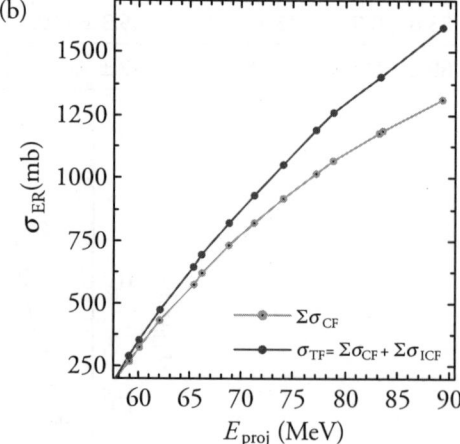

Figure 4.13 (a) Total fusion cross-section ($\sigma_{TF} = \Sigma\sigma_{CF} + \Sigma\sigma_{ICF}$) along with the sum of all CF channels ($\Sigma\sigma_{CF}$) and ICF channels ($\Sigma\sigma_{ICF}$) as a function of projectile energy and (b) the zoom of $\Sigma\sigma_{CF}$ and σ_{TF} comparison on a linear scale for easy visualization of increasing ICF strength with projectile energy. Solid curves represent best fit to the data points

4.1.1.5 ^{12}C+^{175}Lu *System*

The measurement and analysis of excitation functions (EFs) may be used to study the reaction mechanism involved in the production of reaction residues. In the present section, the EFs for ^{183}Ir (4n), ^{182}Ir (5n), ^{182}Os (p4n), ^{181}Re (α2n), ^{179}Re (α4n) and ^{176}Ta (2α3n) radio-nuclides produced in the interaction of the ^{12}C+^{175}Lu system at energies starting from the threshold to $\approx$ 7 MeV/A have been reported. The measured cross-sections for these reactions are tabulated in Table 4.7. The measured EFs have been analysed within the framework of statistical model code PACE4. The production possibilities of Ir, Os, Re and Ta isotopes via different reaction modes and decay routes are discussed here. The higher charge isobar precursor contribution in the population of ^{182}Os (p4n) has been estimated from the cumulative cross-section.

Table 4.7 Measured cross-section values at different energies for the various residues identified in the system ^{12}C+^{175}Lu

System: ^{12}C + ^{175}Lu						
Energy (MeV)	^{183}**Ir** σ (mb)	^{182}**Ir** σ (mb)	^{176}**Os** ^{ind} σ (mb)	^{181}**Re** σ (mb)	^{179}**Re** σ (mb)	^{176}**Ta** σ (mb)
58.42 ± 0.95	73.19 ± 10.97	–	–	7.34 ± 1.66	–	–
60.33 ± 1.18	109.25 ± 16.3	–	–	9.73 ± 2.04	–	–
64.11 ± 0.8	345.75 ± 51.8	–	–	19.92 ± 3.16	–	–

66.23 ± 0.77	493.58 ± 59.0	–	–	25.46 ± 5.59		–
69.30 ± 0.7	557.78 ± 53.6	30.14 ± 4.52	0.635 ± 0.0	31.60 ± 4.74	20.1 ± 3.6	–
71.17 ± 1.31	497.00 ± 59.5	119.33 ± 17.9	2.28 ± 0.3	29.70 ± 4.45	25.6 ± 4.6	–
73.49 ± 1.02	510.99 ± 76.6	334.23 ± 50.1	12.85 ± 1.9	25.30 ± 3.79	30.3 ± 5.4	9.98 ± 1.7
76.40 ± 0.9	379.34 ± 56.9	478.11 ± 71.7	27.3 ± 4.0	21.20 ± 3.18	54.3 ± 11.5	15.4 ± 1.0
78.36 ± 1.08	289.47 ± 43.4	528.87 ± 79.3	40.20 ± 6.0	28.41 ± 4.26	70.3 ± 15.7	18.8 ± 1.2
80.70 ± 1.3	196.86 ± 29.5	667.59 ± 100.1	55.30 ± 8.2	40.38 ± 6.05	77.5±13.9	13.6 ± 2.4
83.20 ± 0.8	159.85 ± 23.9	731.50 ± 109.7	67.00 ± 10.0	65.23 ± 9.78	106 ± 19.0	11.6 ± 2.1

The measured EFs for the residues ^{183}Ir (4n), ^{182}Ir (5n) and ^{182}Os (p4n) in interactions of ^{12}C+^{175}Lu system are shown in Figure 4.14 (a–c), along with the predictions of PACE4. The notations used in these figures are self explanatory.

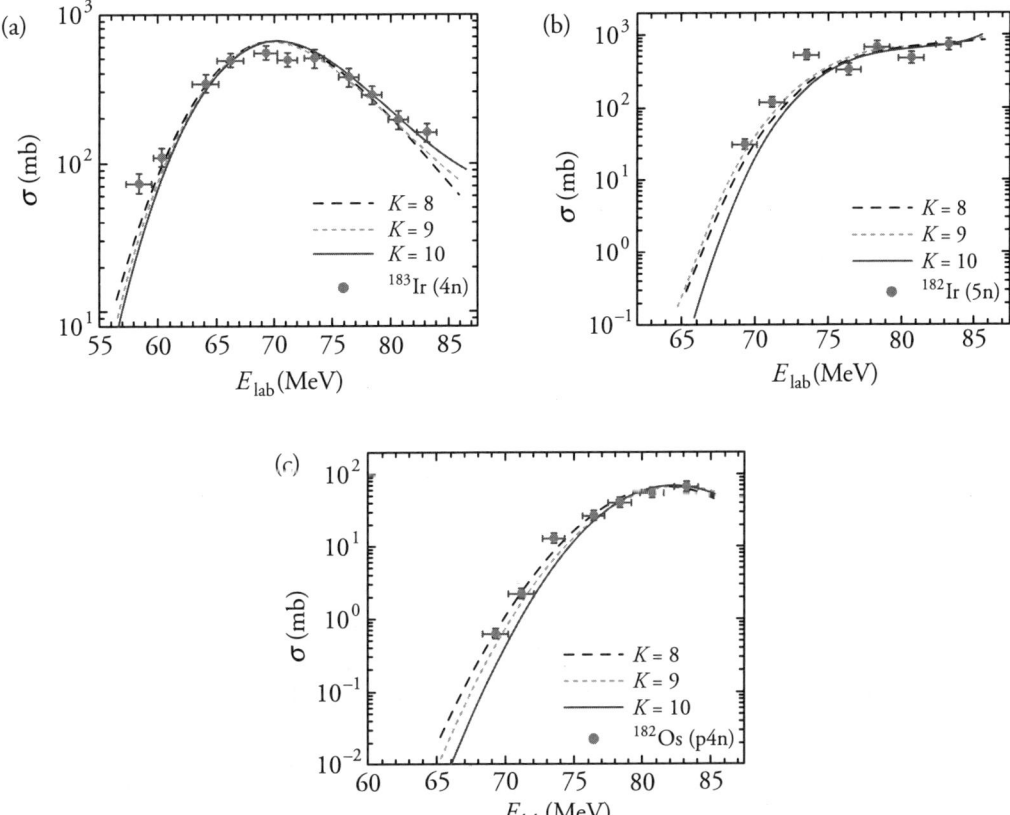

Figure 4.14 Experimental EFs for xn (x = 4 and 5) and p4n-channels populated in the ^{12}C+^{175}Lu system

In PACE4 calculations, the level density parameter ($a = A/K$) is an important parameter and by varying the parameter K, one can reproduce the experimentally measured EFs. In the present work, the experimental data has been tested using three different values of free parameters, viz., $K = 8, 9$ and 10. It may, however, be pointed that a value of $K > 10$, may give rise to anomalous effects in particle multiplicity, and CN temperature[26]. It may be observed from Figure 4.14 that theoretical calculations done using the value $K = 10$ reproduces the experimental data satisfactorily. Hence, in Figure 4.15, the sum of experimentally measured EFs of all xn and pxn channels (i.e., $\Sigma\sigma_{xn+pxn}^{exp}$) has been compared with the corresponding statistical model code PACE4 calculations at K=10. As shown in this figure, the $\Sigma\sigma_{xn+pxn}^{exp}$ is reproduced reasonably well by the PACE4 calculations using the level density parameter 'a' = A/10. This indicates that these residues are populated through the de-excitation of fully equilibrated compound nucleus formed in the CF reaction process.

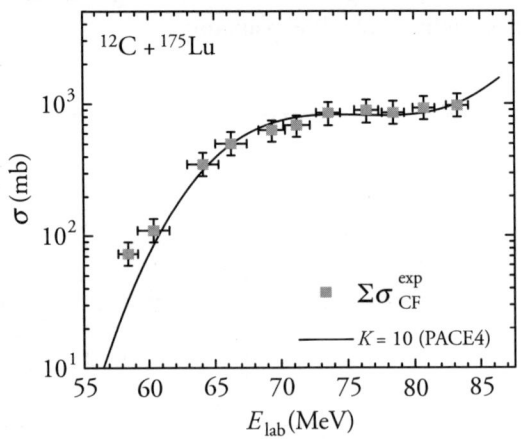

Figure 4.15 Sum of experimentally measured EFs of all xn and pxn-channels ($\Sigma\sigma_{xn+pxn}^{exp}$) along with the values predicted by the PACE4 code at $K = 10$

As has already been pointed out, ICF is not taken into consideration in the theoretical calculations done by using code PACE4; hence, if there is any enhancement in the experimentally measured production cross-section as compared to the calculations, it may be attributed to the ICF processes. In Figure 4.16 (a–c), the experimentally measured and theoretically calculated EFs of ^{181}Re ($\alpha 2n$), ^{179}Re ($\alpha 4n$) and ^{176}Ta ($2\alpha 3n$) residues identified in system ^{12}C+^{175}Lu at energies $\approx$ 4–7 MeV/A have been shown. Although, it is possible to explain all the excitation functions with different values of parameters of the code for individual channels, from the physics point of view, it is quite unreasonable. As such, all calculations have been performed consistently using the same set of parameters for all the channels as used for reproducing the CF channels. As may be seen from Figure 4.16, experimentally measured EFs for all the presently measured α-emitting channels are underestimated by PACE4 calculations done with the same set of parameters. Due to involvement of α-emission in the exit channel, these residues are expected to be populated both via CF and/or ICF processes. In case of CF, the incident projectile entirely fuses with the target nucleus (^{175}Lu) to form a fully equilibrated

CN, which may eventually decay via αxn channels. However, in case of ICF, only a part of the incident projectile (i.e., $^{12}C \rightarrow {}^{8}Be + \alpha$) fuses with the target nucleus to form an incompletely fused composite system; the remnant α particle or ^{8}Be goes on moving in the forward cone as a spectator. In Figure 4.16 (a–c), solid curves are the PACE4 predictions with level density parameter $K = 10$. The observed higher production cross-section over the PACE4 predictions may be because the ICF processes and its contribution are distributed over the full energy range of the EFs, with different behavior depending on the residue. Particularly interesting is the trend of the EF measured through the $\alpha 2n$ channel, which appears to reflect the interplay between CF and ICF processes through three different decay channels.

i. CF 1: The CF of ^{12}C with ^{175}Lu leads to an excited compound nucleus $^{187}Ir^{*}$, which may decay via two protons and four neutrons (2p4n) emission, as;

$$^{12}C + {}^{175}Lu \rightarrow {}^{187}Ir^{*} \rightarrow {}^{181}Re + 2p4n$$

$$Q \text{ value} = -55.52 \text{ MeV}$$

$$E_{\text{threshold}} = 59.34 \text{ MeV}$$

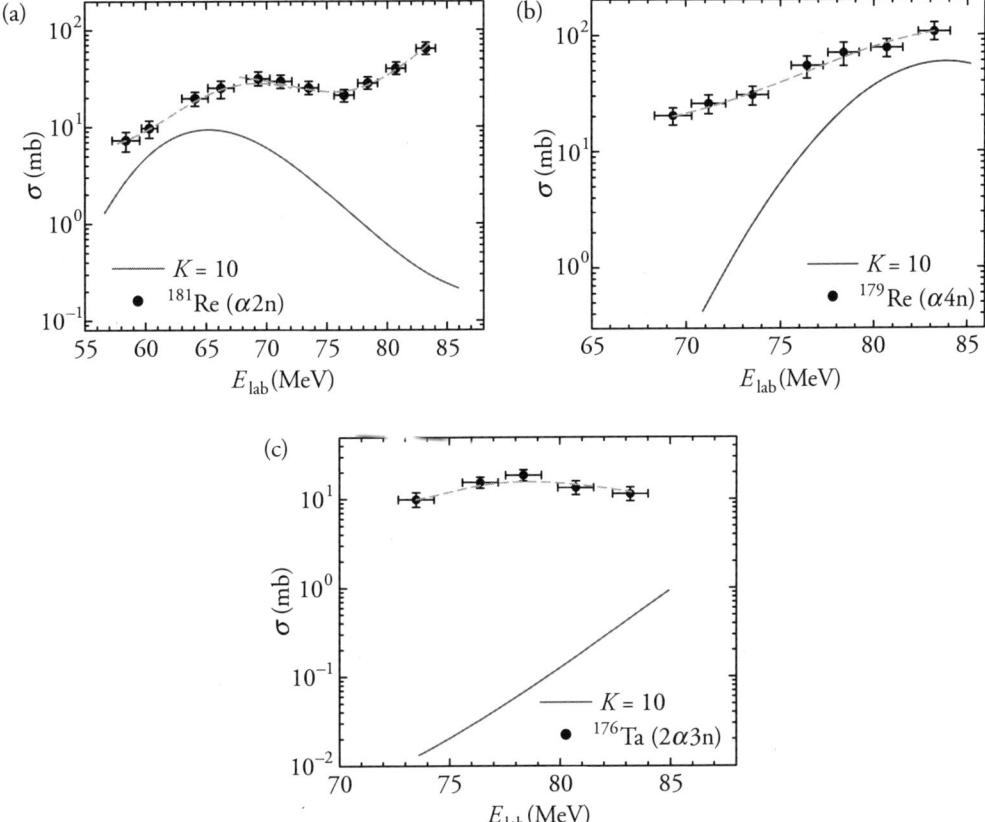

Figure 4.16 Experimentally measured EFs of the $\alpha 2n$, $\alpha 4n$ and $2\alpha 3n$ channels populated in $^{12}C+{}^{175}Lu$ system; the PACE4 predictions at $K = 10$ are shown by solid curves

ii.   CF 2: The excited compound nucleus (^{187}Ir*) formed in a CF reaction may decay through an α cluster and two neutrons ($\alpha 2n$ channel) as

$$^{12}C + {}^{175}Lu \rightarrow {}^{187}Ir^* \rightarrow {}^{181}Re + \alpha 2n$$

Q value = -27.23 MeV

$E_{\text{threshold}} = 29.09$ MeV

iii. ICF: Only a part of the projectile ^{12}C (i.e., ^{8}Be) fuses with ^{175}Lu to form an incompletely fused composite system (^{183}Re*), while an α particle flows in the forward direction as a spectator. The excited ^{183}Re* may then decay by the emission of two neutrons (2n) as

$$^{12}C\,(^{8}Be + \alpha) \rightarrow {}^{8}Be + {}^{175}Lu \rightarrow {}^{183}Re^* \rightarrow {}^{181}Re + \alpha + 2n$$

(α particle as a spectator)

Q value = -19.86 MeV

$E_{\text{threshold}} = 20.77$ MeV

Figure 4.16 (a–c) shows the experimentally measured excitation functions for $\alpha 2n$, $\alpha 4n$ and $2\alpha 3n$ channels respectively. The dotted line through the experimental data points is just to guide the eyes. Further, the PACE4 calculations for each channel are shown by the continuous line. The contributions of the two complete fusion channels (i.e., CF 1 and CF 2) mentioned earlier have already been taken into account in PACE4 calculations. As such, the difference between the PACE4 excitation functions and the experimentally observed excitation functions for each channel may then be attributed to the ICF processes.

The ^{182}Os (p4n) evaporation residues have been found to be strongly fed from their higher charge isobar precursor ^{182}Ir populated via the 5n channel through β^+-emission. The half-life of the precursor, i.e., ^{182}Ir, is 15 min., and is considerably smaller than the half-life of the daughter nuclei, i.e., ^{182}Os $\rightarrow t_{1/2}^{d} = 21.6$ h. As demonstrated by Cavinato et al.[25], the independent production cross-section (σ_{ind}) of the daughter nucleus may be defined in terms of the cumulative (σ_{cum}) and precursor (σ_{pre}) cross-section as follows

$$\sigma_{\text{ind}} = \sigma_{\text{cum}} - F_{\text{pre}} \cdot P_{\text{pre}}$$

Here, F_{pre} is the precursor coefficient, which depends on the branching ratio of the precursor decay (P_{pre}) to the final nucleus as

$$F_{\text{pre}} = P_{\text{pre}} \frac{t_{1/2}^{d}}{(t_{1/2}^{d} - t_{1/2}^{\text{pre}})}$$

Here, $t_{1/2}^{d}$ and $t_{1/2}^{\text{pre}}$ are the half-lives of the daughter and precursor nuclei. P_{pre} is the branching ratio of the precursor to its daughter nuclei. After the inclusion of these, the independent production cross-section (σ_{ind}) can be written as;

$$\sigma_{\text{ind}} = \sigma_{\text{cum}} - P_{\text{pre}} \frac{t_{1/2}^{d}}{(t_{1/2}^{d} - t_{1/2}^{\text{pre}})} \sigma_{\text{pre}}$$

As has already been mentioned, the production of ^{182}Os is substantially fed from its pre-cursor, and the value obtained for F_{pre} is 1.0117, providing;

$$\sigma_{ind}\,(^{182}Os) = \sigma_{cum} - 1.0117\,.\,\sigma_{pre}$$

The value of σ_{ind} is plotted in Figure 4.14 (c) as an independent production cross-section of ^{182}Os (p4n-channel).

The experimentally measured cross-section for the α-emitting channels was underestimated by the PACE4 calculations; hence, significant ICF contributions are expected in α-emitting channels. As such, in order to have a better description of the ICF contribution in the α-emitting channels, the sum of the cross-sections of all identified α-channels ($\Sigma\sigma^{exp}_{\alpha xn+2\alpha xn}$) is compared with that estimated by the statistical model code PACE4 ($\Sigma\sigma^{PACE4}_{\alpha xn+2\alpha xn}$), in Figure 4.17. As can be seen from this figure, the sum of all experimentally measured EFs of α-channels is significantly higher than PACE4 calculations. It may be seen from the figure that ICF starts contributing at ≈ 59 MeV. Further, the enhancement in experimentally measured cross-sections over PACE4 values is relatively higher at higher energies as expected. The contribution of the ICF has been defined as $\sigma_{ICF} = \sigma^{exp}_{\alpha xn+2\alpha xn} - \sum\sigma^{PACE4}_{\alpha xn+2\alpha xn}$. It may be pertinent to mention that the channels viz., ^{180}Re (α3n), ^{178}Re (α5n), ^{177}Re (α6n) and ^{179}W (αp3n) could not be identified in the present work due to their short half-lives; hence, PACE4 values for these α-emitting channels have been corrected for the missing channels. The deduced values of the σ_{ICF} are plotted in

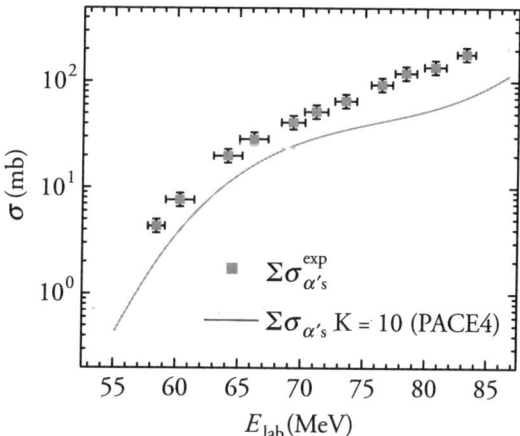

Figure 4.17 Comparison of the ($\Sigma\sigma^{exp}_{\alpha xn+2\alpha xn}$) in the ^{12}C+^{175}Lu system with corresponding PACE4 prediction at $K = 10$

Figure 4.18. As can be seen from the figure, ICF contribution increases rapidly with energy, as expected. Further, the incomplete fusion fraction percentage (% F_{ICF}) has also been estimated from the experimentally measured production cross-section.

$$\% F_{ICF} = \frac{\sum \sigma_{ICF}}{\sigma_{TF}} \times 100$$

The values of F_{ICF} deduced here are plotted as a function of the reduced incident projectile energy (E_{Lab}/V_b) in Figure 4.19. Figure 4.19 defines the empirical probability of ICF at different incident projectile energies. As mentioned earlier, $\sum \sigma_{TF}$ has been corrected for the missing channels that could not be measured in the present work. As such, the F_{ICF} values may be taken as the lower limits of the ICF contribution.

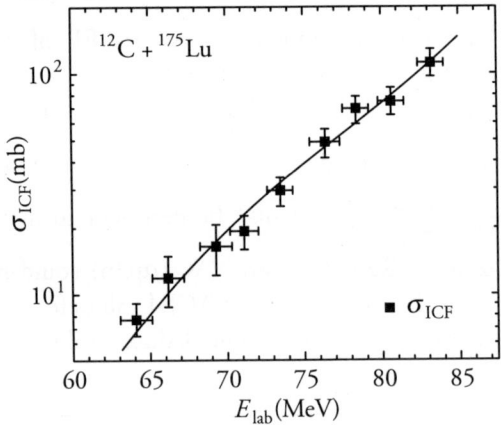

Figure 4.18 Deduced values of σ_{ICF} plotted as a function of projectile energy

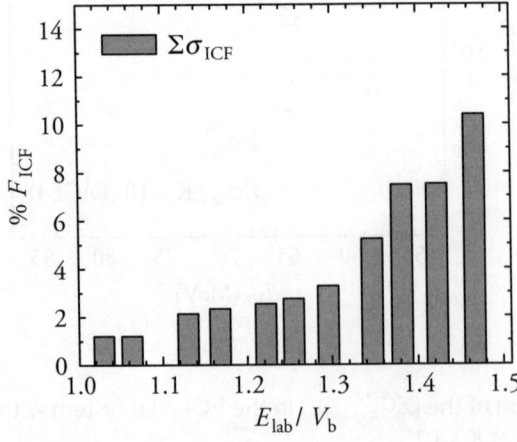

Figure 4.19 A histogram of deduced %F_{ICF} values as a function of normalized laboratory energy

4.1.2 Reactions initiated by ^{13}C beam

4.1.2.1 ^{13}C+^{169}Tm system

Most of the experiments used to study ICF reactions have been carried out using alpha cluster beams like ^{12}C and ^{16}O. However, in order to see if non-alpha cluster beams also give rise to ICF reactions, an experiment has been carried out using the ^{13}C beam on a ^{169}Tm target[27]. The ^{13}C is loosely bound as compared to ^{12}C, because it has one extra neutron in it. As such, this reaction may be considered to be extension of our earlier work to study the effect of a one neutron (1n) excess projectile on the onset and strength of the ICF. Here, the excitation functions (EFs) for the reaction residues populated in the interaction of ^{13}C with ^{169}Tm have been measured at energies ≈4.4–6.5 MeV/nucleon. Similar to the other systems, the measured EFs have been analysed within the framework of statistical model code PACE4, to estimate the percentage fraction of ICF contribution, if present. The excitation functions for all the measured reaction residues in the ^{13}C+^{169}Tm interaction are tabulated in Table 4.8.

Table 4.8 Experimentally measured production cross-sections at different energies for the residues populated via CF/ICF processes in system ^{13}C+^{169}Tm

System: ^{13}C + ^{169}Tm					
Energy (MeV)	^{179}Re σ (mb)	^{178}Re σ (mb)	^{177}Re σ (mb)	^{176}Re σ (mb)	^{179}Re σ (mb)
59.60 ±1.0	125.1 ± 18.7	202.4 ± 31.8	–	–	–
64.05 ± 0.9	80.21 ± 15.5	473.6 ± 76.0	3.21 ± 0.5	–	–
66.82 ± 1.1	33.45 ± 5.01	530.1 ± 73.8	28.4 ± 9.6	–	0.53 ± 0.09
69.18 ± 0.8	25.65 ± 3.8	552.1 ± 82.8	140.2 ± 24.0	–	6.23 ± 1.08
71.55 ± 0.7	13.63 ± 2.04	506.7 ± 91.0	280.5 ± 35.3	–	5.09 ± 0.90
75.14 ± 0.7	–	303.1 ± 47.0	603.1 ± 84.2	–	25.52 ± 4.50
75.49 ± 1.3	–	280.6 ± 45.1	600.2 ± 87	–	23.60 ± 4.23
77.30 ± 1.0	–	176.9 ± 26.5	765.4 ± 122.4	–	38.22 ± 6.84
80.31 ± 0.9	–	94.2 ± 16.2	790.8 ±111.7	92.87 ± 14.8	59.10 ± 10.63
80.92 ± 1.0	–	75.9 ± 9.4	810.6 ±121.3	135.33 ± 21.6	62.32 ± 11.21
82.7 ± 1.3	–	52.1 ± 7.8	800.2 ±135.8	279.75 ± 28.7	66.34 ± 11.88
85.2 ± 0.8	–	23.5 ± 3.5	695.9 ±111.1	438.62 ± 54.1	76.42 ± 13.75
Energy (MeV)	^{175}Ta σ (mb)	^{173}Ta σ (mb)	^{172}Ta σ (mb)	^{172}Lu σ (mb)	^{171}Lu σ (mb)
64.05 ± 0.9	21.22 ± 3.19	12.65 ± 1.39	1.95 ± 0.29	1.95 ± 0.21	–
66.82 ± 1.1	45.61 ± 6.84	14.01 ± 1.54	1.86 ± 0.27	2.01 ± 0.22	–

69.18 ± 0.8	65.98 ± 9.75	10.11 ± 1.11	2.54 ± 0.37	3.22 ± 0.33	–
71.55 ± 0.7	80.32 ± 13.6	13.28 ± 1.46	4.35 ± 0.65	3.53 ± 0.38	–
75.14 ± 0.7	88.35 ± 14.75	30.27 ± 3.34	7.16 ± 1.07	4.91 ± 0.55	–
75.49 ± 1.3	84.35 ± 14.12	36.95 ± 4.01	10.69 ± 2.20	4.91 ± 0.53	–
77.30 ± 1.0	90.12 ± 16.61	43.19 ± 4.7	14.91 ± 3.4	6.30 ± 0.69	–
80.31 ± 0.9	62.92 ± 8.38	70.32 ± 7.73	26.92 ± 3.2	7.12 ± 0.78	2.67 ± 0.41
80.92 ± 1.0	63.41 ± 9.06	72.77 ± 8.00	25.40 ± 3.51	7.59 ± 0.83	3.1 ± 0.49
82.7 ± 1.3	65.56 ± 9.75	75.46 ± 8.30	27.77 ± 5.56	6.41 ± 0.70	15.44 ± 2.31
85.2 ± 0.8	67.17 ± 10.67	85.12 ± 9.36	39.04 ± 7.35	9.61 ± 1.05	12.88 ± 1.93

In Figure 4.20(a) and (b), the ratios of the individual cross-sections for xn channels to the sum of all measured neutron emission channels (σ_{xn}) are shown. It may be pointed out that this representation of data has been taken to observe the typical behaviour of only neutron-emitting channels with respect to the total fusion cross-section. In the plot shown in Figure 4.20(a), the PACE4 calculations for different level density parameters are shown for ^{178}Re residues populated via the 4n channel. As may be observed from this, a value of $K = 10$ in

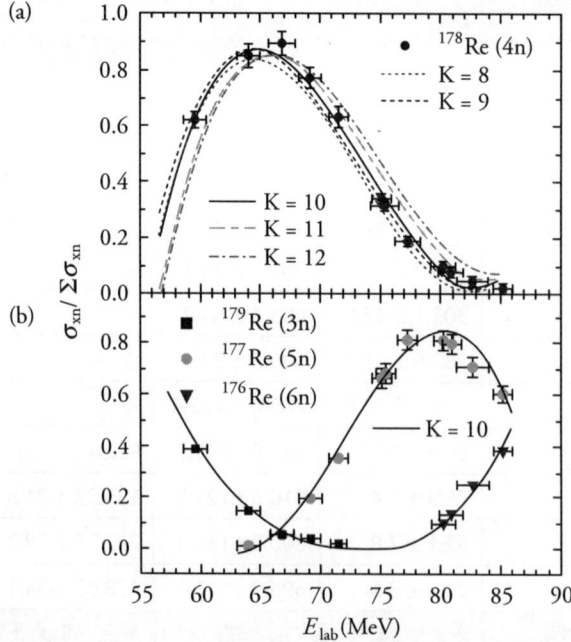

Figure 4.20 Ratio of the individual channel cross-section σ_{xn} to the total channel cross-section $\Sigma\sigma_{xn}$ as a function of laboratory energy for 4n channel along with PACE4 predictions (for $K = 8$ to 12); (b) measured EFs for all xn (x = 3, 5 and 6) channels along with PACE4 calculations

PACE4 calculations satisfactorily reproduce the EFs for the population of ^{178}Re residues. This indicates that these residues are populated via 4n emission from the excited composite system ^{182}Re* via complete fusion. It may also be seen from Figure 4.20(b) that the same value of $K = 10$ nicely reproduce the measured EFs for 3n, 5n and 6n channels, which further confirms that these channels are populated by the complete fusion process.

In Figure 4.21, the measured EF for the ^{177}W (p4n) channel is compared with that predicted by PACE4 code. As may be observed in Figure 4.21, the calculations underestimate the measurements. The residues ^{177}W may be populated via two channels: (i) through the β^+ - decay from ^{177}Re nuclei and also (ii) independently via the de-excitation of the ^{182}Re* compound nucleus by emitting p4n. The two different feeding paths for the population of ^{177}W may be shown by the following equations and also with the help of Figure 4.22

    a. ^{13}C+^{169}Tm $\rightarrow$ ^{182}Re* $\rightarrow$ ^{177}W +p+4n

    b. ^{13}C+^{169}Tm $\rightarrow$ ^{182}Re* $\rightarrow$ ^{177}Re +5n

^{177}Re undergoes β^+/EC decay giving rise to the production of ^{177}W.

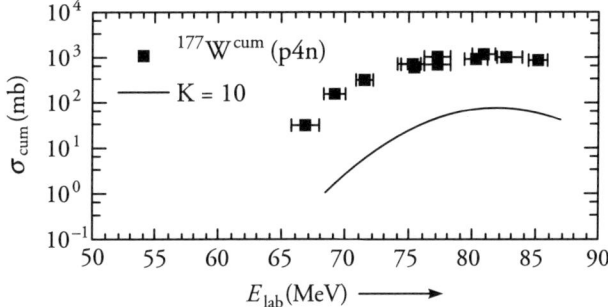

Figure 4.21 Experimentally measured and theoretically calculated excitation functions for ^{177}W residue populated via p4n channel

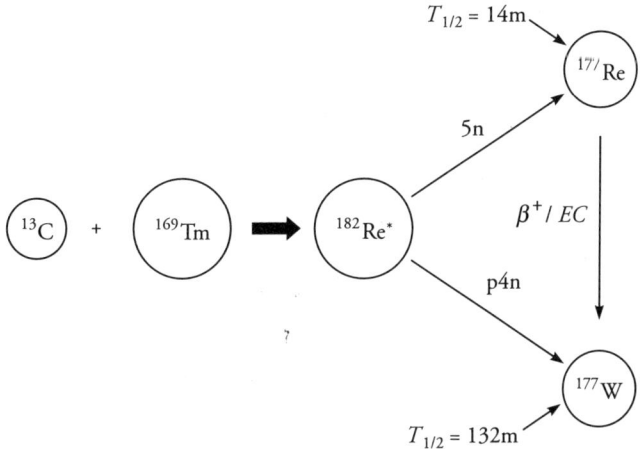

Figure 4.22 An example of the residue ^{177}W being populated via two different routes

The independent cross-section (σ_{ind}) of ^{177}W has been estimated from the cumulative cross-section (σ_{cum}) using the prescriptions of Cavinato et al[25]. The value of σ_{ind} for ^{177}W deduced in this way has been compared with the PACE4 predictions in Figure 4.23(b). Since the measured cross-section values and those calculated using code PACE4 for the population of residue ^{177}W agree well with each other (see Figure 4.23(b)), it may be concluded that this residue is fed through the complete fusion route. The contribution of precursors is also plotted in Figure 4.23(c) to show that on account of its short half-life, the precursor decays rapidly via β^+/EC decay giving a significant contribution to the cumulative cross-section of ^{177}W. The Coulomb barrier is also indicated in this figure.

In Figure 4.23(d), the sum of all CF channels ($\Sigma\sigma_{CF}^{exp}$) is compared with corresponding theoretical PACE4 predictions ($\Sigma\sigma_{CF}^{th}$). The data of $\Sigma\sigma_{CF}^{exp}$ at various energies has been found to

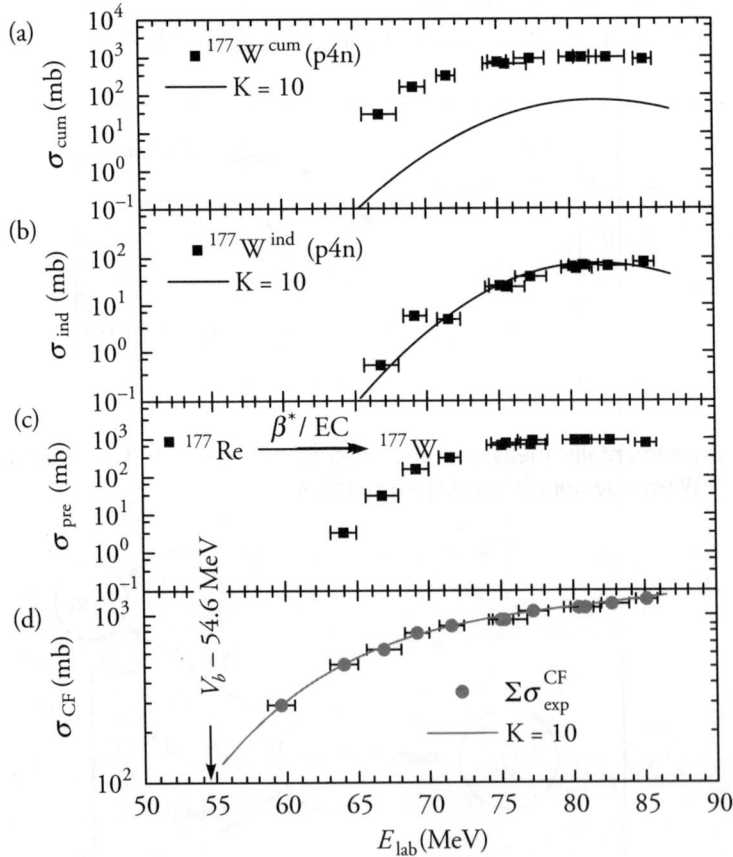

Figure 4.23 Experimentally measured EF for the residues ^{177}W(p4n channel) compared with PACE4 predictions: (a) cumulative cross-section; (b) independent cross-section. (c) Experimentally measured and theoretically predicted EFs of all xn and pxn channels. The solid lines are the PACE4 predictions with $K = 10$. In Figure 4.23(d), V_b(= 54.6 MeV) is the Coulomb barrier in laboratory frame

be in good agreement with that predicted by the PACE4 code. It may be pointed out that this excellent agreement between the two values enhances our confidence in the choice of statistical model parameters used for calculations involving code PACE4. As such, the same set of parameters has been used to compare the data for all alpha-emitting channels in this system.

In the system $^{13}C+^{169}Tm$, the residues $^{175, 174, 173}Ta$ and $^{172, 171}Lu$ are populated via emitting channels $\alpha3n$, $\alpha4n$, $\alpha5n$, $2\alpha2n$ and $2\alpha3n$ respectively. The population of these residues may be due to both CF as well as ICF processes. As already mentioned, ICF is not taken into consideration in PACE4 code. As such, the results of calculations using this code may indicate the underlying physical effects of alpha-emitting channels. In order to make a comparison, the measured EFs for these alpha-emitting channels are presented in Figures 4.24(a)–(d). As shown in these figures, the measured EFs for the residues $^{173,174,175}Ta$ are quite large compared to PACE4 predictions in the entire range of energy under investigation. This difference between the measured EFs and the theoretical predictions of PACE4 may be due to incomplete fusion processes. As an example, the residues ^{174}Ta may be produced via both CF and/or ICF reactions via the following possible channels.

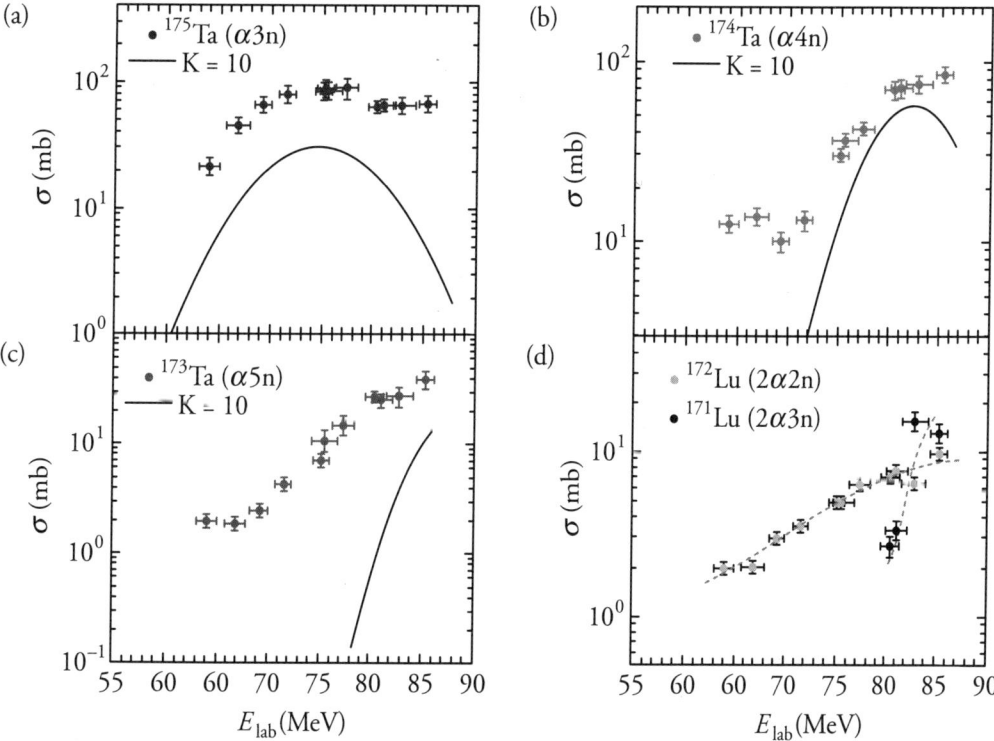

Figure 4.24(a)–(d) Experimentally measured EFs for the residues ^{175}Ta, ^{174}Ta, ^{173}Ta, ^{172}Lu and ^{171}Lu compared with the PACE4 predictions respectively. Solid curves represent theoretical calculations. In panel (d), the dotted lines through the data points are drawn to show the trend of the excitation functions

a. CF → Decay of ^{182}Re* via two proton and six neutrons (2p6n channel)

^{13}C+^{169}Tm → ^{182}Re* → ^{174}Ta +2p6n

Q value ≈ −69.15 MeV and the threshold energy ≈ 74.47 MeV

b. CF → Decay of ^{182}Re* via an α particle and four neutrons (α4n channel)

^{13}C+^{169}Tm → ^{182}Re* → ^{174}Ta + α4n

Q value ≈ −40.86 MeV and the threshold energy ≈ 43.99 MeV

c. ICF → ^{13}C breaks up as (^{9}Be + α), where ^{9}Be fuses with ^{169}Tm forming ^{178}Ta* composite nucleus; the alpha particle moves in the forward direction as a spectator. The remaining excited compound nucleus ^{178}Ta* decays via the emission of four neutrons

^{13}C(^{9}Be+α) → ^{9}Be +^{169}Tm → ^{178}Ta* (α as a spectator)

→ ^{178}Ta* → ^{174}Ta + 4n

Q value ≈ −30.21 MeV and the threshold energy ≈ 31.82 MeV

The EFs for the population of residues ^{171,172}Lu are shown in Figure 4.24(d), where the code PACE4 predicts negligible cross-sections; this indicates that these residues are populated largely through ICF process. It may, therefore, be inferred that ICF contributes significantly to the production of ^{173,174,175}Ta and ^{171,172}Lu isotopes. In order to have a relatively better visualization of increase in ICF values of cross-sections with beam energy, the sums of all experimentally measured cross-sections for α and 2α emitting channels are compared with PACE4 predictions in Figure 4.25(a). As may be seen in this figure, the values $\Sigma\sigma_{exp}^{\alpha's}$ are

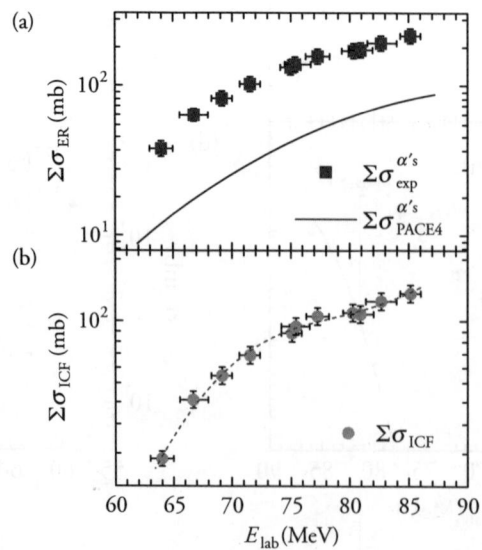

Figure 4.25 (a) Sums of all experimentally measured cross-sections for α and 2α emitting channels are compared with corresponding PACE4 predictions. The values $\Sigma\sigma_{exp}^{\alpha's}$ are significantly higher than that predicted by PACE4. (b) Deduced σ_{ICF} is plotted as a function of beam energy. The dashed line through the data points in Figure 4.25(b) is drawn just to guide the eyes

significantly higher than that predicted by PACE4. The values of $\Sigma\sigma_{ICF}$ have been plotted in Figure 4.25(b)—here an increasing trend has been obtained, which indicates strong energy dependence of the ICF fraction.

It may also be mentioned here that the contribution shown in Figure 4.25(b) gives a lower limit of the incomplete fusion contributions as all the expected reactions leading to alpha particles in the exit channel could not be measured.

4.1.3 Reactions initiated by ^{14}N beam

4.1.3.1 ^{14}N+^{128}Te *system*

The excitation functions for the reactions ^{128}Te(^{14}N, 4n)^{138m}Pr, ^{128}Te(^{14}N,5n)^{137}Pr, ^{128}Te(^{14}N, p4n)^{137g}Ce, ^{128}Te(^{14}N,α5n)^{133}La, ^{128}Te(^{14}N,α6n)^{132g}La,^{128}Te(^{14}N,α2pn)^{135m}Cs, ^{128}Te (^{14}N, 2α2pn)^{131}I and ^{128}Te(^{14}N, 3α)^{130g}I have been measured in the beam energy range $\approx$ 64–90 MeV. A list of reaction residues and their measured cross-sections are tabulated in Table 4.9.

Table 4.9 Experimentally measured cross-sections at different energies for the reaction residues identified in the system ^{14}N+^{128}Te

System: ^{14}N + ^{128}Te								
Energy	^{138m} Pr	^{137}Pr	^{137g} Ce	^{133}La	^{132g}La	^{135m} Cs	^{131}I	^{130g} I
(MeV)	σ(mb)	σ(mb)	σ(mb)	σ(mb)	σ(mb)	σ(mb)	σ(mb)	σ(mb)
64 ± 0.8	255 ± 55	–	–	–	–	–	–	–
71 ± 0.8	381 ± 42	–	–	–	–	–	–	–
76 ± 0.7	178 ± 20	456 ± 65	144 ± 19	–	–	46 ± 4	–	2 ± 0.5
81 ± 0.6	62 ± 8	463 ± 60	194 ± 21	–	–	6 ± 0.6	–	28 ± 4
86 ± 0.5	66 ± 7	577 ± 62	220 ± 35	376 ± 38	18 ± 2	1 ± 0.1	98 ± 32	12 ± 2
90 ± 0.5	48 ± 5	611 ± 72	263 ± 50	1435 ± 144	478 ± 48	141 ± 40	36 ± 4	2 ± 0.3

Analysis of the data has been carried out using the computer code ALICE-91[14] and PACE4[23]. Details of these codes are already given in Chapter 2. Though the details of code ALICE are given elsewhere, for the sake of completeness, some information is also included here for ready reference. This code has been developed by M. Blann and is based on the Weisskopf–Ewing model[18] for CN calculations and the hybrid model[19] for simulating PE emissions. The code assumes equipartition of energy among the initially excited particles and holes. It also uses Gove mass tables[28] or the Myers Swiatecki/Lysekil mass formula[29]. The option that substitutes Gove's table for Myers Swiatecki/Lysekil mass formula including shell corrections was used in the present calculations. In order to calculate inverse cross-sections, the optical model subroutine with the parameters of Becchetti and Greenlees[16] was employed. In this code, there are three important parameters, viz., the level density parameter a, initial exciton number n_0 and the mean free path multiplier COST. The level density parameter mainly affects the equilibrium component and is given by $a = A/K$, where A is the mass number of the nucleus and K is a constant, which may be varied to match the experimental data. In this work, the effect of variation of K on the calculated excitation functions has also been studied. The value of K was varied from 9 to 18. As a typical example, the calculated excitation functions for different values of K are shown in Figure 4.26(a); it was found that $K = 18$ gives the best reproduction of the experimental data for all the channels presently measured.

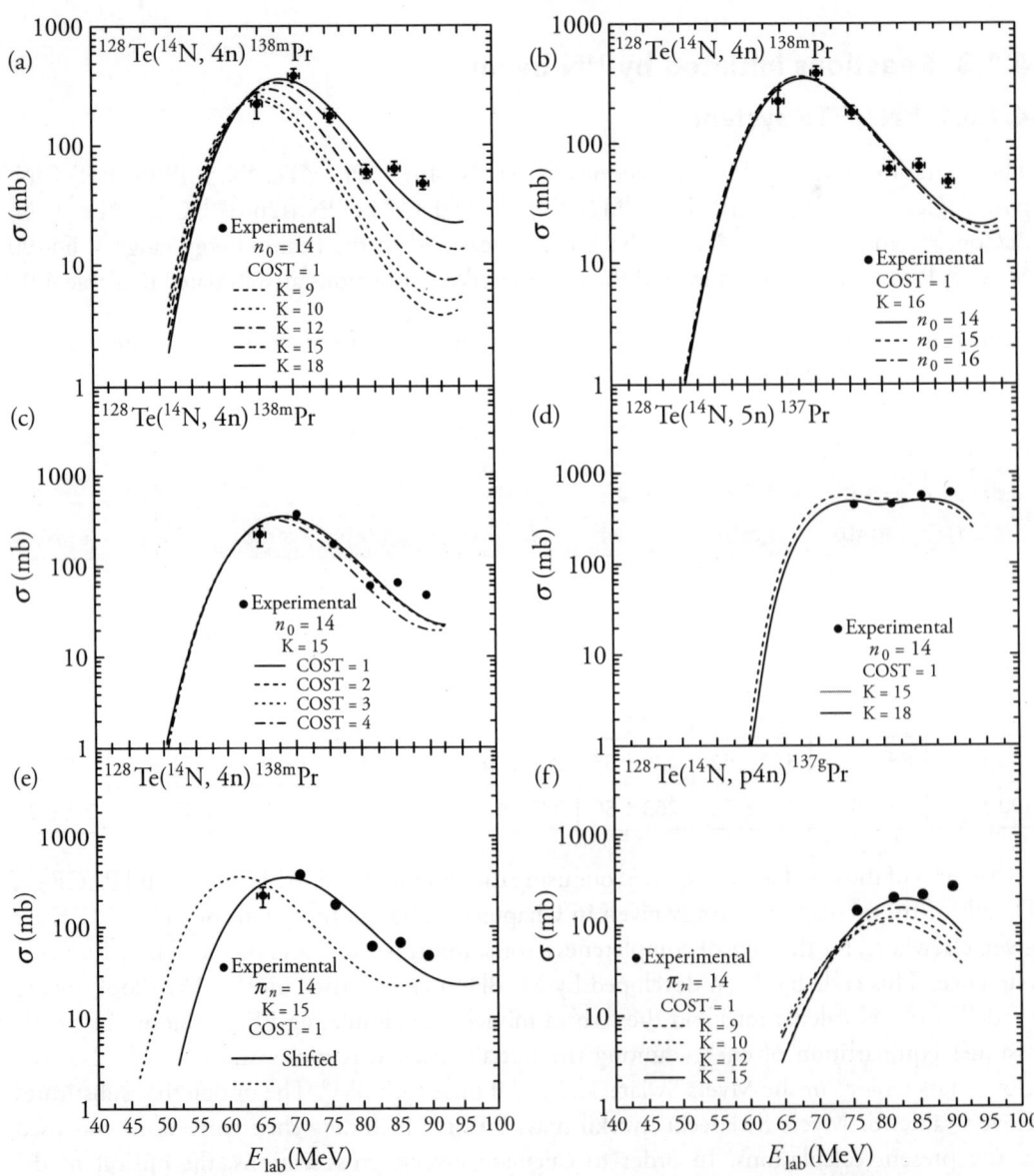

Figure 4.26 Excitation functions for reactions ^{128}Te(^{14}N, 4n)^{138m}Pr, ^{128}Te(^{14}N, 5n)^{137}Pr and ^{128}Te(^{14}N, p4n)^{137g}Ce. The symbols with dark circles represent the experimental data. Various curves correspond to the theoretical predictions of the ALICE-91 code

The parameters n_0 and COST mainly govern the pre-equilibrium component. The initial exciton number decides the initial configuration, where a small value of n_0 means that that initial state is less complex and is far from the equilibrium. As such, in this case (small value of n_0), a large PE contribution is expected. On the other hand, a larger value of n_0 means that the initial system is nearer to the equilibrium stage and therefore, a smaller component of PE emission is expected. To see the effect of variation of n_0 on calculated EFs, the value of n_0 was varied from 14 to 16. Such calculations for ^{128}Te(^{14}N,4n) reaction are presented in Figure 4.26(b). As can be seen from this figure, a value of $n_0 = 14$ is well suited for the experimental data. This value of $n_0 = 14$ may also be justified by considering that the projectile ^{14}N breaks up in the nuclear field of the target nucleus creating 14 excitons.

The parameter COST, which is generally used to adjust the mean free path for two-body residual interactions inside the nuclear matter, was varied from 1 to 4 and its effect on calculated excitation functions for the reaction ^{128}Te(^{14}N,4n) is shown in Figure 4.26(c). It may be remarked that in general, the set with $K = 18$, $n_0 = 14$ and COST = 1 gives a quite satisfactory reproduction of the experimental data. In the same way, the excitation function for the ^{128}Te(^{14}N, 5n) reaction is shown in Figure 4.26(d), which is also nicely reproduced in terms of the magnitude by the selected set of parameters. During calculations, it has been observed that the calculated excitation functions for all the presently studied reactions with the aforementioned set of parameters have their peaks shifted towards the lower energies, as compared to the experimental data. This shifting is expected in HI induced reactions because the projectile imparts a large angular momentum to the composite system. This higher input angular momentum imparted to the composite system may inhibit particle emission in the last stages of the de-excitation. As such, the peaks of the experimental excitation functions corresponding to a particle emission mode are likely to be shifted towards higher energies. One may have an estimate of the shift from the rotational energy E_{rot}, which may be approximated using the relation $E_{rot} = (m/M)E_{lab}$, where m and M are the projectile and target masses while E_{lab} is the incident energy. In the range of energies used in the present work, the rotational energy shift (E_{rot}) has been found to vary in the range $\approx$ 7–9 MeV. Further, since the angular momentum effects have not been considered in the Weisskopf–Ewing, calculations the calculated EFs were shifted by an amount $\approx F_{rot}$ towards the higher energy side. A satisfactory agreement between experimental and theoretical EFs has, in general, been found. As an example, the effect of rotational energy on calculated EF for the reaction ^{128}Te(^{14}N,4n) is presented in Figure 4.26(e). In Figure 4.26(f), the measured cross-sections for the reaction ^{128}Te(^{14}N,p4n) channel are compared with the theoretically predicted values. It may be observed from this figure that even by changing the level density parameter constant K (= 9 – 18), the measured cross-sections are underestimated by the calculated values, particularly at higher energies. These enhanced values of the cross-sections at higher energies may be due to contribution from the precursor decay. As during the irradiation of the sample, the residual nuclei ^{137}Ce may be produced via two different routes—through the reaction ^{128}Te(^{14}N,p4n) and also from the β^+ decay of the residual nucleus ^{137}Pr formed via the reaction ^{128}Te(^{14}N,5n). In this way, the measured cross-sections of the reaction ^{128}Te(^{14}N,p4n) will also have contribution from the β^+ decay of the higher charge isobar precursor. Though, it is quite possible to estimate the contribution from the precursor decay using the Batemann equations, this could not be done in the present case as the isomeric state of ^{137}Ce ($t_{1/2} = 34.4$ hrs) could not be observed.

Further, in Figure 4.27(a), the measured EF for the reaction ^{128}Te(^{14}N,4n) has been compared with the theoretical predictions based on the CN model as well as by including the pre-compound component to it. The Weisskopf–Ewing model calculations are shown by dashed line. As may be seen from this figure, the CN calculations do not match the experimental data at higher energies, where the pre-compound emission is more likely. As such, the calculations were also done using the hybrid model option of the code ALICE-91, which includes the pre-compound emission as well. It may be observed from Figure 4.27(a) that the high energy portion of the data, referred to as the tails portion, is very near to the calculated values when pre-compound emission is included in the calculations. As such, it may be concluded that there is a significant pre-compound contribution in the reaction ^{128}Te(^{14}N,4n) at relatively higher energies, as expected. Further, it may be observed from Figure 4.27(b) and 4.27(c) respectively, that theoretical excitation functions for the reactions ^{128}Te(^{14}N,α5n) and ^{128}Te(^{14}N,α6n) calculated employing the code ALICE-91 are quite small compared to the measured values. This difference in the measured cross-sections from the theoretical calculations may be explained in terms of the contribution from the incomplete fusion of the ^{14}N ions. It may be assumed that when ^{14}N comes near the nuclear field of the target nucleus, it breaks up into two fragments ^{10}B and ^{4}He. One of the fragments ^{10}B fuses with the target nucleus (while ^{4}He moves along the beam direction as a spectator) forming the excited composite nucleus ^{138}La*. This excited nucleus may then emit 5n/6n leading to the formation of residual nuclei ^{133}La and ^{132}La respectively. The theoretical calculations done using code ALICE-91 do not include such ICF processes and as such, the enhancement may be attributed to incomplete fusion process. On the other hand, in the case of ^{128}Te(^{14}N,α2pn), ^{128}Te(^{14}N,2α2pn) and ^{128}Te(^{14}N,3α) reactions, as shown in Figures 4.27(d)–(f), the calculated values from code ALICE-91 give negligible cross-sections though the measured cross-sections are quite substantial. This discrepancy may again be explained in terms of the contribution from incomplete fusion process. For the ^{128}Te(^{14}N,α2pn) reaction, the higher cross-section may be explained by the fact that ^{10}B (if ^{14}N breaks up into ^{10}B and ^{4}He) fuses with the target nucleus and emits two protons and a neutron. In the same way, the reaction ^{128}Te(^{14}N,2α2pn) may be explained by assuming that ^{14}N breaks up into ^{6}Li and two alpha particles, where ^{6}Li fuses with the target nucleus emitting two protons and a neutron. In the ^{128}Te(^{14}N,3α) reaction, it may be considered that ^{14}N breaks up into three alpha particles and a deuteron, where the fusion of deuteron takes place with the target nucleus resulting in the formation of a residual nucleus ^{130}I, which may decay by emitting gamma radiations. It may be pointed out that the theoretical calculations done using code ALICE-91 do not take the incomplete fusion process into account as such a relatively large part of the reaction may be considered to proceed through the ICF process.

The calculations for the system ^{14}N+^{128}Te have also been done using the code PACE4, which is based on the statistical approach. A Monte Carlo approach is used to consider the de-excitation of the compound nucleus. The angular momentum projections are calculated at each stage of de-excitation, which enables the determination of the angular distribution of

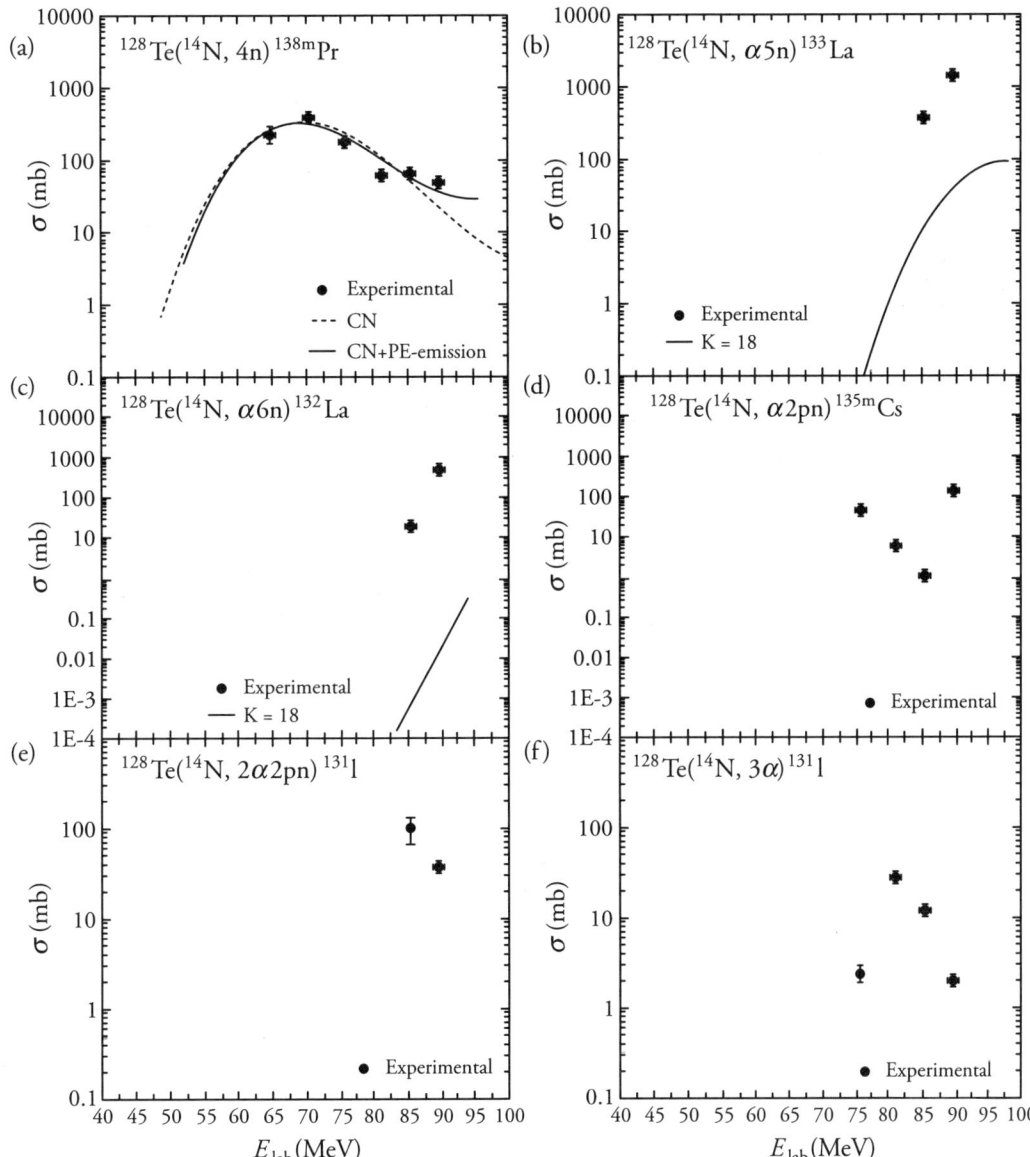

Figure 4.27 Excitation functions for the reactions ^{128}Te(^{14}N, 4n)^{138m}Pr, ^{128}Te(^{14}N, α5n)^{133}La, ^{128}Te(^{14}N, α6n)^{132g}La, ^{128}Te(^{14}N, α2pn)^{135m}Cs, ^{128}Te(^{14}N, 2α2pn)^{131}I and ^{128}Te(^{14}N, 3α)^{130g}I. The filled circles represent the experimental data. The solid and dotted lines correspond to the theoretical predictions of the ALICE-91 code

the emitted particles. As already mentioned, the level density is an important parameter and may be varied to match the measured excitation function data. In Figure 4.28(a)–(c), the effect of variation of level density parameter K(= 8–11) on calculated excitation functions for the reactions ^{128}Te(^{14}N,4n), ^{128}Te(^{14}N,5n) and ^{128}Te(^{14}N,p4n) are shown. It may be seen from these figures that a value of $K = 11$ reproduces the experimental cross-sections satisfactorily, in general, for the reactions $^{128}\mathrm{Te}(^{14}\mathrm{N},4n)$ and $^{128}\mathrm{Te}(^{14}\mathrm{N},5n)$, which are populated via complete fusion channels.

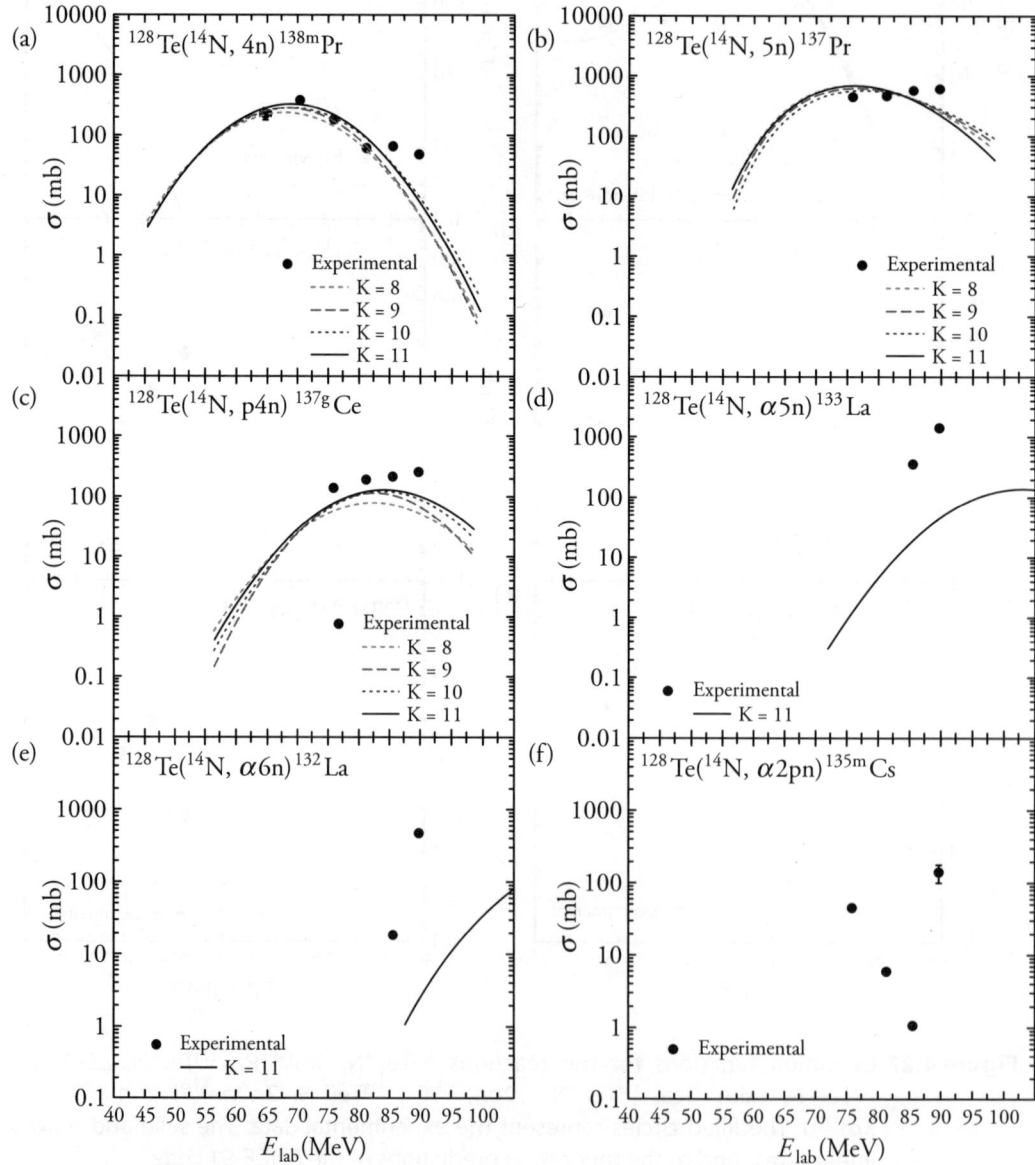

Figure 4.28 Excitation functions for the reactions $^{128}\mathrm{Te}(^{14}\mathrm{N},4n)^{138m}\mathrm{Pr}$, $^{128}\mathrm{Te}(^{14}\mathrm{N},5n)^{137}\mathrm{Pr}$, $^{128}\mathrm{Te}(^{14}\mathrm{N},p4n)^{137g}\mathrm{Ce}$, $^{128}\mathrm{Te}(^{14}\mathrm{N},\alpha5n)^{133}\mathrm{La}$, $^{128}\mathrm{Te}(^{14}\mathrm{N},\alpha6n)^{132g}\mathrm{La}$ and $^{128}\mathrm{Te}(^{14}\mathrm{N},\alpha2pn)^{135m}\mathrm{Cs}$. The filled dark circles represent the experimental data. Various curves correspond to the theoretical predictions of the PACE code

From Figure 4.28(c), it may be observed that measured cross-sections are higher than their theoretical predictions. This is due to the contribution from precursor decay. It has already been mentioned that the precursor contributions could not be separated since the metastable state of the ^{137}Ce nucleus could not be observed in the present measurements. It may be pointed out that since the incomplete fusion is not considered in these calculations, the enhancement of the measured EFs for the reactions ^{128}Te(^{14}N,α5n)^{133}La and ^{128}Te(^{14}N, α6n)^{132g}La may be attributed to the fact that apart from the production of these reactions through CF, they may also be produced via ICF process. In the case of ^{128}Te(^{14}N, α2pn)^{135m}Cs reactions (Figure 4.28(f)), a significant contribution may come from the production of these isotopes from ICF channels, which are not considered in these calculations.

4.1.4 Reactions initiated by ^{16}O beam

4.1.4.1 ^{16}O+^{27}Al system

To study the reaction dynamics of heavy ion interaction in the lower mass region, the cross-sections for reaction residues populated in the system ^{16}O+^{27}Al have been measured over a wide range of energies. Most of the studies where the ICF have been studied are carried out generally with heavier target nuclei; there are very few studies with lower mass target nuclei. One of the advantages of using the lighter system is that the possibility of fission may be avoided, which is generally a competing mode of reaction in heavy ion interaction. With a view to study the contribution of ICF in light mass target nuclei, the measurement and analysis of cross-sections for the reaction channels ^{27}Al(^{16}O,2αn), ^{27}Al(^{16}O,3α3p), ^{27}Al(^{16}O,3α3pn), ^{27}Al(^{16}O,4α2pn) and ^{27}Al(^{16}O,4α3p), covering the energy range from nearly 58 MeV to 94 MeV have been done. The measured cross-sections for these reactions are presented in Table 4.10.

Table 4.10 Measured values of cross-sections at different energies for residues populated via different channels and identified in system ^{16}O+^{27}Al

System: ^{16}O + ^{27}Al					
Energy (MeV)	^{34}Cl σ(mb)	^{20}Mg σ(mb)	^{77}Mg σ(mb)	^{24}Ne σ(mb)	^{24}Na σ(mb)
58.0 ± 1.3	9.4 ± 1.69	–	–	–	0.46 ± 0.08
58.7 ± 1.3	14.57 ± 2.62	–	–	–	1 ± 0.18
66.5 ± 1.2	57.7 ± 10.3	–	–	–	0.94 ± 0.17
68 ± 1.2	62.7 ± 11.2	–	–	–	0.96 ± 0.17
68.2 ± 1.2	154.0 ± 27.7	–	–	–	0.275 ± 0.05
71.6 ± 1.1	115.4 ± 20.7	–	–	–	0.63 ± 0.11
75.4 ± 1.1	169.5 ± 30.5	–	–	–	0.89 ± 0.16
76.2 ± 1.1	100.9 ± 18.1	0.08 ± 0.0	–	–	1.35 ± 0.24
77.1 ± 1.0	126.4 ± 22.7	0.09 ± 0.0	–	–	1.17 ± 0.21
78.8 ± 1.0	95.3 ± 17.1	–	–	–	1.2 ± 0.22

81.8 ± 1.0	81.9 ± 14.7	2.48 ± 0.4	–	–	2.35 ± 0.42
82.0 ± 0.96	120.9 ± 21.7	–	–	–	1.37 ± 0.25
85.5 ± 0.6	84.7 ± 15.2	2.21 ± 0.4	–	0.12 ± 0.03	7.88 ± 1.41
85.9 ± 0.9	140.7 ± 28.2	3.11 ± 0.5	–	0.11 ± 0.02	5.15 ± 0.9
88.2 ± 0.6	15.5 ± 2.8	1.53 ± 0.26	0.22 ± 0.05	0.36 ± 0.09	1.03 ± 0.18
88.5 ± 0.8	–	0.42 ± 0.05	0.2 ± 0.05	–	1.58 ± 0.28
91.4 ± 0.6	5.43 ± 0.9	–	0.08 ± 0.02	0.22 ± 0.05	1.15 ± 0.20
93.4 ± 0.8	8.26 ± 1.4	0.4 ± 0.0	0.11 ± 0.02	0.12 ± 0.05	1.5 ± 0.27
94.4 ± 0.6	4.7 ± 0.8	0.2 ± 0.0	0.1 ± 0.03	0.1 ± 0.03	1.33 ± 0.24

Some data on cross-sections in system $^{16}O+^{27}Al$ is available in literature[30]. However, these measurements were done using a low resolution NaI(Tl) scintillator and/or gas flow beta counter. As such, the spread in energy and experimental errors are considerably large. The literature data on cross-sections was explained on the basis of compound reaction (CN) mechanism. In the present work, cross-section measurements are more precise, being done using a high resolution HP(Ge) detector and experimental errors are considerably low. In order to get unambiguous information regarding the reaction mechanism, presently measured cross-section data were compared with the calculations done using three computer codes viz., CASCADE[13], PACE[23] and ALICE[14]. As already mentioned, the CASCADE code is based on the Hauser–Feshback theory and is generally used to get the theoretical estimates of the cross-sections using the CN mechanism. It may be pointed out that it does not take the ICF and PE emission into consideration. In this code, the level density parameter constant K and the ratio of actual moment of inertia to the rigid body moment of inertia of the excited system F_θ are the two important parameters, which may be varied to match the measured excitation functions. It may be remarked that in calculations, a value of $K > 10$ may give rise to an anomalous effect in the particle multiplicity. In the present work, we have calculated the cross-sections consistently using the same set of parameters that are consistently used in our recent publications. Here the calculations have been performed taking a value of $K = 8$. In case of the reactions $^{27}Al(^{16}O,2\alpha n)$, the residue ^{34}Cl is produced, which has metastable as well as ground states. In the present work, the metastable state ^{34m}Cl was observed through a 146.3 keV gamma ray of intensity 40.5%. The intensities of the gamma lines of the ground state of ^{34g}Cl are quite low; therefore, it could not be observed. However, the production cross-section of the residues ^{34m}Cl was used to deduce the total cross-section using the standard radioactive decay method. Since the code CASCADE gives the total production cross-section of the residues, it is reasonable to compare the total cross-section of the residues ^{34}Cl with the calculations. The experimentally measured and theoretically calculated excitation functions for the reaction $^{27}Al(^{16}O,2\alpha n)^{34}Cl$ are shown in Figure 4.29. The measured values of the cross-sections for the residues ^{34}Cl by Landenbauer-Bellis et al.[30], which have some contribution from the residue ^{38}Cl, are also shown in this figure. The literature values of the cross-sections shown in this figure have large uncertainties in the energy, as mentioned earlier. However,

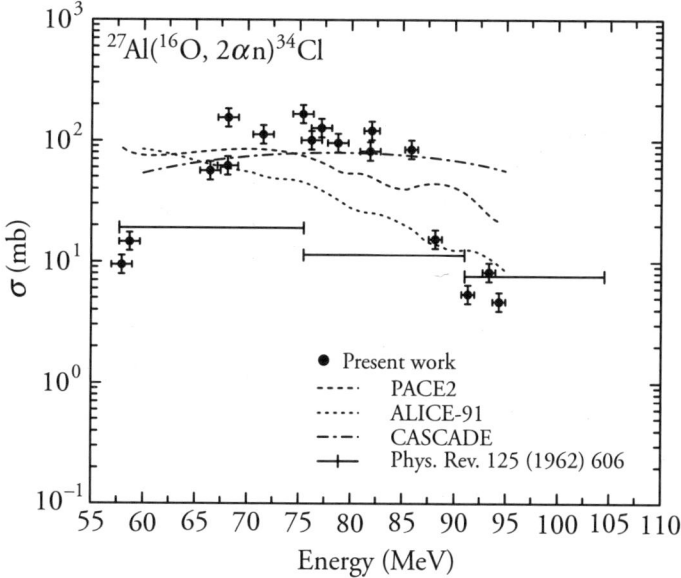

Figure 4.29 Experimentally measured and theoretically calculated excitation functions for the ^{27}Al(^{16}O,2αn) reaction. Literature values are also shown in the figure

in the present work, the energy uncertainties due to finite thickness of the sample are much smaller. Further, as may be seen in the figure, the values provided by Landenbaurer-Bellis et al. have effectively only three data points, while the present work contains 19 data points, indicating a precise measurement at very close energy intervals. Since the code CASCADE does not consider the ICF contribution, the difference in the measured cross-sections for the reaction ^{27}Al(^{16}O,2αn)^{34}Cl may be attributed to the ICF process.

In Figure 4.30(a)–(d), the experimentally measured excitation functions for the reactions ^{27}Al(^{16}O,3α3p)^{28}Mg, ^{27}Al(^{16}O,3α3pn)^{27}Mg, ^{27}Al(^{16}O,4α2pn)^{24}Na and ^{27}Al(^{16}O,4α3p)^{24}Ne are shown. In these figures, the solid curves guide the eyes through the data points by curve fitting.

Based on the trend of these curves, Landenbaurer-Bellis et al. have concluded that this reaction proceeds through the compound nucleus mechanism. Since the calculated values of the EFs using the code CASCADE for these reactions are negligibly small, they are not shown in these figures. Thus, the observed enhancement by several orders of magnitude over the negligible theoretical predictions for these channels may be attributed to the fact that these reactions proceed predominantly via any reaction process other than the compound nucleus process. The theoretical calculations of the cross-sections for these reactions have also been attempted using the code PACE2, which is based on the statistical approach. For all the alpha-emitting channels, the calculations give very small cross-sections indicating that these reactions proceed predominantly through the incomplete fusion process.

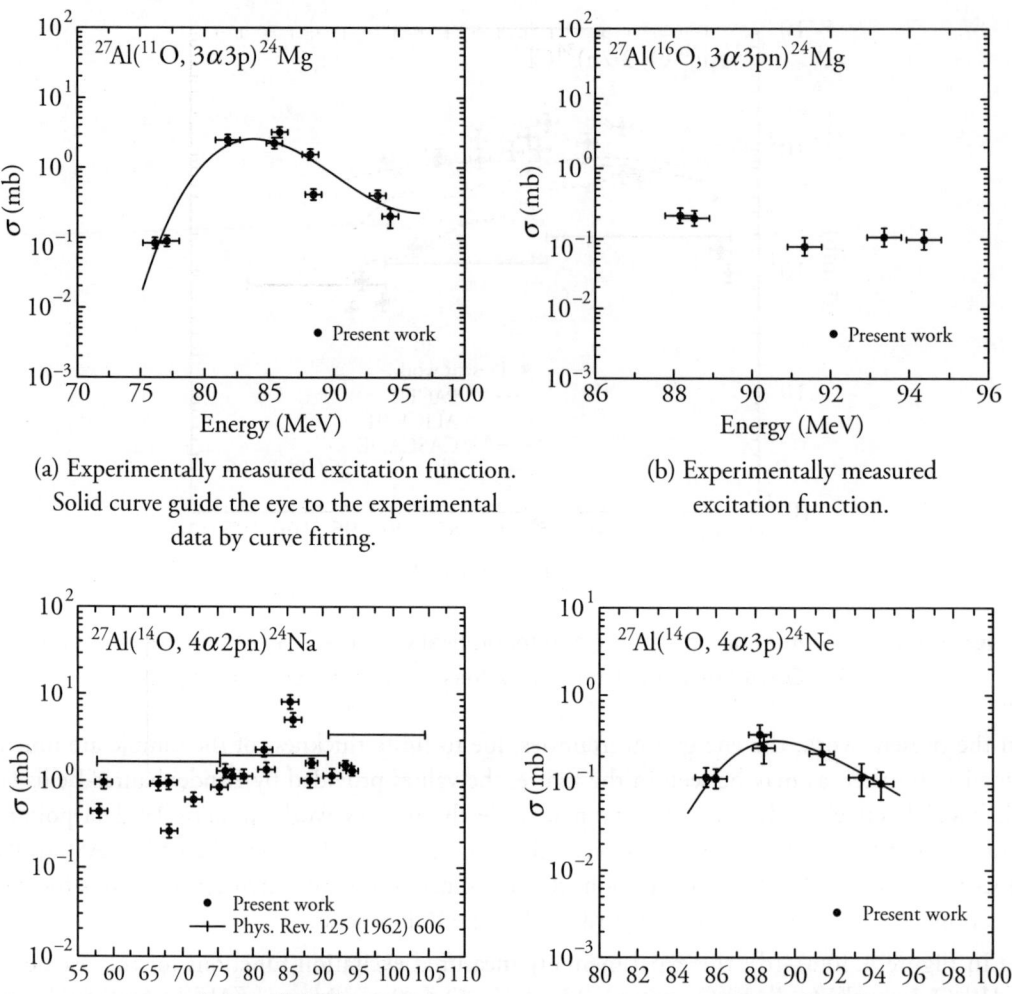

(a) Experimentally measured excitation function. Solid curve guide the eye to the experimental data by curve fitting.

(b) Experimentally measured excitation function.

(c) Experimentally measured excitation function. Literature values are also shown.

(d) Experimentally measured excitation function. Solid curve guide the eye to the experimental data.

Figure 4.30 Experimentally measured excitation functions for the reactions (a) $^{27}Al(^{16}O, 3\alpha3p)$ ^{28}Mg, (b) $^{27}Al(^{16}O,3\alpha3pn)^{27}Mg$, (c) $^{27}Al(^{16}O,4\alpha2pn)^{24}Na$ and (d) $^{27}Al(^{16}O, 4\alpha3p)^{24}Ne$

4.1.4.2 $^{16}O+^{159}Tb$ system

In the system $^{16}O+^{159}Tb$, the excitation functions for $^{159}Tb(^{16}O,3n)^{172}Ta$, $^{159}Tb(^{16}O,4n)^{171}Ta$, $^{159}Tb(^{16}O,5n)^{170}Ta$, $^{159}Tb(^{16}O,p3n)^{171}Hf$, $^{159}Tb(^{16}O,p4n)^{170}Hf$, $^{159}Tb(^{16}O,2p2n)^{171}Lu$, ^{159}Tb $(^{16}O,\alpha n)^{170}Lu$, $^{159}Tb(^{16}O,\alpha 2n)^{169}Lu$ and $^{159}Tb(^{16}O,2\alpha 2n)^{165}Tm$ have been measured in the energy range ≈70–95 MeV and are tabulated in Table 4.11.

Table 4.11 Measured cross-sections at different energies for residues identified in the system $^{16}O+^{159}Tb$

System: $^{16}O + ^{159}Tb$						
Energy (MeV)	^{172}Ta σ(mb)	^{171}Ta σ(mb)	^{170}Ta σ(mb)	$^{171}Hf^{cum}$ σ(mb)	$^{171}Hf^{ind}$ σ(mb)	$^{170}Hf^{cum}$ σ(mb)
69.4 ± 1.0	0.3 ± 0.0	–	–	–	–	–
75 ± 0.9	19.8 ± 3.3	–	–	13.6 ± 5.6	–	–
78.7 ± 0.9	34.8 ± 6.8	67.0 ± 11.8	–	80.2 ± 19	11.01 ± 5	–
83.2 ± 0.8	48.14 ± 5.5	280.5 ± 47	64.9 ± 6.6	431.2 ± 86.2	141.4 ± 56.6	84.6 ± 10
87.2 ± 0.8	12.6 ± 2.2	244 ± 43.3	100.2 ± 15.4	322 ± 74.4	69.9 ± 29.9	177.3 ± 20
90 ± 1.0	11.8 ± 2.4	228.3 ± 39.2	263.4 ± 38.7	340.1 ± 73.4	104.3 ± 41.7	574.3 ± 68.8
95 ± 0.4	8.53 ± 1.5	157.9 ± 28.5	390.1 ± 45.5	205.4 ± 46.2	42.1 ± 18	427.3 ± 55
Energy (MeV)	$^{170}Hf^{ind}$ σ(mb)	^{171}Lu σ(mb)	^{170}Lu σ(mb)	^{169}Lu σ(mb)	^{165}Tm σ(mb)	
78.7 ± 0.9	–	29.5 ± 3.1	–	4.2 ± 0.4	–	
83.2 ± 0.8	19.2 ± 2.3	61.6 ± 6.5	6.5 ± 1.0	76.8 ± 11.2	5.53 ± 0.58	
87.2 ± 0.8	76.3 ± 11.8	1167.62 ± 170.6	49.8 ± 6.7	43.4 ± 4.5	7.58 ± 0.8	
90 ± 1.0	308.2 ± 36.7	611.8 ± 64	31.3 ± 5.4	80.65 ± 9.4	10.36 ± 1.3	
95 ± 0.4	34.8 ± 10.8	255.8 ± 26.3	31.3 ± 5.4	34.4 ± 4.5	5.76 ± 1.1	

These excitation functions have been plotted in Figures 4.31 and 4.32. As is obvious, the residues $^{(175-x)}Ta$ (x = 3 – 5) and $^{(175-x)}Hf$ (x = 4 – 5) are populated only via CF process. However, the residues $^{(175-x)}Lu$ (x = 4 – 6) and the residue ^{165}Tm may have contributions not only from CF but also from ICF reactions. When ^{159}Tb interacts with the projectile ^{16}O, some of the residues may be produced directly as an independent yield while some of them may be produced through beta positive emission and/or EC decay. In these cases, the cumulative cross-sections have been measured. The separation of the independent yield from the cumulative yield has been obtained using the successive radioactive decay formulations given by Cavinato et al.[25]. The analysis of the data has been done using the code PACE2, which is based on a statistical approach while the de-excitation of the CN is analysed using the Monte Carlo procedure.

As can be seen from Figure 4.31, the 3n, 4n and 5n channels are nicely reproduced by the theoretical calculations done using code PACE that is essentially based on the complete fusion model. The enhancement shown in the experimental data for the p3n channel is mainly because of the contribution from higher charge isobar precursors due to β^+ decay. However, once the contribution of the decay from the precursors was separated, the independent cross-section excitation function data agrees reasonably well with the calculations, indicating its population from complete fusion process only.

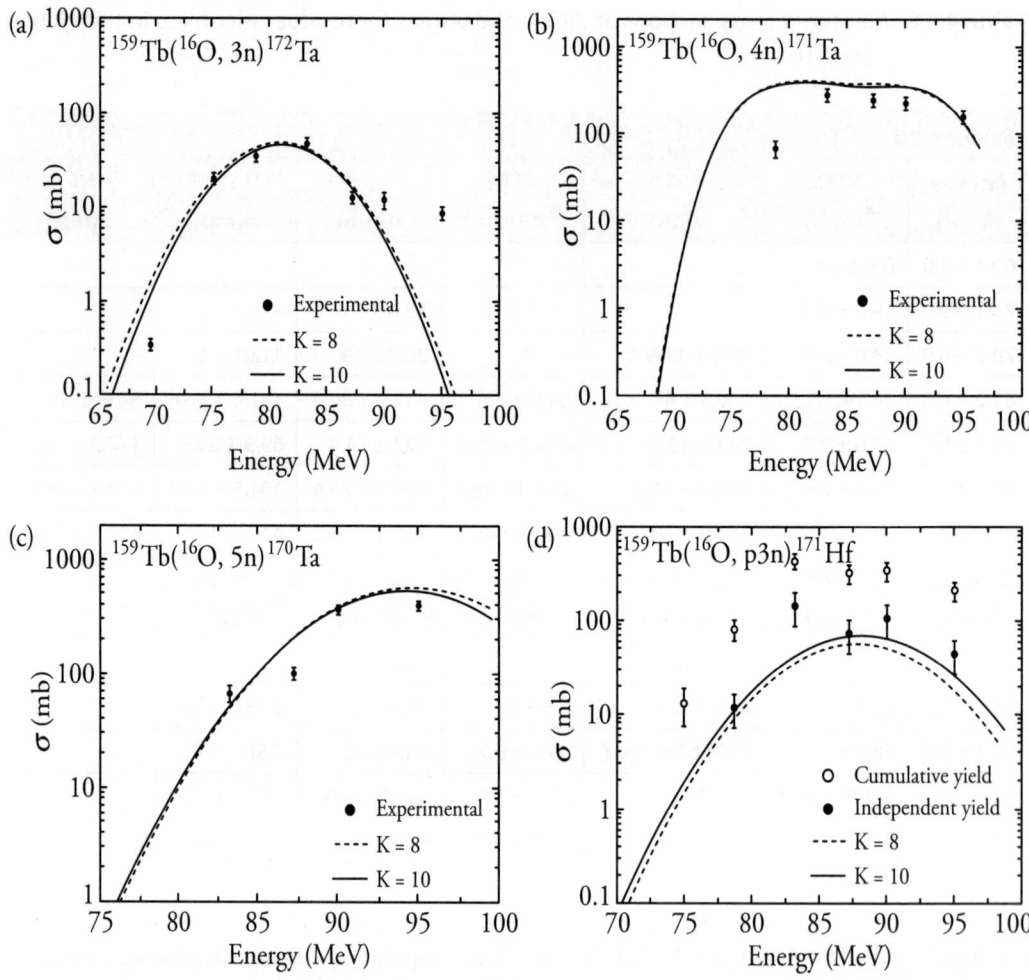

Figure 4.31 Experimentally measured and theoretically calculated EFs using code PACE2. The effect of variation of parameter *K* is also shown in these figures. In Figure 4.31(d), the open circles represent the cumulative yield for the production of the residue ^{171}Hf, while the solid circles represent its independent yield

In the reactions ^{159}Tb(^{16}O,2p2n)^{171}Lu, ^{159}Tb(^{16}O,αn)^{170}Lu, ^{159}Tb(^{16}O,α2n)^{169}Lu and ^{159}Tb (^{16}O,2α2n)^{165}Tm, the calculated values are smaller than the measured cross-sections as shown in Figure 4.32. The difference between the measured cross-sections and the theoretical predictions may be due to the ICF process. A detailed description for the measurement and analysis for this system is given by Gupta et al.[32]

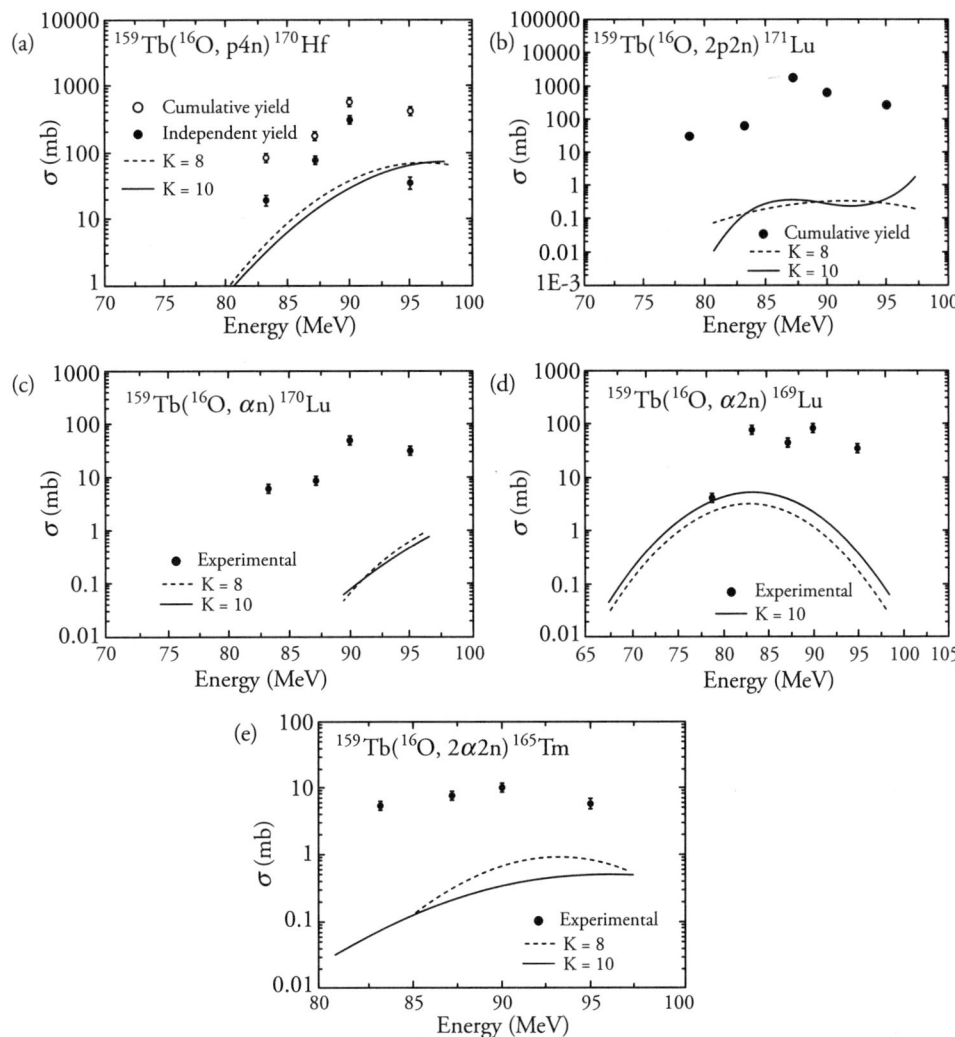

Figure 4.32 Experimentally measured and theoretically calculated EFs using code PACE. The effect of variation of parameter *K* is also shown in these figures. In Figure 4.32(a), the open circles represent the cumulative yield for the production of the residue ^{170}Hf, while the solid circles represent its independent yield

4.1.4.3 ^{16}O+^{169}Tm system

As a part of our study of complete fusion (CF) and incomplete fusion (ICF) dynamics in heavy ion interactions at low energies, the excitation functions for eight reactions in the system ^{16}O+^{169}Tm have been measured[6] in the energy range ≈70–95 MeV. The stacked foil activation technique has been used for the measurements. Excitation functions for the reactions ^{169}Tm(^{16}O,3n)^{182}Ir, ^{169}Tm(^{16}O,4n)^{181}Ir, ^{169}Tm(^{16}O,p2n)^{182}Os, ^{169}Tm(^{16}O,p3n)^{181g}Os, ^{169}Tm(^{16}O,2p2n)^{181}Re,

^{169}Tm(^{16}O,α3n)^{178}Re, ^{169}Tm(^{16}O,2αpn)^{175}Hf and ^{169}Tm(^{16}O,3αn)^{172}Lu have been measured and are listed in Table 4.12.

Table 4.12 Measured cross-sections at different energies for the residues identified in the system ^{16}O+^{169}Tm

System: ^{16}O + ^{169}Tm					
Energy (MeV)	^{182}Ir σ(mb)	^{181}Ir σ(mb)	^{182}Os^{cum} σ(mb)	^{182}Os^{ind} σ(mb)	^{181}Os^{cum} σ(mb)
71.7 ± 1.0	3.28 ± 0.7	–	4.58 ± 1.4	1.26 ± 0.6	2.72 ± 0.4
74.9 ± 0.9	42.3 ± 7.2	28.7 ± 12.5	82.7 ± 10	39.98 ± 5.1	4.81 ± 1.2
78.7 ± 0.9	59.8 ± 13.9	110 ± 15.6	139.4 ± 22.9	78.8 ± 13.8	32.81 ± 4.3
82 ± 0.8	86.4 ± 14.8	170.4 ± 28.4	155.6 ± 20.7	68.15 ± 8.3	129 ± 16.3
85.8 ± 0.8	47.6 ± 7.7	250.1 ± 67.7	107.4 ± 14.4	59.2 ± 7.5	198.02 ± 23.3
88.9 ± 1.0	35.2 ± 3.9	316.7 ± 34.85	71.5 ± 9.9	35.84 ± 6.2	250.9 ± 31.3
91.6 ± 0.4	13.7 ± 3.2	229.8 ± 39.6	29.3 ± 4.1	15.45 ± 4.4	153.8 ± 17.8
94.6 ± 0.4	8.47 ± 1.4	183.8 ± 27.2	18.4 ± 3.7	9.85 ± 2.7	173.1 ± 22.8
Energy (MeV)	^{181}Re σ(mb)	^{178}Re σ(mb)	^{175}Hf σ(mb)	^{182}Lu σ(mb)	
71.7 ± 1.0	2.6 ± 0.7	–	–	–	
74.9 ± 0.9	5.35 ± 0.7	–	–	–	
78.7 ± 0.9	137.3 ± 28.7	1.74 ± 0.2	–	14.9 ± 2.1	
82 ± 0.8	391.4 ± 83	5.2 ± 0.8	–	20.5 ± 2.6	
85.8 ± 0.8	594 ± 90.6	9.02 ± 1.2	0.57 ± 0.1	31.3 ± 3.9	
88.9 ± 1.0	607.9 ± 86.7	27.34 ± 5.3	2.53 ± 0.4	30.2 ± 3.7	
91.6 ± 0.4	526.2 ± 78.8	32.14 ± 3.7	2.96 ± 0.5	28.3 ± 3.1	
94.6 ± 0.4	441.9 ± 66.9	34.31 ± 0.6	4.62 ± 0.6	28.1 ± 3.3	

Measured excitation functions (EFs) have been analysed using the statistical codes ALICE-91, CASCADE and PACE2. Though the details of the code are already given in Chapter 2, some salient features of the codes are mentioned here. The code ALICE-91, developed by M. Blann, accounts for the CN and PE emission in nuclear reactions. The CN calculations in this code are performed using the hybrid/geometry dependent hybrid model. The initial exciton number n_0, which is the sum of the initially excited particles and holes, is one of the (rather) free parameters that may be varied to reproduce the experimental data. The mean free path multiplier COST is another important parameter, which accounts for the actual mean free path inside the nuclear matter because the actual mean free path may be quite different from the one calculated using free nucleon–nucleon scattering data. In order to compensate for this

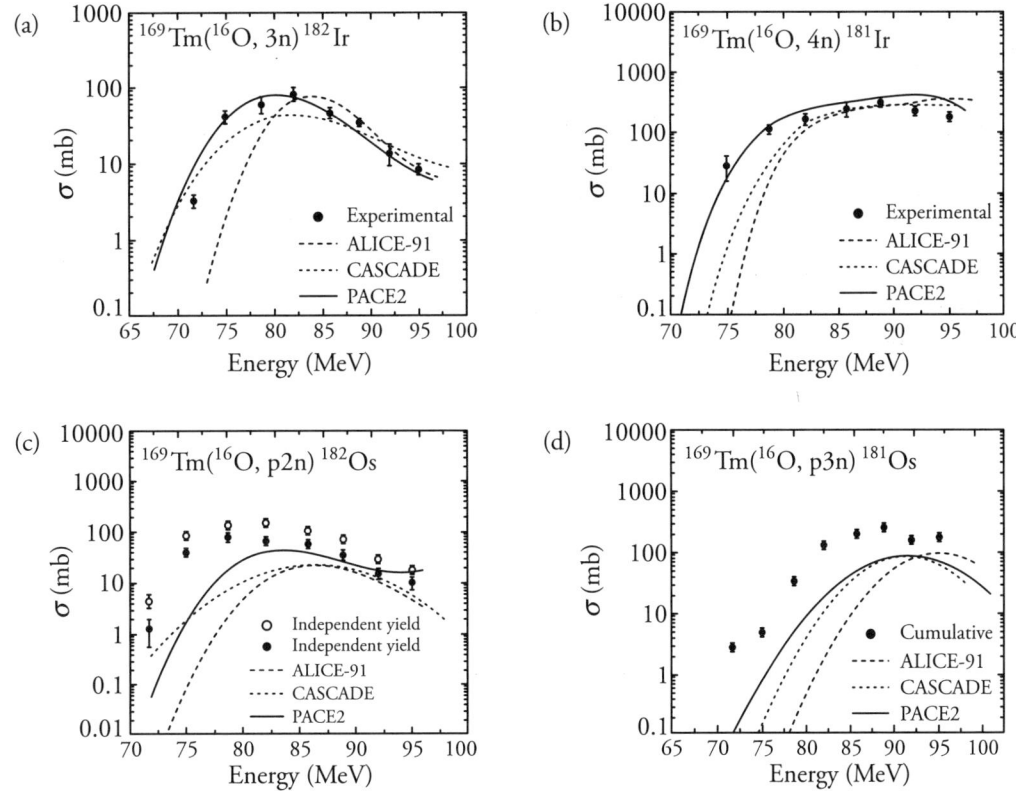

Figure 4.33 Experimentally measured and theoretically calculated EFs using codes ALICE-91, CASCADE and PACE2. In Figure 4.33(c), the open circles represent the cumulative yield for the production of the residue ^{182}Os, while dark circles represent its independent yield

difference, the parameter COST is provided in the code ALICE. The level density parameter a mainly affects the equilibrium component and is given by $a = A/K$, where A is the mass number of the nucleus and K is the free parameter. In this work, the same values of parameters have been used to fit the experimental data on xn channels. Further, the code CASCADE is based on the Hauser–Feshbach theory and does not take the pre-equilibrium emission and the incomplete fusion into account. In this code, the level density parameter and the ratio of the actual moment of inertia to the rigid body moment of inertia of the compound system F_θ are the two important parameters that are generally used to fit the experimental data. The level density parameter a_f at the saddle point, another free parameter of the code, may be obtained using the relations $a_f = A/D_{AF}$, where A is the mass number of the nucleus and D_{AF} is a parameter that may be varied to match the experimental data. The CASCADE calculations give satisfactory reproduction of the data for the values $K = 14$, $D_{AF} = 14$ and $F_\theta = 0.85$, as shown in Figure 4.33. Similarly, PACE2 calculations are shown in this figure for the reactions ^{169}Tm(^{16}O,3n)^{182}Ir, ^{169}Tm(^{16}O,4n)^{181}Ir, ^{169}Tm(^{16}O,p2n)^{182}Os and ^{169}Tm(^{16}O,p3n)^{181g}Os. As can be seen from this figure, 3n and 4n channels are satisfactorily reproduced by the calculations but the p2n, and p3n channels are underestimated by calculations. The reason for this may be that the production of residues resulting in these reactions may also have contributions from their higher charge precursors. The independent production cross-sections for residues ^{182}Os has been obtained after subtracting the contributions from the precursor (Figure 4.33 (c)).

Further, the same input parameters have been used to calculate the EFs for the alpha-emitting channels as shown in Figure 4.34. Theoretical calculations using all the three codes give negligible cross-sections for the production of residues ^{176}Hf and ^{172}Lu, and therefore, they are not shown in Figure 4.34(c) and (d). As can be seen from these figures, the calculations give quite small cross-sections at each energy as compared to the measured values, except for the reaction ^{169}Tm(^{16}O,α3n)^{178}Re. Since the ICF is not considered in these calculations done using different codes, the enhancement in cross-section values may be attributed to the incomplete fusion process.

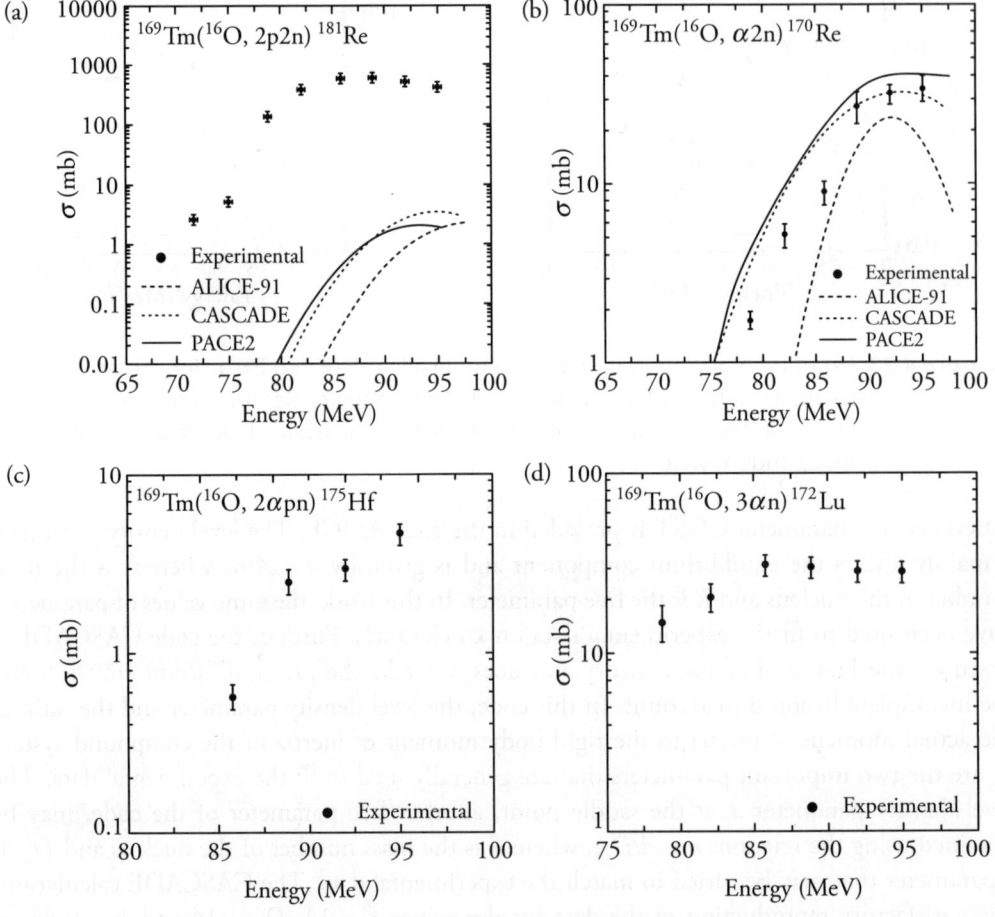

Figure 4.34 Experimentally measured and theoretically calculated EFs for alpha-emitting channels. Theoretical calculations have been done using codes ALICE-91, CASCADE and PACE2. In Figure 4.34(c) and (d), calculated EFs are not shown

4.1.4.4 $^{16}O+^{103}Rh$ *system*

Experimentally measured cross-sections at different energies for the feeding of different identified residues in the system $^{16}O+^{103}Rh$ are presented in Table 4.13. The detailed analysis and measurement descriptions for this system are given by Gupta et al.[32]

Table 4.13 Experimentally measured cross-sections at different energies for the population of different identified residues in the system $^{16}O+^{103}Rh$

System: $^{16}O + ^{103}Rh$							
Energy (MeV)	^{117}Te cum σ(mb)	^{117}Te ind σ(mb)	^{116}Te cum σ(mb)	^{116}Te ind σ(mb)	^{115}Te cum σ(mb)	^{115}Te ind σ(mb)	^{114}Te σ(mb)
51.8 ± 3.2	6.3 ± 0.6	2.1 ± 0.2	–	–	–	–	–
57.8 ± 3.5	13.8 ± 1.3	5.9 ± 0.6	102.3 ± 10.2	68.6 ± 6.8	3 ± 0.3	2.9 ± 0.3	–
65.8 ± 2.3	5.6 ± 0.5	2.3 ± 0.2	227.3 ± 22.7	161.2 ± 16.1	27.2 ± 2.7	24.3 ± 0.2	–
70.8 ± 3.1	3.4 ± 0.3	1.1 ± 0.1	212.7 ± 21.3	154.2 ± 15.4	42.5 ± 4.3	39.1 ± 0.4	–
77.1 ± 2.6	0.8 ± 0.1	0.8 ± 0.1	150.7 ± 15	127.6 ± 12.8	118.2 ± 11.8	95.1 ± 9.5	4.1 ± 0.4
82.8 ± 2.6	–	–	136.4 ± 13.6	116.8 ± 12	160.7 ± 16.1	138.4 ± 13.8	15.3 ± 1.5
Energy (MeV)	^{117}Sb cum σ(mb)	^{117}Sb ind σ(mb)	^{115}Sb cum σ(mb)	^{115}Sb ind σ(mb)	^{111}In σ(mb)	^{110g}In σ(mb)	^{110m}In σ(mb)
51.8 ± 3.2	14.4 ± 2	1.5 ± 0.15	–	–	–	–	–
57.8 ± 3.5	19.6 ± 2	3 ± 0.3	2.3 ± 0.2	–	–	0.8 ± 0.1	0.9 ± 0.1
65.8 ± 2.3	24.6 ± 3	2.1 ± 0.2	51.8 ± 5.2	–	0.5 ± 0.1	2.7 ± 0.3	18.1 ± 2
70.8 ± 3.1	8.7 ± 0.8	0.4 ± 0.1	146.4 ± 14.6	16.7 ± 1.7	2.8 ± 0.3	1.2 ± 0.4	51.2 ± 5.1
77.1 ± 2.6	4 ± 0.4	–	269.4 ± 27	57.5 ± 5.7	4.6 ± 0.5	2.4 ± 0.3	93.1 ± 9.3
82.8 ± 2.6	2.1 ± 0.2	–	387.6 ± 38.7	111.4 ± 11.1	18.2 ± 1.8	3.7 ± 0.4	131.6 ± 13.2
Energy (MeV)	^{109}In σ(mb)	^{108g}In σ(mb)	^{108m}In σ(mb)	^{106m}Ag σ(mb)	^{104g}Ag σ(mb)	^{103g}Ag σ(mb)	
51.8 ± 3.2		–	–	–	–	–	
57.8 ± 3.5		–	–	–	–	–	
65.8 ± 2.3	0.5 ± 0.1	–	–	9.6 ± 1.0	–	5.1 ± 0.5	

70.8 ± 3.1	1.8 ± 0.2	0.4 ± 0.04	3.6 ± 0.4	12.4 ± 1.2	–	16.2 ± 1.6	
77.1 ± 2.6	10 ± 1	1.6 ± 0.2	4.9 ± 0.5	18.9 ± 1.9	2.7 ± 0.3	28.3 ± 2.8	
82.8 ± 2.6	19.6 ± 2	2.5 ± 0.3	9.5 ± 1.0	29.1 ± 0.3	5.1 ± 0.5	–	

4.1.4.5 $^{16}O+^{181}Ta$ system

To study the incomplete fusion reaction dynamics within the framework of the break-up fusion model, at energies near and above the Coulomb barrier, another experiment[33] was carried out for the system $^{16}O+^{181}Ta$. Details of the experimental setup and more are already given in Chapter 3. The measured cross-section for reaction residues in this system are given in Table 4.14.

Table 4.14   Experimentally measured cross-sections for the reaction residues populated in the system $^{16}O+^{181}Ta$

System: $^{16}O + ^{181}Ta$						
Energy (MeV)	^{194m}Tl σ(mb)	^{194g}Tl σ(mb)	^{193m}Tl σ(mb)	^{193g}Tl σ(mb)	^{192m}Tl σ(mb)	^{192g}Tl σ(mb)	^{193g}Hg σ(mb)
76 ± 1.1	2 ± 0.2	2 ± 0.2	0.1 ±0.0	26 ± 3.8	–	–	23 ± 3.5
80 ± 1.5	6 ± 0.8	6 ± 0.8	0.2 ± 0.0	45 ± 6.8	22 ± 3.2	22 ± 3.2	47 ± 7
85 ± 1.2	4 ± 0.5	4 ± 0.5	0.2 ± 0.0	68 ± 10.2	61 ± 9.1	61 ± 9.1	60 ± 8.9
87 ± 1.0	3 ± 0.4	3 ± 0.4	0.3 ± 0.0	46 ± 6.9	44 ± 6.5	44 ± 6.5	49 ± 7.4
88 ± 1.6	2 ± 0.2	2 ± 0.2	0.1 ± 0.0	44 ± 6.5	91 ± 13.7	91 ± 13.7	42 ± 6.2
93 ± 1.1	2.5 ± 0.3	2 ± 0.3	0.1 ± 0.0	35 ± 5.2	184 ± 27.6	184 ± 27.6	29 ± 4.4
97 ± 1.0	2 ± 0.3	1.5 ± 0.2	0.1 ± 0.0	15 ± 2.3	171 ± 25.5	171 ± 25.5	12 ± 1.7
99 ± 0.9	1 ± 0.1	1 ± 0.1	0.1 ± 0.0	17 ±2.5	222 ± 33.3	222 ± 33.3	10 ± 1.5
Energy (MeV)	^{193m}Hg σ(mb)	^{192}Hg σ(mb)	^{191g}Hg σ(mb)	^{191m}Hg σ(mb)	^{192g}Au σ(mb)	^{191g}Au σ(mb)	^{190g}Au σ(mb)
80 ± 1.5	21 ± 2.1	4 ± 0.5	–	–	2 ± 0.2	–	–
85 ± 1.2	30 ± 3.0	40 ± 6.0	–	–	10 ± 1.5	–	8 ± 1.3
87 ± 1.0	22 ± 2.2	36 ± 5.5	–	–	12 ± 1.8	2 ± 0.3	6 ± 0.8
88 ± 1.6	24 ± 2.3	65 ± 9.8	3 ± 0.5	0.3 ± 0.0	31 ± 4.6	2 ± 0.3	23 ± 3.5
93 ± 1.1	13 ± 1.3	121 ± 18.2	5 ± 0.7	3 ± 0.5	46 ± 6.9	3 ± 0.5	20 ± 2.9
97 ± 1.0	8 ± 0.7	131 ±6	7 ± 0.9	8 ± 1.2	63 ± 9.5	14 ± 2.1	40 ± 5.9
99 ± 0.9	6 ± 0.5	154 ± 23.2	14 ± 2.1	18 2.7	50 ± 7.5	22 ± 3.2	21 ± 3.2

The experimentally measured excitation functions for the residues populated via the xn channel are shown in Figure 4.35(a). It may be pointed out that for 3n and 5n channels, the metastable and ground states of the respective residues decay with gamma rays of nearly the same energy and half-life. Hence, the observed composite decay curve gives the sum of both the states in each case. The individual cross-sections were obtained by dividing the measured

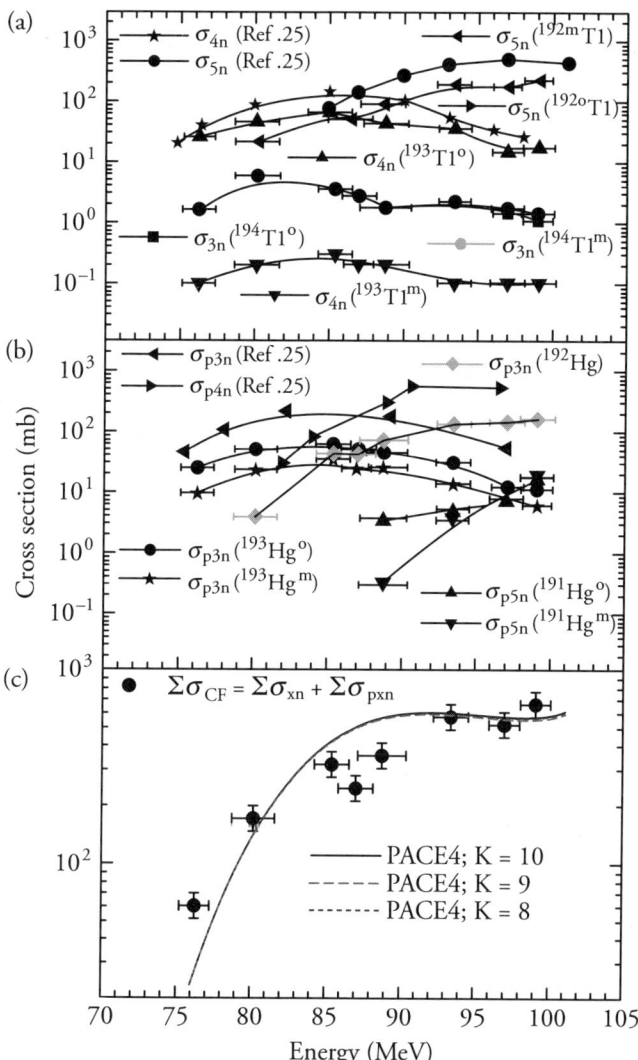

Figure 4.35 Measured EFs for (a) xn (x = 3, 4 and 5) channels and (b) pxn (x = 3, 4 and 5) channels. In panels (a) and (b), the lines joining the experimental data points are just to guide the eyes. Panel (c) shows the sum of cross-sections for the xn and pxn channels. The effect of the variation of the choice of the level density parameter $K = 8$, 9 and 10 (dotted, dashed and solid lines, respectively) on calculated $\Sigma\sigma_{CF}$ is also shown

composite cross-sections according to the ratio of their gamma ray intensities[25]. In case of the ^{193g,m}Tl residue populated via 4n channel, the metastable state has a half-life of nearly 2 min; it then decays to the ground state, which has a half-life of nearly 22 min. Since the counting of the samples was done after about 10 minutes from the stop of irradiation, the measured cross-sections for the ground state contain about 0.38% contributions from the isomeric state. Hence, an upper limit for the cross-section for the independent production of the metastable state has been deduced and is given Table 4.14. The total cross-section ($\Sigma\sigma_{xn}$) for all the residues produced via xn (x = 3, 4 and 5) channels is also given in Figure 4.35(a), which indicates the initial rise in $\Sigma\sigma_{xn}$ values and then saturation at relatively higher energies. It may be pointed out that in case of pxn channels, there is no likelihood of incomplete fusion and hence, these channels are populated only by CF, like the xn channels.

The product residues formed via p3n, p4n and p5n channels have been identified as discussed earlier. For p3n and p5n channels, both the metastable as well as the ground state are populated while for the p4n channel, only one state is formed. All the residues of the pxn channel decay independently with their respective half-lives. The cross-section for these channels are plotted in Figure 4.35(b) and are tabulated in Table 4.14. It is worth mentioning that the residues populated via the pxn channel may be populated both independently, and also by the decay of higher charge isobar precursor, as shown here:

$$^{16}O + {}^{181}Ta \Rightarrow {}^{197}Tl^* \Rightarrow {}^{193}Tl^* + 4n$$

$$^{193}Tl^* \Rightarrow {}^{193}Hg + \beta^+/EC \text{ (precursor decay)}$$

$$^{16}O + {}^{181}Ta \Rightarrow {}^{197}Tl^* \Rightarrow {}^{193}Hg + p3n \text{ (independent decay)}$$

Similarly, the population of residues ^{192}Hg and ^{191}Hg may also be expected via the independent decay as well as the precursor decay of the type:

$$^{16}O + {}^{181}Ta \Rightarrow {}^{197}Tl^* \Rightarrow {}^{192}Tl^* + 5n$$

$$^{192}Tl^* \Rightarrow {}^{192}Hg + \beta^+/EC \text{ (precursor decay)}$$

$$^{16}O + {}^{181}Ta \Rightarrow {}^{197}Tl^* \Rightarrow {}^{192}Hg + p4n \text{ (independent decay)}$$

and

$$^{16}O + {}^{181}Ta \Rightarrow {}^{197}Tl^* \Rightarrow {}^{191}Tl^* + 6n$$

$$^{191}Tl^* \Rightarrow {}^{191}Hg + \beta^+/EC \text{ (precursor decay)}$$

$$^{16}O + {}^{181}Ta \Rightarrow {}^{197}Tl^* \Rightarrow {}^{191}Hg + p5n \text{ (independent decay)}$$

It may, however, be pointed out that in case of the p5n channel, the precursor ^{191}Tl, which may be produced via the 6n channel, is not likely to be populated in the present case because of its higher threshold which is around 100 MeV. Further, in case of the p3n and p4n channels, it was not possible to determine the precursor contributions because of either the incomplete decay or the unknown decay characteristics of the precursor. As an example, in case of the p4n channel, the probability for the independent decay of the precursor formed by the 5n channel determined from its characteristic gamma lines are found to be higher than the cross-sections for residues ^{192}Hg populated via the p4n channel. It is likely in case the pre-cursor does not feed the residue ^{192}Hg formed via the p4n channel. It may be remarked that the decay schemes of ^{192}Hg and ^{193}Hg available at present are not complete and need further investigation. The cross-section values given in Table 4.14 for these reactions also contain the contribution from precursor decay, if any, in the case of p3n and p4n channels. Further, in Figure 4.35(b), the cross-sections for all presently measured pxn channels are shown along with the literature values. To get the total measured fusion cross-sections, $\Sigma\sigma_{CF}$, shown in Figure 4.35(c), cross-sections for all the xn and pxn channels have been added and compared with $\Sigma\sigma_{CF}$ obtained using the code PACE4. Here the calculations for different level density parameters are also shown. As can be seen from this figure, the measured $\Sigma\sigma_{CF}$ agree reasonably well with the theoretical cross-sections for $\Sigma\sigma_{CF}$. This agreement in the measured and theoretical total fusion cross-sections indicates that the choice of statistical parameters used in theoretical calculations is satisfactory.

The measured cross-sections for the population of ^{190,191,192}Au isotopes produced via αxn channels are shown in Figure 4.36(a). It may be seen that in this case, the residues may be populated in two ways: (i) by the fusion of ^{16}O with the target nucleus ^{181}Ta forming the CN and then the evaporation of neutrons and alpha particle; this is referred to as the complete fusion mode, or (ii) the ^{16}O ion may break into $\alpha + {}^{12}$C and ^{12}C fuses with the target nucleus^{181}Ta leaving an alpha particle as a spectator; here the excited nucleus formed as a result of fusion of ^{12}C may eject neutrons during the de-excitation process; this may be referred to as the incomplete fusion mode. The reactions may be represented with the help of the following equations:

$$^{16}O(^{12}C + \alpha) \Rightarrow {}^{12}C + {}^{181}Ta \Rightarrow {}^{193}Au^* + \alpha; (\alpha \text{ as spectator})$$

The residue ^{192}Au^g may be populated via CF and/or ICF channels as

 i. Complete fusion of ^{16}O, i.e.,

$$^{16}O + {}^{181}Ta \Rightarrow {}^{197}Tl^* \Rightarrow {}^{192}Au^g + \alpha + n$$

 ii. Incomplete fusion of ^{16}O, i.e.,

$$^{16}O(^{12}C + \alpha) + {}^{181}Ta \Rightarrow {}^{193}Au^* + \alpha, (\alpha \text{ as spectator})$$

$$^{193}Au^* \Rightarrow {}^{192}Au^g + n$$

The residues ^{191}Au^g may be populated via CF and/or ICF channels as

 i. Complete fusion of ^{16}O, i.e.,

$$^{16}O + {}^{181}Ta \Rightarrow {}^{197}Tl^* \Rightarrow {}^{191}Au^g + \alpha + 2n$$

ii.   Incomplete fusion of ^{16}O, i.e.,

$$^{16}O(^{12}C + \alpha) + ^{181}Ta \Rightarrow ^{193}Au^{*} + \alpha, (\alpha \text{ as spectator})$$

$$^{193}Au^{*} \Rightarrow ^{191}Au^{g} + 2n$$

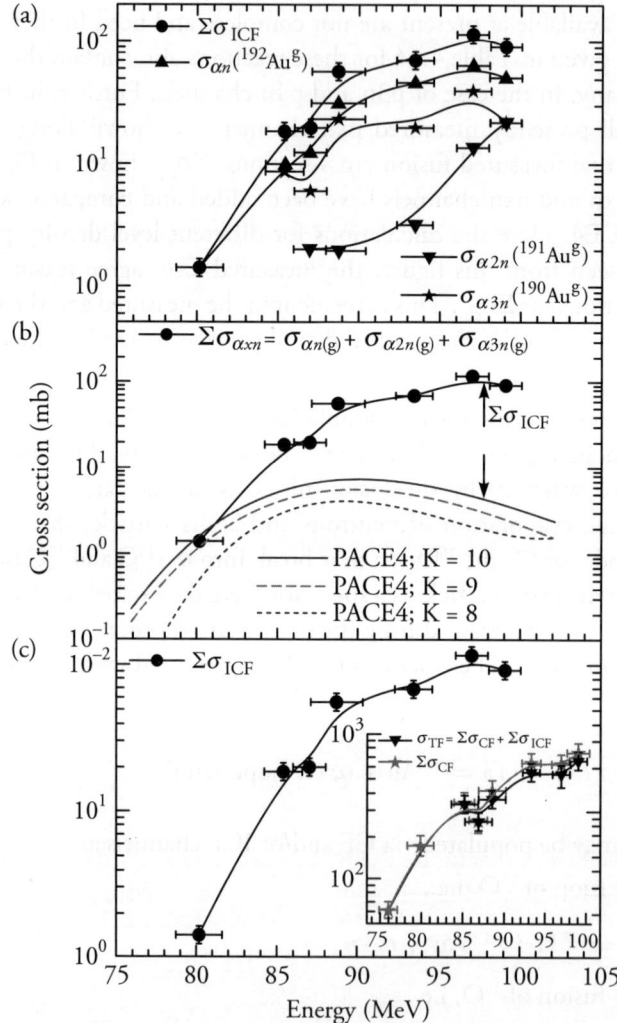

Figure 4.36 *(a) Measured EFs for αxn (x = 1, 2 and 3) channels, (b) sum of the αxn channels, measured as well as calculated using PACE4 for K = 8, 9, 10 (dotted, dashed and solid lines, respectively), and (c) sum of σ_{ICF} (all αxn) channels. In panels (a), (b) and (c), the spline-like lines joining the experimental data points are just to guide the eyes. The inset shows cross-sections for the sum of both CF and ICF channels and for CF channels separately. The increasing difference between the two curves in the inset, with energy indicates the dominance of ICF processes with energy*

Similarly, the residues ^{190}Au^g may be populated via CF and/or ICF channels as

i. Complete fusion of ^{16}O, i.e.,

$$^{16}O + ^{181}Ta \Rightarrow ^{197}Tl^* \Rightarrow ^{190}Au^{g*} + \alpha + 3n$$

ii. Incomplete fusion of ^{16}O, i.e.,

$$^{16}O(^{12}C + \alpha) + ^{181}Ta \Rightarrow ^{193}Au^* + \alpha, (\alpha \text{ as spectator})$$

$$^{193}Au^* \Rightarrow ^{190}Au^{g*} + 3n$$

It may be remarked that the residues ^{192}Au^g may also be populated via the decay of ^{192}Hg through β^+/EC decay. Both these residues, i.e., ^{192}Au^g ($T_{1/2}$ = 4.94 h) and ^{192}Hg ($T_{1/2}$ = 4.85 h) have nearly the same half-lives. Here, it was possible to separate the contribution from the decay of ^{192}Hg, which is populated via the p4n channel, using the decay analysis. It is known from successive radioactive decay that if the daughter nucleus half-life (T_A) and the parent nucleus half-life (T_B) are nearly equal, as in the present case, such that $T_A = T_B(1+\delta)$, where $\delta \ll 1$, then the activity ratio increases approximately linearly with time, so long as $t \ll 2\tau_B/\delta$, where τ_B is the mean lifetime of the parent nucleus. In order to get the cross-section for ^{192}Au^g, a curve between the lapse time and its production cross-section was plotted at different times and also at different energies. To get the independent cross-section at each energy, plots of different lapse times were extrapolated at t = 0 time using a least square linear fitting method. The cross-section at time t = 0 is the independent cross-section for the production of ^{192}Au^g. The cross-section deduced as given here for the independent production of ^{192}Au^g is shown in Figure 4.36(b). In this figure, the sum of cross-sections for all measured αxn channels is also shown and is found to increase linearly with incident energy. To get the contribution from ICF processes to the αxn channels, the measured $\Sigma\sigma_{\alpha xn}$ (expt) has been compared with the corresponding values calculated using the code PACE4, which is based on statistical CN theory. Since the code PACE4 does not take ICF into consideration, the calculated cross-section for $\Sigma\sigma_{\alpha xn}$ have predictions based only on the CF model. A comparison of $\Sigma\sigma_{\alpha xn}$ (expt) has also been made with $\Sigma\sigma_{\alpha xn}$ (Th) in Figure 4.36(b) for level density parameters K = 8, 9 and 10. It may be observed from this figure that theoretical calculations do not reproduce the data at energies above 85 MeV. This difference in cross-section values may be attributed to the ICF reactions. Further, the difference between $\Sigma\sigma_{\alpha xn}$ (expt) and $\Sigma\sigma_{\alpha xn}$ (Th) is found to increase with energy above 80 MeV, indicating the importance of ICF reactions at relatively higher energies; maximum ICF contribution was found at 100 MeV energy. In Figure 4.36(c), the $\Sigma\sigma_{ICF}$ deduced by subtracting $\Sigma\sigma_{ICF}$(Th) from measured $\Sigma\sigma_{\alpha xn}$ has been plotted with energy. One can see from this figure that ICF production increases quite rapidly with increase in energy. In the inset of Figure 4.36(c), the $\Sigma\sigma_{TF}$ (total sum of cross-sections for all measured channels) and $\Sigma\sigma_{CF}$ are compared. It may be observed from this figure that ICF increases with the incident beam energy, as expected.

4.1.5 Reactions initiated by ^{18}O beam

4.1.5.1 ^{18}O+^{159}Tb system

In order to have a better and clear picture of incomplete fusion reaction dynamics at energies ~ 4–7 MeV/nucleon, the excitation function measurements have also been performed for the ^{18}O+^{159}Tb system. Here a non-alpha cluster beam has been used. The measured cross-sections for several *xn* and *αxn* channels are given in Table 4.15. The experimental data have been analysed within the framework of compound nucleus decay. The cross-sections for xn/pxn-channels are found to be well reproduced by PACE4 predictions, which suggest their production via complete fusion process. However, a significant enhancement in the excitation functions of *α*-emitting channels has been observed over the theoretical ones, which has been attributed to the incomplete fusion processes.

Table 4.15 Measured cross-sections at different energies for the residues identified in the system ^{18}O+^{159}Tb

System: ^{18}O + ^{159}Tb				
Energy (MeV)	^{174}Ta σ(mb)	^{173}Ta σ(mb)	^{172}Ta σ(mb)	^{171}Ta σ(mb)
70.35 ± 0.70	4.63 ± 0.46	30.3 ± 3.7	–	–
72.86 ± 0.73	6.75 ± 0.68	150.7 ± 11.1	0.5 ± 0.05	–
77.84 ± 0.78	4.40 ± 0.44	280.2 ± 21.2	80.23 ± 8.9	–
80.36 ± 0.80	4.13 ± 0.41	320.5 ± 31.78	180.89 ± 21.45	–
81.97 ± 0.82	1.25 ± 0.13	250.9 ± 23.8	290.34 ± 30.89	–
84.19 ± 0.84	2.10 ± 0.21	232.1 ± 21.2	410.56 ± 43.64	0.59 ± 0.07
87.04 ± 0.87	0.46 ± 0.05	100.3 ± 10.98	550.91 ± 59.12	15.12 ± 2.25
91.27 ± 0.91	0.17 ± 0.02	32.5 ± 2.89	610.39 ± 58.34	130.96 ± 19.50
93.66 ± 0.94	0.08 ± 0.01	21.1 ± 2.11	500.16 ± 49.55	259.11 ± 38.85
95.57 ± 0.96	0.04 ± 0.001	15.9 ± 1.2	490.31 ± 52.19	290.45 ± 43.50
99.16 ± 0.99	0.01 ± 0.001	6.2 ± 0.6	350.92 ± 38.72	560.55 ± 84.0
Energy (MeV)	^{171}Lu σ(mb)	^{168m}Lu σ(mb)	^{167}Lu σ(mb)	^{167}Yb σ(mb)
70.35 ± 0.70	4.63 ± 0.51	–	–	–
72.86 ± 0.73	6.75 ± 0.78	–	–	–
77.84 ± 0.78	4.40 ± 0.84	–	–	–
80.36 ± 0.80	4.13 ± 0.32	–	–	–
81.97 ± 0.82	1.25 ± 0.26	–	–	3.2 ± 0.87
84.19 ± 0.84	2.10 ± 0.37	12 ± 1.1	0.1 ± 0.87	2.52 ± 0.52

87.04 ± 0.87	0.46 ± 0.05	32 ± 5.2	5 ± 0.69	8.01 ± 1.65
91.27 ± 0.91	4.13 ± 0.02	38 ± 4.9	3 ± 0.71	51.12 ± 7.05
93.66 ± 0.94	1.25 ± 0.09	80 ± 10.83	27 ± 3.78	93.35 ± 14.05
95.57 ± 0.96	4.13 ± 0.73	180 ± 23.12	72 ± 11.41	79.04 ± 12.48
99.16 ± 0.99	1.25 ± 0.30	295 ± 22.78	39 ± 6.75	106.02 ± 17.08

In order to understand the formation mechanism of the residues produced during the $^{18}O+^{159}Tb$ interactions, the experimentally measured excitation functions have been analysed within the framework of the statistical model code PACE4[23], which is based on equilibrated CN decay of the Hauser–Feshbach theory. It may however be pointed out that the ICF and pre-equilibrium emission (PEE) are not taken into consideration in this code. In this code, level density parameter ($a = A/K$) is an important input parameter that affects the CF cross-sections and where K may be varied to match the experimental cross-sections. As a representative case, in Figure 4.37, the EFs for 3n, 4n, 5n and 6n channels have been compared with corresponding PACE4 predictions for three different values of level density parameters and $K = 8$ has been found to reproduce satisfactorily the experimental data, which shows that the production of this residue is via the CF process.

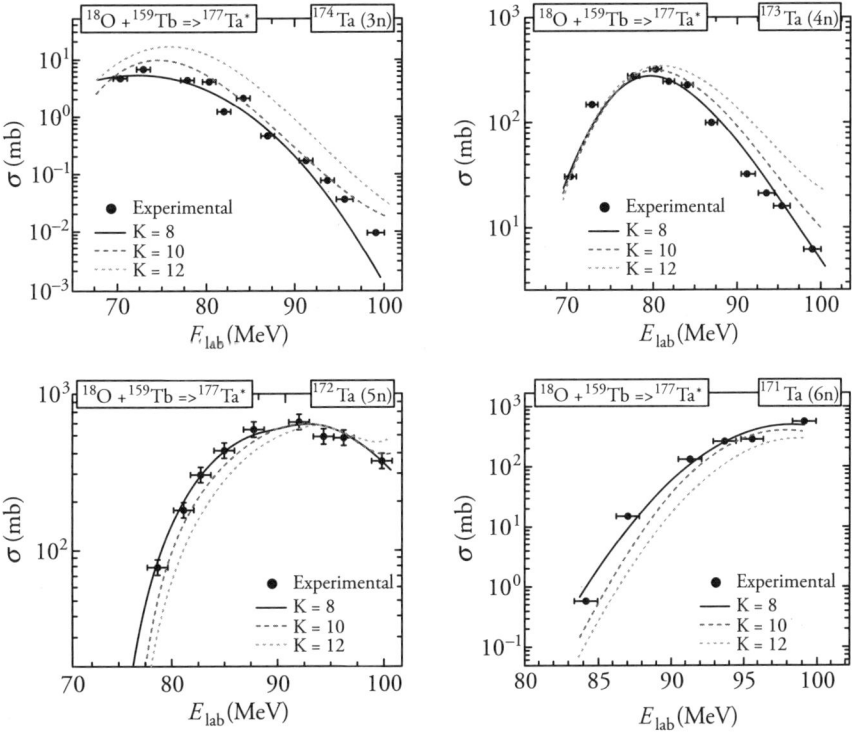

Figure 4.37 Experimental EFs of ^{174}Ta (3n), ^{173}Ta (4n), ^{172}Ta (5n) and ^{171}Ta (6n) have been compared with that predicted by PACE4 for different values of level density parameter ($a = A/K$ MeV^{-1}, where K is varied from 8 to 12)

The same set of parameters has also been used to check the production mechanism of α-emitting channels. As shown in Figure 4.38, the measured cross-sections for ^{171}Lu($\alpha 2n$), ^{168}Lu($\alpha 5n$), ^{167}Lu($\alpha 6n$) and ^{167}Yb($\alpha p5n$) residues are found to be significantly enhanced over their theoretical predictions. It is worth mentioning that the CN reaction theory predicts almost negligible cross-sections for the production of residues ^{171}Lu and ^{167}Y and therefore, are not shown in Figure 4.38.

It has already been mentioned that PACE4 does not take ICF and PEE into account and hence, this enhancement may be assumed to be due to the ICF reaction mechanism. It is evident from the analysis that ICF reactions contribute significantly to the production cross-section of α-emitting channels at the studied energy range. Further, the ICF contribution for individual channels has been deduced by subtracting CF cross-sections (σ_{CF}) at each energy from the corresponding experimentally measured total fusion cross-sections (σ_{TF}), which is plotted in Figure 4.39(a). It may be observed in Figure 4.39(a), that the ICF contribution increases with beam energy. It is not out of place to mention that the σ_{TF} has been corrected for the missing channels (which could not be measured experimentally) by their PACE4 values. Hence, the σ_{ICF} may be taken at least as the lower limit of ICF contribution.

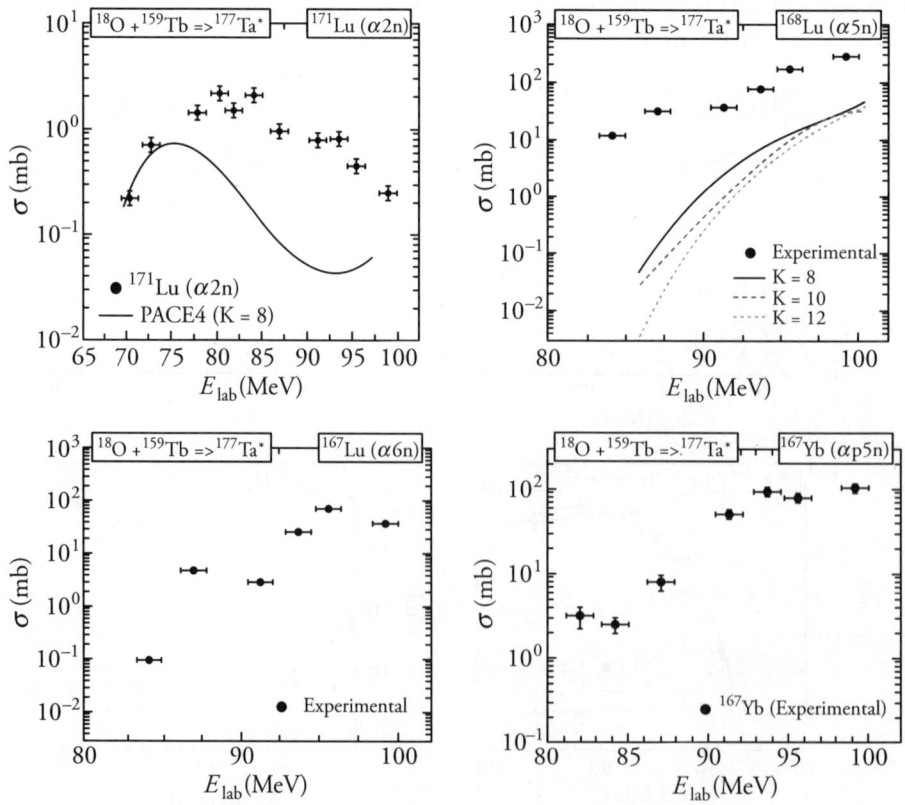

Figure 4.38 Experimental EFs of ^{171}Lu ($\alpha 2n$), ^{168}Lu($\alpha 5n$), ^{167}Lu($\alpha 6n$) and ^{167}Yb($\alpha p5n$) residues have been compared with that predicted by PACE4 for different values of level density parameter ($a = A/K$ MeV), where K is varied from 8 to 12

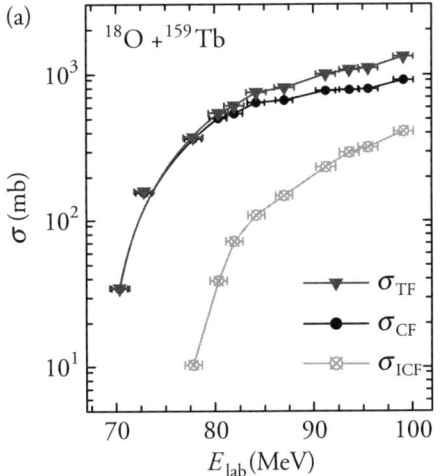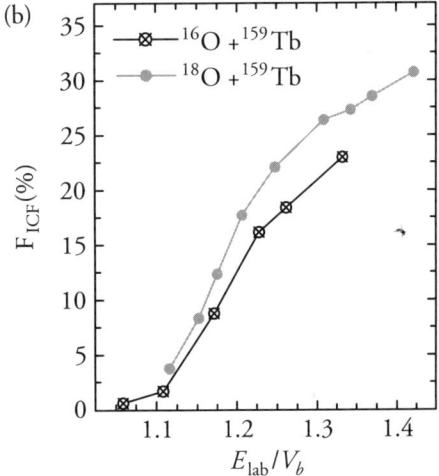

Figure 4.39 (a) Incomplete fusion cross-section along with total and complete fusion cross-section. (b) Comparison of F_{ICF} for ^{18,16}O projectiles on the same target

4.1.6 Reactions initiated by ^{19}F beam

4.1.6.1 ^{19}F+^{159}Tb system

It may be observed from the earlier description that considerable efforts have been made to understand the reaction dynamics of the heavy ion induced reactions at energies near the Coulomb barrier to well above it. However, most of the measurements have been done using the alpha cluster beams. With a view to study the reaction dynamics of HI interactions in reactions induced by non-alpha cluster beams, experiments were also done[10] using ^{19}F beam and ^{159}Tb and ^{169}Tm targets. Details of the experimental methodology are described in Chapter 3 and are similar to that used in experiments with alpha cluster beams. The measured cross-sections for many residues in the interaction of ^{19}F+^{159}Tb are given in Table 4.16.

Table 4.16 Cross-sections at different energies for the residues identified in the system ^{19}F+^{159}Tb

System: ^{19}F + ^{159}Tb						
Energy (MeV)	^{174}W σ(mb)	^{173}W σ(mb)	^{172}W σ(mb)	^{173}Ta^{ind} σ(mb)	^{171}Hf σ(mb)	^{170}Hf σ(mb)
80.5 ± 1.35	113.8 ± 17	–	–	–	5.2 ± 6.1	0.58 ± 0.08
82.9 ± 1.88	127.3 ± 19.1	97.4 ± 14.6	–	1.5 ± 0.2	17.3 13.7	7.37 ± 1.1
87.5 ± 2.09	110.3 ± 16.5	207.6 ± 31.1	–	12.6 ± 1.8	33.8 ± 22.3	43.7 ± 6.5
90.5 ± 1.41	95.6 ± 14.3	309.9 ± 46.4	–	17.0 ± 2.5	36.7 ± 18.9	74.9 ± 11.2

91.9 ± 1.41	82.1 ± 12.3	382.4 ± 57.3	–	21.8 ± 3.2	32.8 ± 16.1	78.8 ± 11.8
96.7 ± 1.74	38.6 ± 5.8	382.8 ± 57.4	–	46.9 ± 7	28 ± 12.2	125.1 ± 18.7
97.6 ± 1.76	31.5 ± 4.7	372.6 ± 55.8	206.4 ± 30.9	46.8 ± 7	27.7 ± 9.2	128.7 ±19.3
102.6 ± 1.4	11.6 ± 1.7	277.2 ± 41.5	375.6 ± 56.3	61.2 ± 9.1	24.2 ± 4.5	167.2 ± 25
105.4 ± 1.56	–	186.0 ± 27.9	513.7 ± 77	44.6 ± 6.6	23.9 ± 1.8	127.1 ± 19
108.0 ± 1.9	–	126.7 ± 19	526.1 ± 78.9	24.1 ± 3.6	32.6 ± 0.5	121.4 ± 18.2

Energy (MeV)	^{170}Lu σ(mb)	^{167}Yb σ(mb)	^{167}Tm σ(mb)	^{165}Tm σ(mb)		
80.5 ± 1.35	0.36 ± 0.0	1.12 ± 0.16	5.7 ± 0.8	0.8 ± 0.12		
82.9 ± 1.88	4.32 ± 0.6	2.4 ± 0.3	11.7 ± 1.7	3.1 ± 0.4		
87.5 ± 2.09	14.2 ± 2.1	5.0 ± 0.7	14.8 ± 2.2	5.7 ± 0.85		
90.5 ± 1.41	25.5 ± 3.8	7.8 ± 1.1	16 ± 2.4	9.6 ± 1.45		
91.9 ± 1.41	33.2 ± 4.9	7.5 ± 1.1	21.6 ± 3.2	8.2 ± 1.23		
96.7 ± 1.74	33.8 ± 5	13.7 ± 7.7	24.5 ± 3.6	21.7 ± 3.25		
97.6 ± 1.76	43.9 ± 6.5	19.4 ± 2.9	30.4 ± 4.5	22.6 ± 3.39		
102.6 ± 1.4	45.2 ± 6.7	37.1 ± 5.5	46 ± 6.96	30.4 ± 4.5		
105.4 ± 1.56	50.5 ± 7.5	55.3 ± 8.3	49.2 ± 7.3	39.9 ± 5.9		
108.0 ± 1.9	56.4 ± 8.4	61.2 ± 9.1	47.4 ± 7.1	33.2 ± 4.9		

As shown in Table 4.16, excitation functions (EFs) for several reaction residues populated via CF and/or ICF processes in the ^{19}F+^{159}Tb system have been measured in the energy range $\approx$81–110 MeV. These excitation functions have also been analysed within the framework of the statistical model code PACE4. A detailed description and listing of the input parameters of this code have already been presented in the chapter on theoretical codes and also in the descriptions given earlier. It may again be pertinent to point out here that PACE4 computes only the CF events according to the Hauser–Feshbach theory of CN decay; it does not include the transfer and/or ICF channels. As such, if there is any deviation in the experimental cross-sections with respect to the theoretical calculations based on PACE4, it may be attributed to the ICF process. In Figure 4.40, the measured EFs of ^{174}W, ^{173}W, ^{172}W and ^{173}Ta residues populated via 4n, 5n, 6n and p4n channels respectively, are shown and compared with the corresponding PACE4 predictions. It has been observed that the radioisotopes ^{173}Ta(p4n) are strongly fed from its higher charge isobar (precursor hereafter) ^{173}W(5n) through β^+ emission. In this case, the independent cross-sections (σ_{ind}) of ^{173}Ta residues was obtained by using the successive radioactive decay formulations popularly known as the Batemann equations. One can see from this figure that the theoretical predictions of PACE4 code reproduces the experimental data satisfactorily for the level density parameter $a = A/10$ MeV^{-1}. As such, it may be concluded that these channels (4n, 5n, 6n and p4n) are populated via the CF process.

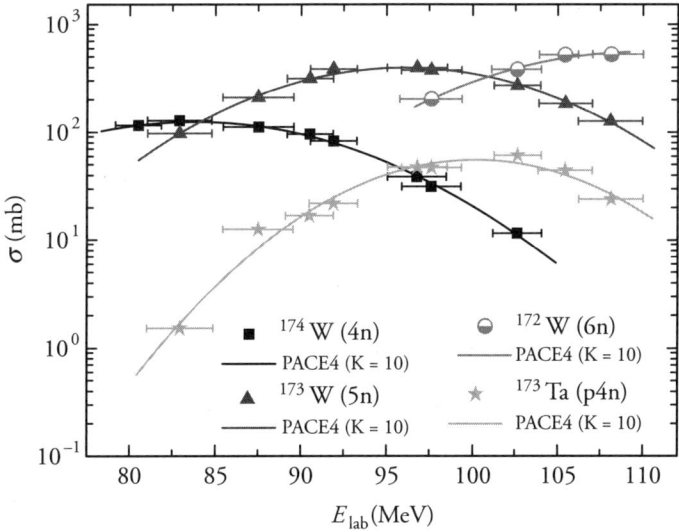

Figure 4.40 Experimentally measured excitation functions for all xn/pxn channels populated in the ^{19}F +^{159}Tb system. The solid lines through the experimental data represent the PACE4 calculations

In order to check if the alpha-emitting channels are populated via CF and/or ICF processes, the sum of experimental excitation functions of all identified alpha-emitting channels for this system was compared with that calculated by corresponding PACE4 values and are presented in Figure 4.41(a). It may be remarked that in these calculations of EFs, the same input parameters that were used for reproducing the EFs of the xn and pxn channels have been utilized. It may be observed from this figure that the experimental EFs for α-emitting channels are quite high compared to the PACE4 predictions. This observed enhancement in the experimental

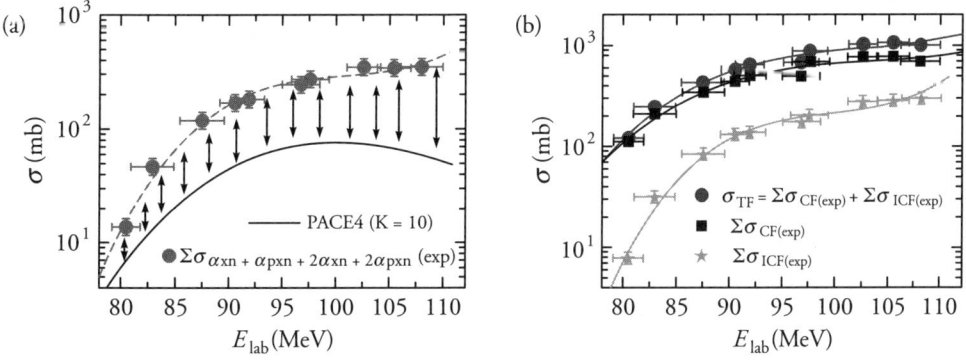

Figure 4.41 (a) Sum of experimentally measured excitation functions of all α-emitting channels compared with the predictions of the statistical model code PACE4. The line through the data points is drawn to guide the eyes. (b) Comparison of σ_{TF}, σ_{CF} and σ_{ICF} cross-sections for the ^{19}F+^{159}Tb system with the incident laboratory energy

EFs of the α-emitting channels may be attributed to the contribution of the ICF process in the population of these residues. An attempt has been made to deduce the ICF cross-section contribution by using the formulation $\sigma_{ICF} = \Sigma\sigma_{exp}^{\alpha's} - \Sigma\sigma_{\alpha's}^{PACE4}$, at each incident energy level. In order to obtain the effect of ICF on the total fusion cross-section, the $\sigma_{TF} = \Sigma\sigma_{CF(exp)} + \Sigma\sigma_{ICF(exp)}$, was obtained and plotted with $\Sigma\sigma_{CF(exp)}$ (sum of all xn and pxn channels) as a function of the projectile energy in Figure 4.41(b). It can be seen from this figure that the onset of ICF takes place at energy as low as 80 MeV (which is around 8% above the barrier energy) and increases as the beam energy increases.

4.1.6.2 ^{19}F+^{169}Tm system

In the system ^{19}F+^{169}Tm, the cross-sections for the residues ^{185}Pt, ^{184}Pt, ^{183}Pt, ^{184}Ir and ^{183}Ir, which are expected to be populated via 3n, 4n, 5n, p3n and p4n emissions, respectively, from the excited ^{188}Pt* CN formed in CF reactions have been measured[11] in the energy range $\approx$4–6 MeV/nucleon. Figure 4.42 (a)–(d) compares the individual experimental EFs of ^{185}Pt(3n), ^{184}Pt(4n), ^{183}Pt(5n) and ^{183}Ir(p4n) residues, respectively with those estimated by PACE4 model predictions.

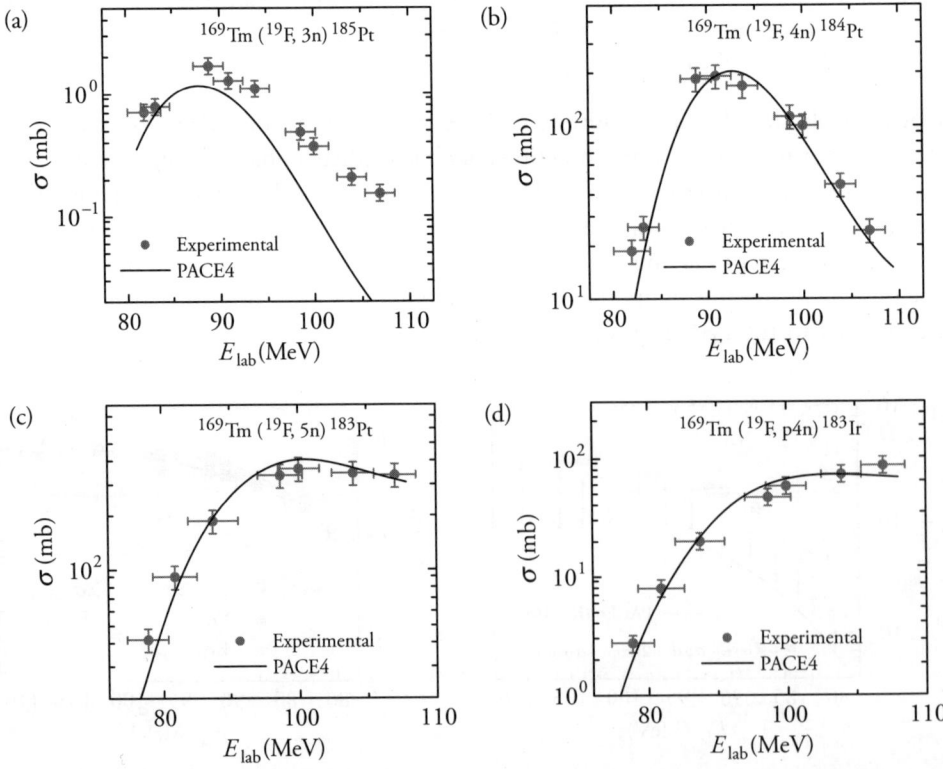

Figure 4.42 (a) Experimentally measured EFs of ^{185}Pt, ^{184}Pt, ^{183}Pt and ^{183}Ir residues populated via 3n, 4n, 5n and p4n in the ^{19}F+^{169}Tm system. The solid lines are the PACE4 calculations done for the level density parameter $a = A/10$ MeV^{-1}. The x error bars on the energy axis indicate the uncertainty in energy due to energy loss of the incident beam in the sample thickness

As can be seen from Figure 4.42(a)–(d), the measured EFs of reaction residues are in good agreement with those predicted by the PACE4 code for the level density parameter a = $A/10$ MeV^{-1} except for ^{185}Pt(3n) residues, which show a significant enhancement in the tail portion of EF as compared to the PACE4 calculations. Since the PACE4 model is based on the Hauser–Feshbach theory of CN decay and does not consider the pre-equilibrium (PE) emission, the discrepancy between the theoretical calculations and the experimentally measured cross-sections for ^{185}Pt(3n) residues may be attributed to the PE emission, which is likely to be a dominant mode of reaction mechanism at relatively higher energies. It may be pointed out that the residues ^{185}Pt cannot be produced by ICF.

Table 4.17 Measured cross-sections at different energies for the residues identified in the system ^{19}F+^{169}Tm

System: ^{19}F + ^{169}Tm					
Energy (MeV)	**^{185}Pt σ(mb)**	**^{184}Pt σ(mb)**	**^{183}Pt σ(mb)**	**^{184}Ir ind σ(mb)**	**^{183}Ir σ(mb)**
80 ± 1.9	0.8 ± 0.1	18.9 ± 2.8	–	0.7 ± 0.1	0.7 ± 0.1
81.6 ± 1.6	0.8 ± 0.1	26 ± 3.96	10.1 ± 1.5	1.2 ± 0.2	0.8 ± 0.1
87.2 1.7	1.7 ± 0.3	185.8 ± 27.9	41.9 ± 6.3	15.7 ± 2.7	2.7 ± 0.4
89.3 ± 1.6	1.3 ± 0.2	193.5 ± 29	92.2 ± 13.8	17.9 ± 2.7	8 ± 1.2
92 ± 1.8	1.1 ± 0.2	171.3 ± 25.7	186.2 ± 27.9	20.5 ± 3	19.7 ± 2.9
97 ± 1.6	0.5 ± 0.1	112.9 ± 16.9	331.1 ± 49.7	13 ± 1.9	46.9 ± 7
98.5 ± 1.5	0.4 ± 0.1	100.2 ± 15	359.4 ± 53.9	11.2 ± 1.7	58.2 ± 8.8
102.5 ± 1.5	0.2 ± 0.1	45.5 ± 6.8	338.4 ± 50.8	5.3 ± 0.8	73.5 ± 11.4
105.4 ± 1.6	0.2 ± 0.1	24.5 ± 3.7	330 ± 49.5	2.9 ± 0.4	85.5 ± 12.8
Energy (MeV)	**^{183}Os σ(mb)**	**^{182}Os σ(mb)**	**^{181}Os σ(mb)**	**^{179}Os σ(mb)**	**^{177}W σ(mb)**
80 ± 1.9	--	0.7 ± 0.1	1.0 ± 0.1	–	–
81.6 ± 1.6	0.3 ± 0.0	2.2 ± 0.3	2.3 ± 0.3	0.7 ± 0.1	0.2 ± 0.0
87.2 ± 1.7	2.5 ± 0.4	4.5 ± 0.7	36.2 ± 5.4	1.0 ± 0.2	1.2 ± 0.2
89.3 ± 1.6	2.5 ± 0.4	7.1 ± 1.0	56.1 ± 8.4	2.1 ± 0.3	1.6 ± 0.2
92 ± 1.8	6.4 ± 1.0	6.6 ± 1.0	62.2 ± 9.3	3.1 ± 0.5	3.5 ± 0.5
97 ± 1.6	10.1 ± 1.5	6.2 ± 0.9	78.8 ± 11.8	8.2 ± 1.2	6.4 ± 1.0
98.5 ± 1.5	14.9 ± 2.2	4.7 ± 0.7	70.7 ± 10.6	13.4 ± 2.0	9.3 ± 1.4
102.5 ± 1.5	17.5 ± 2.6	8.1 ± 1.2	38.7 ± 5.8	26.7 ± 4	10.3 ± 1.5
105.4 ± 1.6	16.3 ± 2.3	12.4 ± 1.9	23.1 ± 3.5	71.7 ± 10.8	12.1 ±1.8

Energy (MeV)	^{185}W σ(mb)	^{174}W σ(mb)	^{176}Ta σ(mb)	^{175}Ta σ(mb)
87.2 ± 1.7	0.3 ± 0.0	2.1 ± 0.3	2.0 ± 0.3	0.3 ± 0.0
89.3 ± 1.6	1.1 ± 0.2	2.3 ± 0.3	2.6 ± 0.4	2.3 ± 0.3
92 ± 1.8	3.2 ± 0.5	5.4 ± 0.8	3.3 ± 0.5	6.7 ± 1.0
97 ± 1.6	6.8 ± 1.0	8.8 ± 1.3	4.6 ± 0.7	13.4 ± 2.0
98.5 ± 1.5	7.3 ± 1.1	9.3 ± 1.4	5.6 ± 0.8	14.3 ± 2.1
102.5 ± 1.5	9.7 ± 1.5	15.5 ± 2.3	9.7 ± 1.4	19.8 ± 2.7
105.4 ± 1.6	11 ± 1.6	17.1 ± 2.65	16.1 ± 2.4	24.3 ± 3.6

Figure 4.44(a) shows the EF of residues ^{184}Ir populated via the p3n channel; it is found that the measured EF is enhanced as compared to the PACE4 model predictions. It has also been observed that the residues ^{184}Ir(p3n) are strongly fed from their higher charge isobar ^{184}Pt(4n) through β^+ emission. The different possible decay routes of ^{184}Ir population shown in Figure 4.43 are as follows.

i. ^{19}F+^{169}Tm $\Rightarrow^{188}$Pt* $\Rightarrow^{184}$Ir + p+ 3n

ii. ^{19}F+^{169}Tm $\Rightarrow^{188}$Pt* $\Rightarrow^{184}$Pt + 4n

i.e., ^{184}Pt $\Rightarrow$ undergoes β^+/EC $\Rightarrow$ gives ^{184}Ir

These routes can be pictorially represented as shown in Figure 4.43.

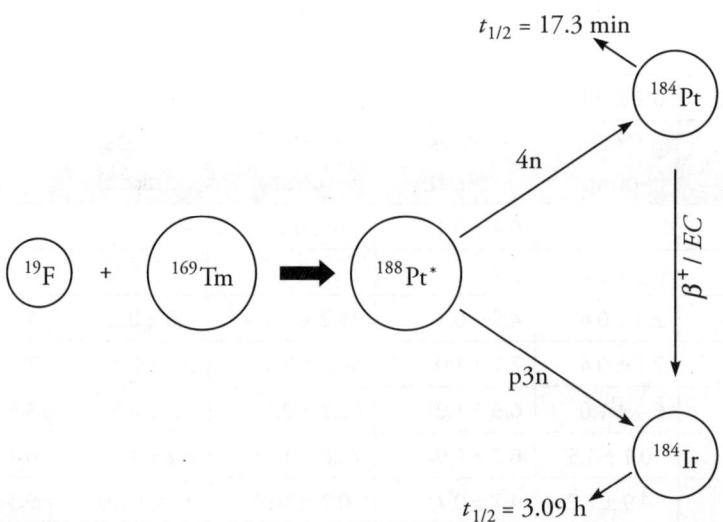

Figure 4.43 An example showing the two different routes for the population of the residues ^{184}Ir

The half-life of the daughter nuclei (i.e., ^{184}Ir $\rightarrow t^{d}_{1/2}$ = 3.09 hr) is quite large compared to the half-life of the precursor nucleus (i.e., ^{184}Pt $\rightarrow t^{d}_{1/2}$ = 17.3 m). Thus, the independent production cross-section (σ_{ind}) of residues ^{184}Ir^{Ind}(p3n) has been deduced from the cumulative cross-section (σ_{cum}), i.e., ^{184}Ir^{cum}(p3n) using the successive radioactive decay formulations presented by Cavinato et al. [25] based on standard Batemann equations [21].

$$\sigma_{ind} = \sigma_{cum} - P_{pre} \frac{t^{d}_{1/2}}{(t^{d}_{1/2} - t^{pre}_{1/2})} \sigma_{pre}$$

where, σ_{pre} is the cross-section of the parent nuclei, and $t^{d}_{1/2}$ and $t^{pre}_{1/2}$ are the half-life of the daughter and precursor nuclei respectively. The term P_{pre} is the branching ratio of the precursor to its daughter nuclei. The independent cross-sections deduced using this formulation for residues ^{184}Ir(p3n) are compared with those estimated by the PACE4 code and the result is shown in Figure 4.44(b). As can be seen from this figure, the EF of ^{184}Ir(p3n) residues is very well matched with PACE4 calculations for the level density parameter a = $A/10$ MeV^{-1} and confirms the production of ^{184}Ir residues solely via CF mode. The satisfactory reproduction of the cross-sections for xn and pxn channels by the theoretical calculations based on the CN model indicates the production of these channels via the CF process. In the same way, the residues ^{183}Ir populated via the p4n channel are also likely to be populated via the precursor decay through β^+ emission from the reaction residues ^{183}Pt, populated in the 5n channel. It can be seen from Figure 4.42(d) that PACE4 predictions match the measured cross-section for the reaction ^{169}Tm(^{19}F, p4n) ^{183}Ir over the entire range of energy. As such, it may be assumed that the precursor contribution to this channel is negligibly small.

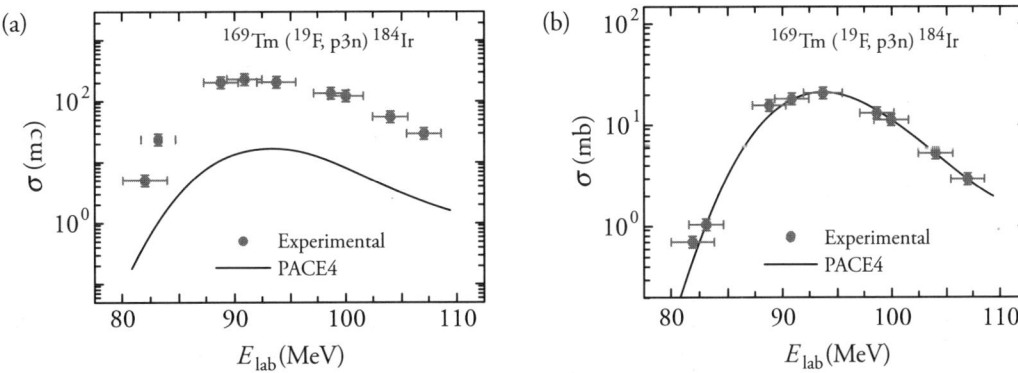

Figure 4.44 Experimentally measured EF of ^{184}Ir(p3n) channel compared with PACE4 calculations. (a) Cumulative cross-section; (b) independent cross-section. The solid line is the PACE4 calculations at K = 10

The individual EFs of residues ^{183,182,181,179}Os (αxn), where x = 1, 2, 3 and 5; ^{177,175,174}W (2αxn), where x = 3, 5 and 6; and ^{176,175}Ta (2αpxn), where x = 3 and 4, are shown in Figures

4.45 and 4.46 respectively. As a matter of fact, these reactions involve α particle(s) in the outgoing channel. Hence, these reactions may be assumed to be produced both from CF and ICF processes. In case of CF, after the entire fusion of the projectile (^{19}F) with the target nucleus (^{169}Tm), the excited CN (^{188}Pt*) decays by emitting the α particle(s). However, in the ICF process, only a part of the projectile (^{19}F $\rightarrow$ ^{15}N + α) fuses with the target nucleus (^{169}Tm); the remaining α particle or ^{15}N nucleus moves forward as a spectator. The experimental EFs of these α-emitting channels are compared with those calculated by the corresponding PACE4 code, which is based on the statistical CN decay.

The PACE4 calculations have been done with the same set of input parameters that have been used to reproduce the EFs of xn and pxn channels produced via CF mode. As already discussed, the PACE4 code does not calculate the ICF cross-sections, and therefore any enhancement in the experimental EFs over the PACE4 predictions may be attributed to the contribution from ICF processes. As can be seen from Figures 4.45(a)–(d) and 4.46(a), the

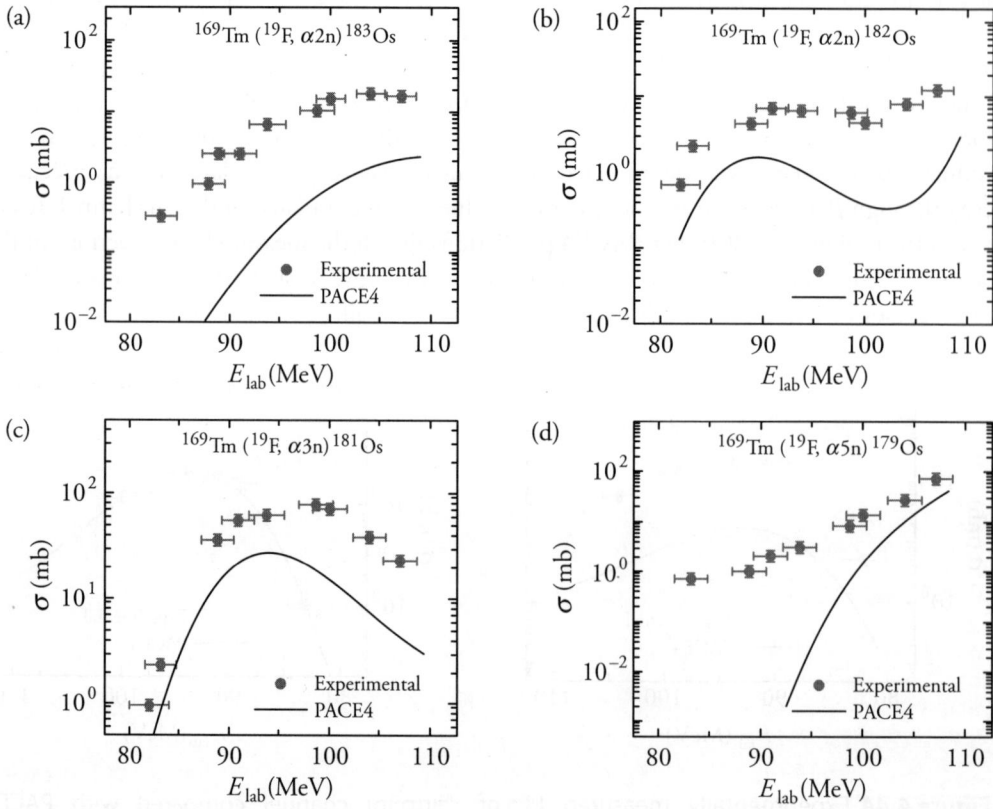

Figure 4.45 Experimentally measured EFs of ^{183,182,181,179}Os (axn, where x = 1, 2, 3 and 5) residues populated in the ^{19}F+^{169}Tm system and compared with those calculated by the PACE4 model. Solid lines are the PACE4 calculations done for the level density parameter $a = A/10$ MeV^{-1}

experimental EFs of ^{183,182,181,179}Os and ^{177}W residues show a significant enhancement as compared to the PACE4 calculations. The present enhancement in the experimental cross-sections of the α-emitting channels may be attributed to the contribution of ICF processes at the energy range of interest.

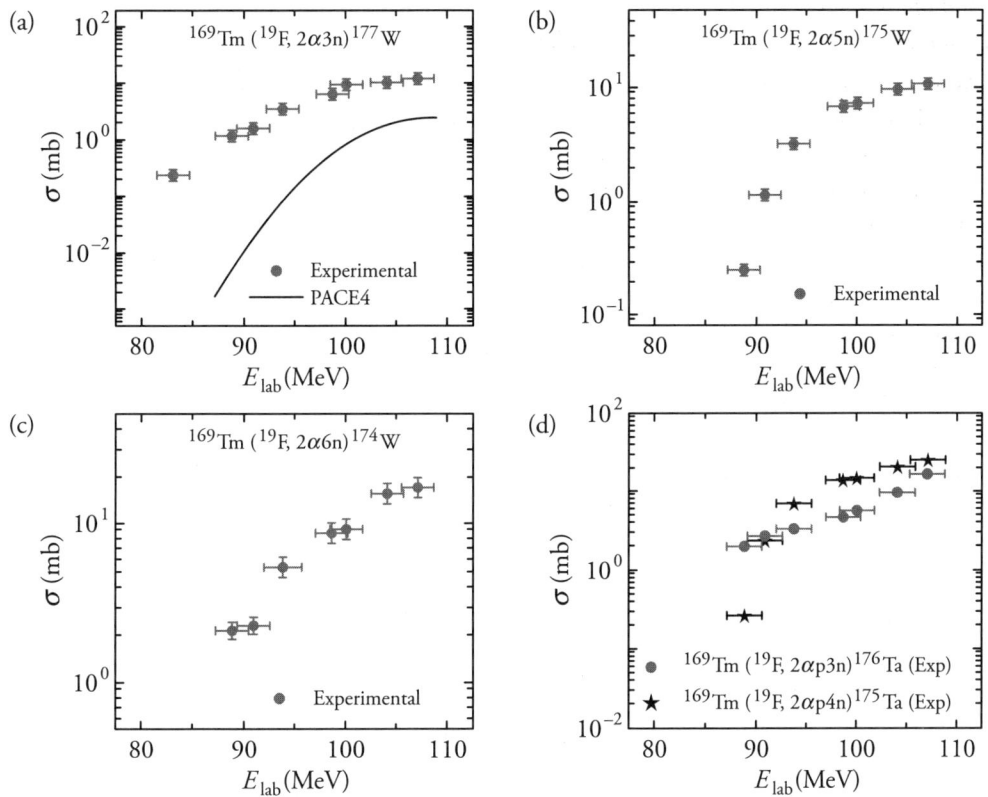

Figure 4.46 Experimentally measured EFs of ^{177,175,174}W (2αxn, where x = 3, 5 and 6) and ^{176,175}Ta(2αpxn, where x = 3 and 4) residues populated in the ^{19}F+^{169}Tm system. The solid line is the PACE4 calculations for the level density parameter $a = A/10$ MeV^{-1}

However, Figure 4.46(b), (c) and (d) show the experimental EFs of ^{175,174}W and ^{176,175}Ta residues, where the corresponding PACE4 code predicts negligible cross-sections; this indicates that the production of these residues are solely via ICF processes.

4.2 Measurement of Recoil Range Distributions (RRD) and their Analysis

In a complete fusion event, the incident heavy ion imparts all its linear momentum to the fused composite system, which recoils in the forward direction with the total linear momentum.

However, in the case of incomplete fusion, where only a part of the incident ion fuses with the target, the fused system gets only a fraction of the incident linear momentum and, therefore, it recoils with a reduced linear momentum in the forward direction. If these recoiling residues are made to pass through an absorbing medium, say a bunch of aluminium foils, the complete fusion residues will have a longer range in the absorbing medium as compared to the incompletely fused residues. Thus, recoil range distribution of residues may be used to disentangle complete fusion and incomplete fusion residues.

With a view to disentangle the complete and incomplete fusion events in heavy ion interactions, the forward recoil ranges of residues populated via CF and/or ICF have been measured at several energies, in different systems. It may be pointed out that the measurement and analysis of the forward recoil ranges is one of the most direct and irrefutable method to disentangle CF and ICF events. In the present measurements, the recoil catcher technique followed by off-line gamma-ray spectroscopy has been employed. The various identified peaks, which have been used for recoil range analysis, are assigned to the different reaction products on the basis of their characteristic gamma ray energies and measured half-lives. The measured intensities of the characteristic gamma radiations were used to obtain the production cross-sections for various residues. In order to obtain the normalised yield of the residues as a function of cumulative thickness of the Al-catcher foils, the cross-section of the reaction products in each catcher foil was divided by the respective catcher foil thickness. The resulting normalised yield of reaction products have been plotted against the cumulative catcher foil thickness to obtain the differential recoil range distributions. A list of systems in which the forward recoil range distributions have been measured is given in Table 4.18.

The details of the techniques employed are already given in Chapter 3. The complete and incomplete fusion events have been tagged by full and partial linear momentum transfer components respectively. An attempt has been made to separate the relative contributions of complete and incomplete fusion events using the measured recoil range distributions. A case by case study of RRD for different systems is presented here.

Table 4.18 List of systems for which the recoil range distributions have been measured along with the values of incident energies at which measurements have been done and the Coulomb barrier for the system

S.No	System	Energies (MeV)	Coulomb barrier (MeV)
1	^{12}C + ^{159}Tb	74, 80 and 86	≈ 48.53
3	^{16}O + ^{159}Tb	90	≈ 63.69
4	^{16}O + ^{169}Tm	86.6	≈ 66.99
5.	^{16}O + ^{181}Ta	81, 90 and 96	≈ 70.09

4.2.1 Recoil range distribution for the system ^{12}C+^{159}Tb

As already mentioned, the analysis of RRD may be used to disentangle CF and ICF events. RRD experiments for the system ^{12}C+^{159}Tb were done[8] at three incident beam energies ≈ 74, 80 and 86 MeV. Here, an attempt was made to get detailed information about the ICF reactions. It may be mentioned that these measurements are in continuation and complimentary to the measurement of excitation functions for the system $^{12}C+^{159}Tb$, where an attempt was made to look for the role of break up processes. The experiments were done using the ^{12}C ion beam delivered from the Pelletron accelerator of the IUAC, New Delhi. Though, the experimental methodology is similar to the previous case, for the sake of completeness, a few details are given here. The three different stacks consisting of a ^{159}Tb target (abundance 100%) followed by a series of thin aluminium catcher foils to trap the recoiling residues, have been irradiated separately by ^{12}C beam at the aforementioned three energies. The targets were prepared by the rolling technique, as mentioned in Chapter 3, while the thin Al-catcher foils were made by the vacuum evaporation technique. The thickness of each target and catcher foil was measured by the α transmission method. In the present case, the thickness of the ^{159}Tb target was ≈ 190 $\mu g/cm^2$; the thicknesses of the Al catchers was $\approx 15–50$ $\mu g/cm^2$. After the irradiation, the catcher foils were counted and the RRD was measured as already discussed. Depending upon the degree of linear momentum transfer from the projectile to the target nucleus, one can disentangle the CF and ICF events. It may be pointed out that the velocity distribution of a given type of reaction products is symmetric about v_0, having a width that depends on the mass of the evaporated particles from the compound nucleus. The mean velocity v_0, may be presented as;

$$v_0 = v_{CN} = \frac{\sqrt{(2M_p E)}}{(M_{P+T})}$$

where, M_p stands for the mass of the incident ion, M_{P+T} is the mass of the composite system and E is the energy of the incident ion. As such, the degree of linear momentum transfer may be given as;

$$\rho_{LMT} = \frac{P_{frac}}{P_{proj}}$$

where, P_{frac} is the linear momentum of the fused part of the projectile while P_{proj} is the total linear momentum of the projectile. It may be mentioned again that ρ_{LMT} is proportional to the fused mass of the incident particle, where the maximum LMT may give rise to the maximum recoil velocity to the reaction products. In a complete fusion process, the maximum LMT is transferred to the target nucleus, while in an ICF process, only partial linear momentum is transferred. Thus, the measured recoil range distributions of the reaction products may give information of considerable value regarding the reaction mechanism involved.

In order to obtain the normalised yields as a function of the cumulative depth of the stopping medium Al, the cross-section of the reaction residues in each catcher foil was divided by its thickness. The resulting normalised yields have been plotted as a function of the cumulative catcher foil thickness to get the recoil range distributions of the identified residues viz., $^{169}Lu(3n)$, $^{167}Lu(4n)$, $^{165}Lu(6n)$, $^{167}Yb(p3n)$, $^{165}Tm(\alpha 2n)$, $^{163}Tm(\alpha 4n)$, $^{161}Ho(2\alpha 2n)$, $^{160}Ho^g(2\alpha 3n)$ and $^{160}Ho^m(2\alpha 3n)$. To show the CF and ICF components, as a representative

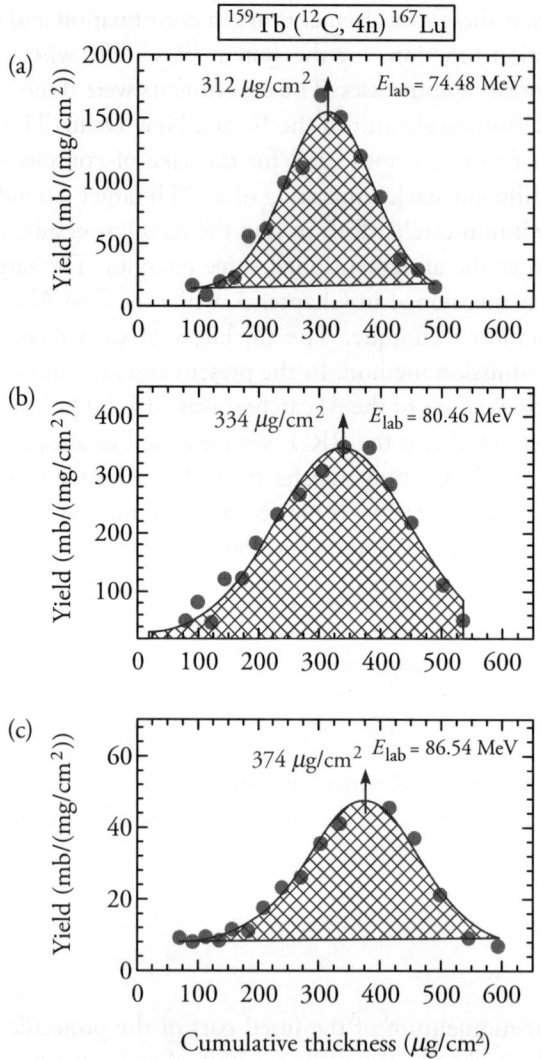

Figure 4.47 Measured recoil range distributions (RRDs) for ^{167}Lu residues populated via the 4n channel respectively (a) at 74.48 MeV, (b) at 80.46 MeV and (c) at 86.54 MeV beam energies

case, the measured recoil range distribution for ^{167}Lu(4n), ^{165}Tm(α4n) and ^{161}Ho(2α2n) have been presented in Figures 4.47–4.49 respectively, at three different incident energies. The size of the circles in these figures indicates the uncertainty in the yield values.

The measured distributions show different linear momentum transfer events depending on the fused mass of the projectile with the target nucleus. For the 4n channel (Figure 4.47), the measured RRDs show only one peak at all the three energies, indicating that only one linear momentum transfer component (a characteristic of the CF process) is involved in the production of ^{167}Lu residues. It may also be observed from this figure that the RRD peak shifts towards the relatively higher cumulative catcher thickness with the increase in the incident beam energy. Moreover, the particle emission, particularly the neutron, from the recoiling residues may change the energy and momentum of the final residues. This is reflected in the width (FWHM) of the measured recoil range distributions. The observed width may also have a contribution from straggling. The identified reaction products and their experimentally measured most probable ranges R_p^{expt} in aluminium, in units of µg/cm² are given in Table 4.19. The most probable theoretical recoil ranges (R_p^{theo}) have also been calculated, assuming that in the case of CF, the incident ions completely fuse with the target nucleus and transfer their total linear momentum to the composite fused system, which then recoils for the conservation of linear momentum.

Table 4.19 Experimentally measured most probable ranges $R_{p(\exp)}$ deduced from RRD data, and theoretically calculated mean ranges $R_{p(\text{the})}$ in aluminium in units of µg/cm² for CF and ICF components using the range energy relation along with the reaction products produced in the interaction of ^{12}C with ^{159}Tb at ≈ 74 MeV.

Residues	$R_{p(\exp)}^{CF}$	$R_{p(\text{the})}^{CF}$	$R_{p(\exp)}^{ICF-\,^8Be}$	$R_{p(\text{the})}^{ICF-\,^8Be}$	$R_{p(\exp)}^{ICF-\,^4He}$	$R_{p(\text{the})}^{ICF-\,^4He}$
^{168}Lu	315 ± 43	321	–	–	–	–
^{167}Lu	312 ± 48	321	–	–	–	–
^{165}Lu	314 ± 52	321	–	–	–	–
^{167}Yb	330 ± 28	321	–	–	–	–
^{165}Tm	340 ± 32	321	163 ± 23	150	–	–
^{163}Tm	333 ± 61	321	158 ± 19	150	–	–
^{161}Ho	334 ± 53	321	150 ± 21	150	22 ± 9	21
^{160}Ho^g	348 ± 32	321	141 ± 28	150	23 ± 8	21
^{160}Ho^m	337 ± 61	321	145 ± 26	150	25 ± 11	21

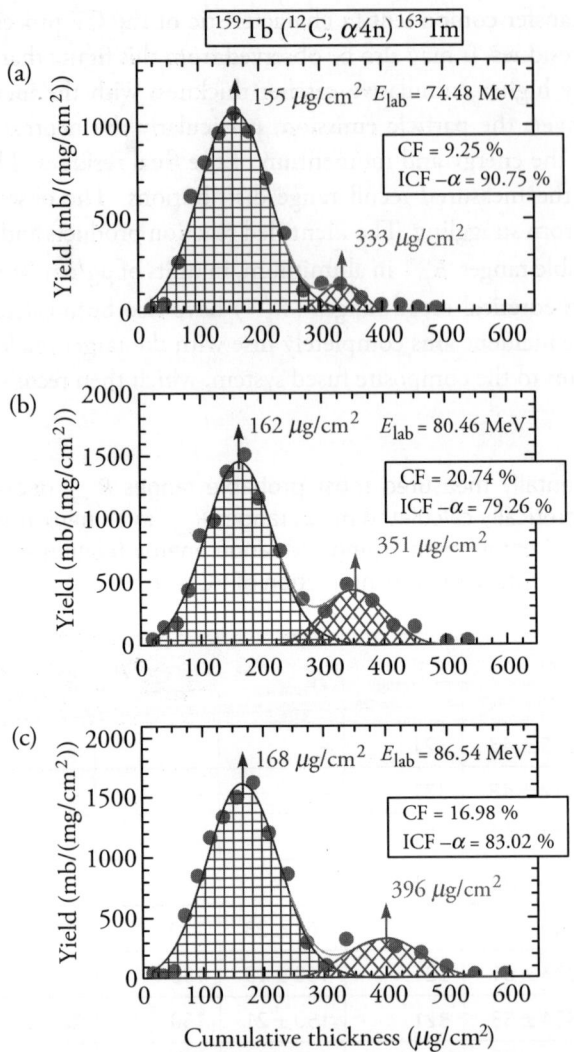

Figure 4.48 Typical RRDs for ^{165}Tm populated via the α4n channel at $\approx$ 74, 80 and 87 MeV beam energies. Two distinct peaks in the RRD distributions corresponding to two momentum transfer components may be seen in the figure

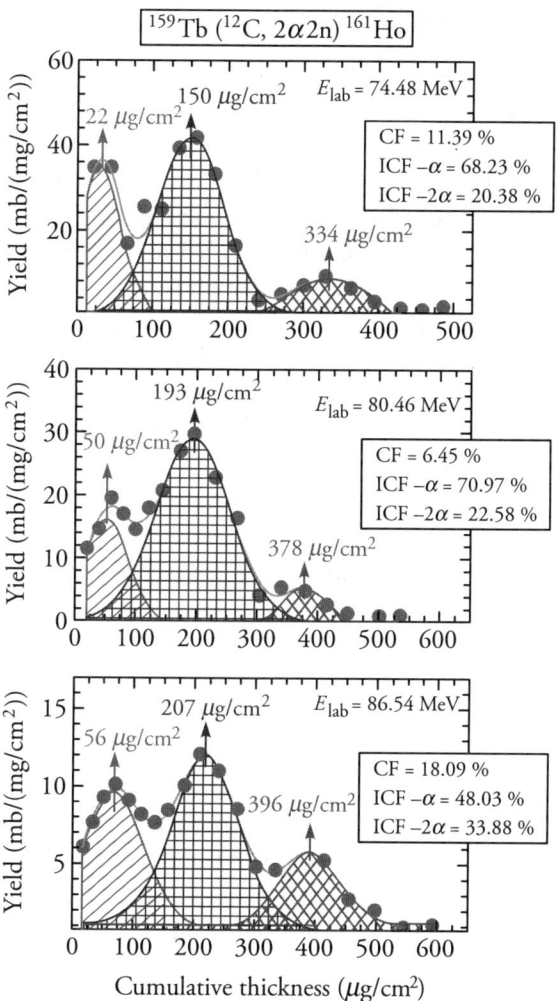

Figure 4.49 Measured FRRDs for ^{161}Ho residues populated via the 2α2n channel at ≈ 74, 80 and 87 MeV beam energies, having three different momentum transfer components

Table 4.20 Comparison of normalised FWHM for various RRD distributions

Residues	≈ 74 MeV CF	≈ 80 MeV CF	≈ 87 MeV CF	≈ 74 MeV ICF-α	≈ 80 MeV ICF-α	≈ 87 MeV ICF-α	≈ 74 MeV ICF-2α	≈ 80 MeV ICF-2α	≈ 87 MeV ICF-2α
^{168}Lu (3n)	0.62	0.77	–	–	–	–	–	–	–
^{167}Lu (4n)	0.56	0.70	0.65	–	–	–	–	–	–
^{165}Lu (6n)	0.67	0.65	0.67	–	–	–	–	–	–
^{167}Yb (p3n)	0.58	0.63	0.71	–	–	–	–	–	–
^{165}Tm(α2n)	0.31	0.25	0.22	0.83	0.72	0.81	–	–	–

^{163}Tm(α4n)	0.20	0.30	0.34	0.87	0.80	0.83	–	–	–
^{161}Ho(2α2n)	0.32	0.20	0.34	0.69	0.75	0.63	1.89	1.66	1.80
^{160}Ho(2α2n)	0.27	0.22	0.26	0.74	0.79	0.73	1.81	1.76	1.90

In order to check the consistency in the FWHM of the measured FRRDs, the normalised FWHM (FWHM/R_p^{expt}) has been deduced and tabulated in Table 4.20 for the observed residues. As may be seen from the table, the FWHM are individually consistent both for the CF and the ICF residues. Moreover, for the alpha-emitting channels, the average peak resolution for CF is ≈ 0.28, while for ICF-α and ICF-2α, the average peak resolution increases to 0.69 and 1.78 respectively, as expected. All these observations clearly demonstrate that the reaction products ^{167}Lu are clearly produced via the 4n channel and are associated with the complete momentum transfer from projectile to the target nucleus. It may be represented as

$$^{12}C + {^{159}Tb} \Rightarrow {^{171}Lu^*} \Rightarrow {^{167}Lu} + 4n$$

Similarly, the RRDs for the residues ^{168}Lu(3n), ^{165}Lu(6n) and ^{167}Yb(p3n) are observed to have one peak associated with the full or complete linear momentum transfer from the projectile to the composite nucleus, indicating the population of these residues only via the complete fusion process. On the other hand, in case of channels where alpha particle emission is involved, the residues ^{165}Tm, ^{163}Tm, ^{161}Ho, ^{160}Ho^g and ^{160}Ho^m are expected to be produced via the α2n, α4n, 2α2n and 2α3n channels. The measured recoil range distributions have been separated into two Gaussian peaks for αxn channels, using the software ORIGIN. In order to represent a typical case, the RRDs for the residues ^{163}Tm (α4n) have been plotted at three different energies in Figure 4.48(a)–(c). It may be observed from this figure that these RRDs may be fitted with two Gaussian peaks, one, respectively at ≈333±32, 351±60 and 396±65 μg/cm^2 in aluminium foils for three energies showing the full momentum transfer events. However, there is another peak at a lower cumulative depth at ≈155±23, 162±40 and 168±45 μg/cm^2, which corresponds to the fusion of ^{8}Be (if ^{12}C is assumed to break up into ^{8}Be + α and ^{8}Be fuses) with ^{159}Tb target nucleus. In the same way, the RRDs for other αxn channels were resolved into two Gaussian peaks, indicating the presence of more than one linear momentum transfer component. It has been observed that the complete as well as incomplete momentum transfer peaks in the RRD spectra are centred at the expected positions indicated by the arrow. It may also be observed from Figure 4.47, that as expected, the mean range R_p^{exp} shifts towards the higher cumulative catcher thickness as the beam energy increases.

The residue ^{163}Tm may be populated through the α4n channel in two different ways;

i.   Fusion of ^{12}C

$$^{12}C + {^{159}Tb} \Rightarrow {^{171}Lu^*} \Rightarrow {^{163}Tm} + \alpha4n \text{ and/or } 2p6n$$

or

ii.   Fusion of ^{8}Be (α as spectator)

$$^{12}C(^8Be + \alpha) \Rightarrow {^8Be} + {^{159}Tb}$$

$\Rightarrow {}^{167}\text{Tm}^* + \alpha$ (as spectator)

$\Rightarrow {}^{167}\text{Tm}^* \Rightarrow {}^{163}\text{Tm} + 4n$

The measured RRDs have been resolved into three Gaussian peaks in case of $2\alpha xn$ channels. These experimentally measured RRDs for the $2\alpha 2n$ channel have been shown in Figure 4.49(a)–(c) at three incident energies. Here, the observation of three peaks may be explained by assuming the breakup of ${}^{12}\text{C}$ into possible α clusters. The peaks at $\approx 334 \pm 53$, 378 ± 47 and 396 ± 67 $\mu g/cm^2$ depths at three energies are attributed to the complete momentum transfer events (i.e., fusion of ${}^{12}\text{C}$ with the target nucleus). On the other hand, the peaks at $\approx 150 \pm 21$, 193 ± 37 and 207 ± 29 $\mu g/cm^2$ for the three energies belongs to the partial linear momentum transfer (i.e., the fusion of ${}^{8}\text{Be}$). Another peak at the lowest depth corresponds to the fusion of an α particle with the target nucleus, ${}^{8}\text{Be}$ being the spectator. It may be mentioned that the residues ${}^{161}\text{Ho}$ populated via $2\alpha 2n$ channel have contribution from both the CF and ICF processes, which may be explained as follows.

i. Fusion of ${}^{12}\text{C}$

$${}^{12}\text{C} + {}^{159}\text{Tb} \Rightarrow {}^{171}\text{Lu}^* \Rightarrow {}^{161}\text{Ho} + 2\alpha 2n \text{ and/or } 4p6n$$

ii. Fusion of ${}^{8}\text{Be}$ (α as spectator)

$${}^{12}\text{C}({}^{8}\text{Be} + \alpha) \Rightarrow {}^{8}\text{Be} + {}^{159}\text{Tb}$$

$$\Rightarrow {}^{167}\text{Tm}^* + \alpha \text{ (as spectator)}$$

$$\Rightarrow {}^{167}\text{Tm}^* \Rightarrow {}^{161}\text{Ho} + \alpha 2n$$

or

iii. Fusion of α (${}^{8}\text{Be}$ as spectator)

$${}^{12}\text{C}({}^{8}\text{Be} + {}^{4}\text{He}) \Rightarrow {}^{4}\text{He} + {}^{159}\text{Tb}$$

$$\Rightarrow {}^{163}\text{Ho}^* + {}^{8}\text{Be} \text{ (as spectator)}$$

$$\Rightarrow {}^{163}\text{Ho}^* \Rightarrow {}^{161}\text{Ho} + 2n$$

It may again be pointed out that this description of the fusion reactions is based on the breakup fusion model of the reactions where it is considered that the ${}^{12}\text{C}$ ion breaks into fragments (e.g., ${}^{8}\text{Be}+\alpha$ or $\alpha+{}^{8}\text{Be}$) as it enters the field of the target nucleus. The fragments of the incident ion so produced are considered to move nearly with the same velocity as that of the incident ion. One of the fragments fuses with the target nucleus forming an incompletely fused composite system, which recoils in the forward direction for the conservation of the linear momentum.

The range integrated yields of the complete and incomplete fusion reactions have been compared with that obtained using the code PACE4, and are presented in Table 4.21. As can be seen from this table, the theoretically calculated values agree well with the experimentally obtained values from the recoil range distribution measurements, at all the energies for the fusion channels. However, the measured values are considerably larger in alpha-emitting channels because ICF is not taken into consideration in PACE4 calculations.

Table 4.21 Experimentally measured recoil range integrated cross-section σ_{exp}^{RRD} (mb) deduced from RRD curves, and theoretically calculated cross-section σ_{theo}^{PACE} at $\approx$ 74, 80 and 87 MeV

Residues	Energy (E) ≈ 74 McV		Energy (E) ≈ 80 MeV		Energy (E) ≈ 87 MeV	
	σ_{exp}^{RRD}	σ_{theo}^{PACE}	σ_{exp}^{RRD}	σ_{theo}^{PACE}	σ_{exp}^{RRD}	σ_{theo}^{PACE}
^{168}Lu (3n)	3.20	3.18	1.10	0.97	–	–
^{167}Lu (4n)	297	314	96	90.2	12.0	15.9
^{165}Lu (6n)	1.4	0.61	120	109	510	298
^{167}Yb (p3n)	33	29	13.3	12.05	2.82	3.07
^{165}Tm (α2n)	6.79	0.83	11.15	0.28	17.32	0.24
^{163}Tm (α4n)	165.32	10.98	260.9	38.05	280.44	53.09
^{161}Tm (2α2n)	7.27	0.20	5.94	0.23	3.34	0.14

The relative contributions of CF and ICF in the population of a particular radioisotope have been calculated by fitting the measured RRDs with the Gaussian distribution employing the ORIGINE software. The yield curves of evaporation residues given by Gaussian RRDs may be given by

$$Y = Y_0 + \frac{A}{\omega_A^2 \sqrt{2\pi}} e^{-(R-R_P)^2/2\pi\omega_A^2},$$

where R_p is the most probable mean range, ω_A is the width parameter (FWHM) of the distribution and A is the area under the RRD peak. The normalised yield Y may be obtained by the χ square fit (χ^2) of the measured range distribution and may be represented as

$$\chi^2 = \frac{1}{(m-p-1)}\{Y(A) - Y_0(A)\}^2$$

The χ square (χ^2) was minimised using the non-linear least square fit routine, keeping the width parameter (ω_A) and the most probable mean range (R_p) in the RRD as a free parameter. As shown in the RRDs for the alpha-emitting channels, there is more than one peak. In such cases, the experimentally measured normalised yields have been fitted employing the multiple-peak fitting option as described earlier. The relative contributions of different fusion components have been obtained by dividing the area under the peak of the corresponding fusion component by the total area obtained from the RRD data. It has been found that the contribution of the CF component reasonably matches with that predicted by the PACE4 code with physically reasonable parameters, which were optimised to explain the residues in the case of CF reactions like xn or pxn channels. It may be pointed out that ICF contribution could not be reproduced by calculations using the same set of parameters mainly due to the fact that PACE code does not take ICF into account.

4.2.2 Recoil range distribution for the system $^{16}O+^{159}Tb$

Another experiment for the measurement and analysis of the recoil range distributions was carried out[31] for the system $^{16}O+^{159}Tb$. It may be remarked that for this system, the excitation functions were already measured and presented; a significant part of the reaction was expected from incomplete fusion processes. However, the relative contribution of complete and incomplete fusion processes could not be obtained there. As such, to get the contribution of ICF, the experiment was carried out to study the RRD of the residues using a 90 MeV ^{16}O beam. As already mentioned, in RRD experiments, the target followed by a stack of thin Al catcher foils was mounted normally to the beam direction in the general purpose irradiation/scattering chamber (GPSC). The recoiling residues were trapped in the stack of Al-foils at different thicknesses depending on the linear momentum transferred. For the sake of completeness, it may be mentioned that a stack of 15 thin Al-catchers prepared using vacuum evaporation technique was used to trap the recoiling residues. Irradiation was continued for about 24 hours with a beam fluence of 355 μC. Attempts were made to maintain a constant beam current. The radio activities in the catchers were recorded using the gamma-ray spectroscopy technique for about two weeks. The yield distribution was obtained as a function of cumulative catcher thickness. For this, cross-section in each catcher was divided by its measured thickness. The measured recoil range distributions have been plotted in Figure 4.50(a) to (k) as a function of cumulative catcher thickness to obtain the recoil range distributions.

In these figures, the solid lines are to guide the eyes to the experimental data. As can be seen from these figures, the recoil range distributions for $^{175-x}Ta$ and $^{175-x}Hf$ isotopes produced via ($^{16}O,xn$) and ($^{16}O,pxn$), x = 3–4 channels respectively have a peak at only one value at the cumulative catcher thickness ≈ 350 $\mu g/cm^2$. The peaks in the range distribution for Ta and Hf isotopes are nearly Gaussian at a depth corresponding to the expected range in aluminium for the compound system ^{175}Ta, which indicates that these residues are populated via complete fusion only. It may be observed from the qualitative inspection of Figure 4.50(e)–(i), that the RRDs for the residues $^{175-x}Lu$, x = 4–6 and $^{167,166}Tm$ have two peaks one at a lower cumulative catcher thickness ($\approx 250 \mu g/cm^2$) and the other at a higher thickness $\approx 350 \mu g/cm^2$. The peak at higher thickness corresponds to the complete fusion process; however, the peak at lower value may be attributed to the fact that these residues are produced via incomplete fusion of ^{16}O, if ^{12}C gets fused with the target and the linear momentum transferred is less than the CF channel. It may also be pointed out that the experimentally measured RRD for the residue ^{171}Lu populated via the reaction channel $^{159}Tb(^{16}O,2n2p)$ has a structure that is a combination of two peaks. The peak at lower value of catcher thickness corresponds to the fraction of residues produced through incomplete fusion process, while the peak at larger value of the catcher thickness corresponds to the fraction of residues populated via complete fusion process. Therefore, this reaction channel may be considered to have contributions from CF as well as ICF processes. Measured recoil range distribution for ^{165}Tm residues (Figure 4.50(j)) populated via $^{159}Tb(^{16}O,2\alpha2n)$ has three distinct peaks at cumulative catcher thicknesses ≈ 400 $\mu g/cm^2$, ≈ 300 $\mu g/cm^2$ and ≈ 200 $\mu g/cm^2$ due to the residues ^{165}Tm produced via three reaction channels i.e., (i) complete fusion of $^{16}O+^{159}Tb$, forming the composite system ^{175}Ta followed by the emission of $2n2\alpha$ particles, (ii) incomplete fusion of ^{16}O, assuming that ^{16}O breaks up into ^{12}C and an α particle, and the fragment ^{12}C fuses with ^{159}Tb forming the composite

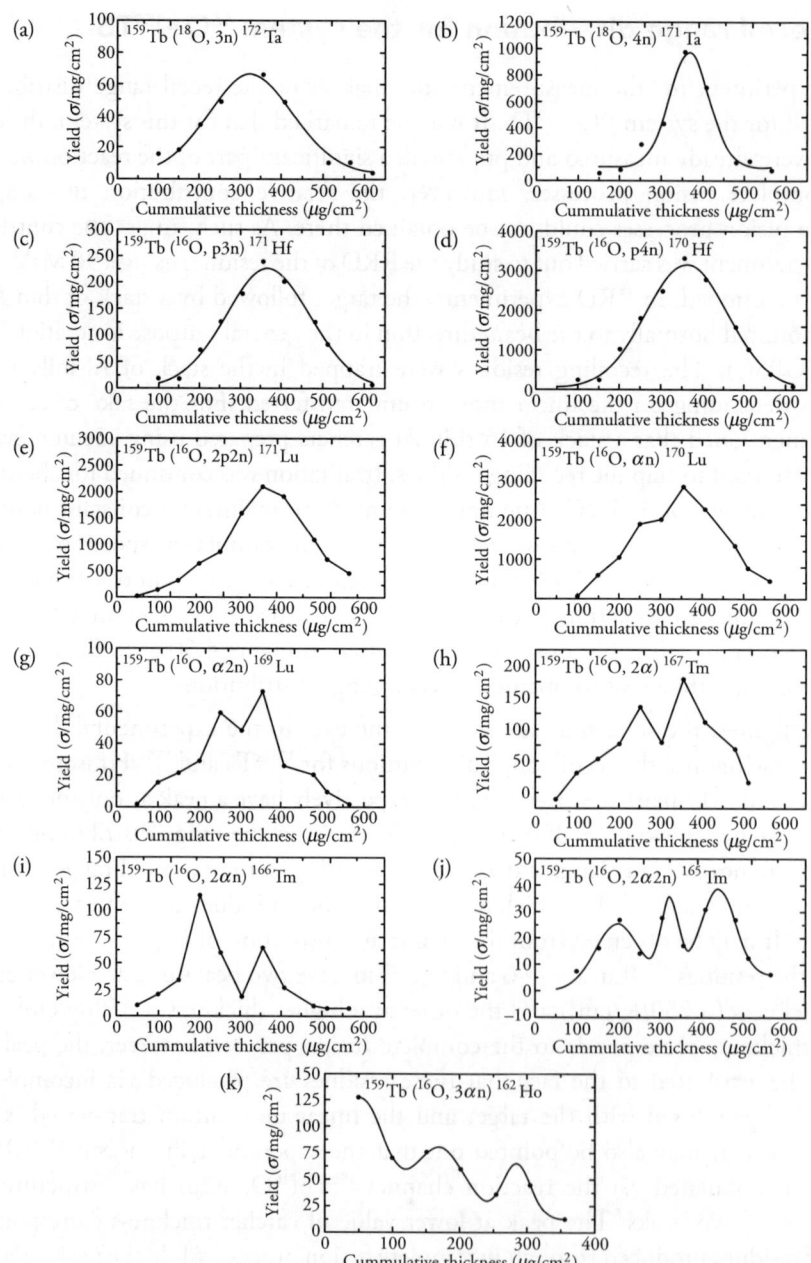

Figure 4.50 Experimentally measured recoil range distributions for various radioactive residues produced in the interaction of ^{16}O beam with ^{159}Tb target at ≈ 90 MeV

nucleus ^{171}Lu followed by the emission of $2n\alpha$ particles, and (iii) incomplete fusion of ^{16}O assuming that ^{16}O breaks up into two ^{8}Be fragments and one of the fragments fuses with the ^{159}Tb forming the composite nucleus ^{167}Tm, followed by 2n emission. In this RRD, the two peaks at the lower cumulative catcher thickness ($\approx 300\mu g/cm^2$ and $\approx 200\mu g/cm^2$) correspond to the two incomplete fusion channels. The experimentally measured recoil range distribution for the reaction channel ^{159}Tb(^{16}O,3αn)^{162}Ho (Figure 4.50(k)) has three peaks at relatively lower cumulative catcher thickness $\approx 75~\mu g/cm^2$, $\approx 150~\mu g/cm^2$ and $\approx 270~\mu g/cm^2$ respectively. In this case, the peak corresponding to the CF channel expected at $\approx 350~\mu g/cm^2$ has not been observed, indicating that this reaction proceeds predominantly via incomplete fusion process. To get the contributions of these CF and ICF processes in these reactions, the measured RRDs have been fitted (Figure 4.51(a)–(c)) with Gaussian distributions employing the ORIGINE software, which required the observed intensity and number of peaks to be fitted, as the input data. The relative contributions of CF and ICF processes are obtained by dividing the area of the corresponding peak by the total area. In Figure 4.51(a)–(c), the relative contributions for the reactions ^{159}Tb(^{16}O,2αn)^{166}Tm, ^{159}Tb(^{16}O,2α2n)^{165}Tm and ^{159}Tb(^{16}O,3αn)^{162}Ho are shown.

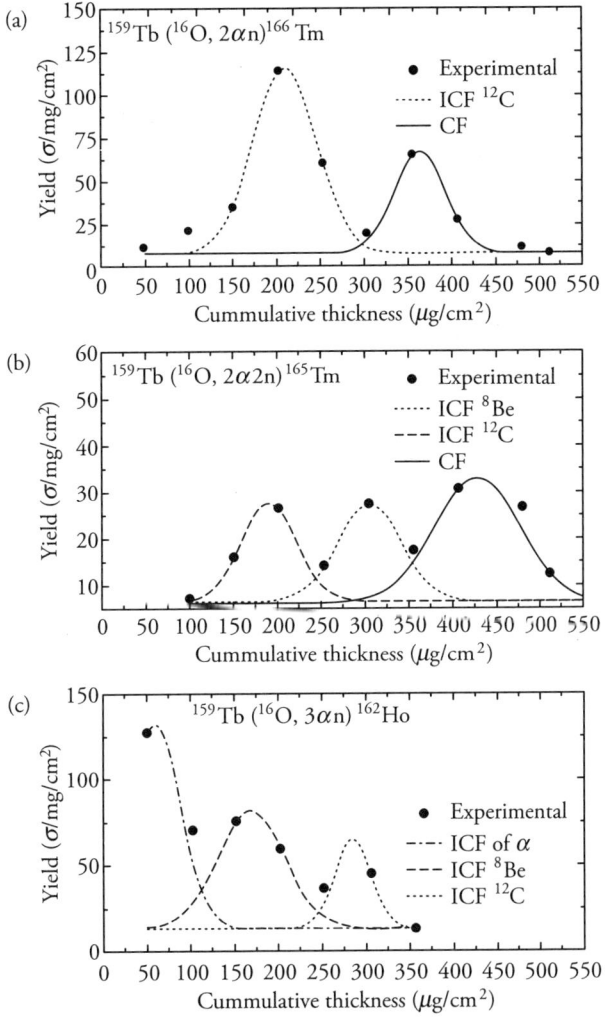

Figure 4.51 The experimental recoil range distributions fitted with Gaussian peaks to disentangle the relative contributions of CF and ICF processes

The relative contribution of incomplete fusion of ^{12}C for reaction ^{159}Tb(^{16}O,2αn)^{166}Tm (Figure 4.51(a)) (shown by dotted curve) is found to be ≈ 70%, while the contribution from CF (shown by solid curve) is ≈ 30%. In the same way, for reaction ^{159}Tb(^{16}O,2α2n)^{165}Tm (Figure 4.51(b)), the contributions of ICF of ^{12}C and ^{8}Be (shown by dotted and dashed curves) are found to be respectively ≈ 22% and ≈ 23%, while the contribution of complete fusion (CF) is ≈ 45%. On the other hand, for reaction ^{159}Tb(^{16}O,3αn)^{162}Ho (Figure 4.51(c)), contributions of incomplete fusion of ^{12}C, ^{8}Be and α particle are found to be ≈ 16%, ≈ 44% and ≈ 40%, respectively.

4.2.3 Recoil range distribution for the system ^{16}O+^{169}Tm

For this system, the recoil range distributions have been measured[6] at the incident energy of 86.6 MeV. The experimental arrangement was same as for the other cases, where the target was mounted in the irradiation chamber with an Al-backing of the thulium sample facing the beam. A catcher stack immediately followed the thulium layer. As a matter of fact, the beam energy on the front Al-backing of the sample surface was 92 MeV, but after an energy loss of nearly 5 MeV in the Al material, the beam energy on the thulium target was 86.6 MeV. In the present experiment, a stack of 19 thin Al-catchers of thickness varying in the range ≈16–45 μg/cm^2 was used to trap the recoiling residues. A list of thickness of various catcher foils is given in Table 4.22. The irradiation was carried out for 18 h with a beam fluence of ≈ 3500 μC. The activities induced in the catcher foils were recorded for about two weeks using the pre-calibrated high-resolution gamma-ray spectrometer. The cross-sections were computed for each catcher foil from the intensity of the activity induced in the foil and the cross-sections so measured were divided by the respective thickness of the catcher foils. The resulting yields are plotted in Figure 4.52(a)–(h) as a function of cumulative catcher thickness to obtain the recoil range distributions.

Table 4.22 A list giving thicknesses of catcher foils used in RRD measurements

S. No	Thickness in μg/cm^2
1	16.8
2	19.6
3	27.4
4	27.8
5	28.6
6	29.5
7	30.2
8	30.6
9	31.3
10	31.9

11	32.1
12	33.2
13	33.9
14	37.1
15	39.9
16	44.2
17	46.1
18	47.0

It may be observed from these figures that the recoil range distributions for ^{182}Ir and ^{181,182}Os isotopes produced through the reactions (^{16}O,3n), (^{16}O,p3n) and (^{16}O,p2n) respectively, have a peak at only one value of cumulative catcher thickness, ≈ 350 $\mu g/cm^2$. These RRDs of Ir and Os isotopes are nearly Gaussian having peak at the depth corresponding to the expected ranges of the compound system ^{185}Ir in the aluminium material, calculated using the stopping power values. This means that these products are formed only by complete fusion (CF) process followed by the evaporation of neutrons and/or protons. On the other hand, for the reactions ^{169}Tm(^{16}O,2p2n)^{181}Re, the RRD has two peaks: one at a relatively lower value of cumulative catcher thickness and the other at ≈ 350 $\mu g/cm^2$, the same as in case of complete fusion process, respectively.

As can be seen from Figure 4.52(d), the maxima at a larger value of cumulative catcher thickness (≈ 350 $\mu g/cm^2$) corresponds to the fraction of residues populated via complete fusion, while the peak at relatively lower ranges of thickness (≈ 250 $\mu g/cm^2$) may be attributed to the fact that the residues ^{181}Re are produced via incomplete fusion of ^{12}C, where the linear momentum transferred is expected to be less than that for complete fusion channel. It may be pointed out that in Figure 4.52(e), the expected data points for peak position of RRD at ≈ 350 $\mu g/cm^2$ for the residues ^{178}Re produced via the (^{16}O,α3n) reaction through CF could not be obtained due to the short half-life (13.3 min) of the residue. However, from the trend, it may be deduced that there may be two peaks – one corresponding to the ICF and the other due to CF channel.

It may be noted that the observed recoil range distribution data (Figure 4.52(f)) for the ^{175}Hf isotope produced via ^{169}Tm(^{16}O,2αpn) reaction have three peaks at cumulative thicknesses ≈ 370 mg/cm^2, ≈ 260 mg/cm^2 and ≈ 150 mg/cm^2 corresponding to the residues ^{175}Hf produced via three different channels, i.e., (a) the complete fusion of ^{16}O with ^{169}Tm, forming the composite nucleus ^{185}Ir, followed by the emission of a proton, a neutron and two α particles; (b) the incomplete fusion of ^{16}O, if it is assumed that ^{16}O breaks up into ^{12}C and an α particle while the fragment ^{12}C fuses with ^{169}Tm, forming the composite nucleus ^{181}Re, which decays by the emission of a proton, a neutron and an α particle; (c) the incomplete fusion of ^{16}O, assuming that ^{16}O breaks up into two ^{8}Be fragments and one of these fragments fuses with ^{169}Tm, forming the composite nucleus ^{177}Ta, followed by the emission of a proton and a neutron. For the reactions, ^{169}Tm(^{16}O,3αn)^{172}Lu and ^{169}Tm(^{16}O,3α2n)^{171}Lu, the measured

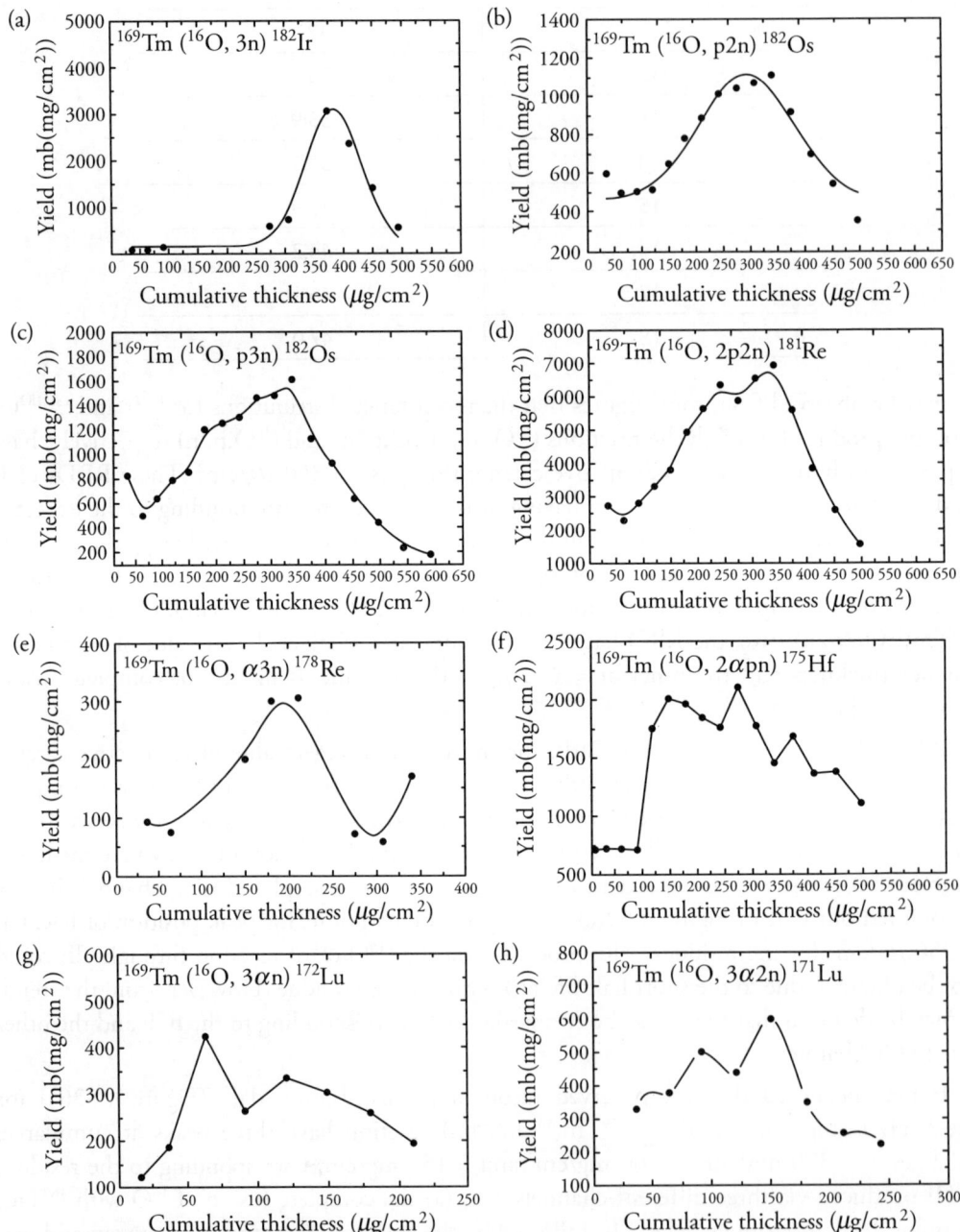

Figure 4.52 Experimentally measured recoil range distributions for various radioactive residues produced in the interaction of ^{16}O beam with ^{169}Tm target at ≈ 87 MeV

RRDs (Figures 4.52(g) and (h)) show two peaks at relatively lower values of cumulative catcher thicknesses, i.e., at ≈ 75 mg/cm^2 and ≈ 150 mg/cm^2, respectively. This indicates that these products are not populated by the complete fusion process but by some other process in which the linear momentum transferred is less than that for the complete fusion process. This is possible when only a part of the projectile fuses with the target (incomplete fusion) and the rest of it move with a velocity nearly equal to the velocity of the projectile. As such, in these reactions, the contribution of complete fusion is expected to be negligible. This may also be confirmed from the fact that the theoretical calculations of EFs for these channels using all the three codes, ALICE-91, CASCADE and PACE2, give negligible cross-section values.

To obtain the relative contributions of complete and incomplete fusion processes in the ^{169}Tm(^{16}O,2p2n)^{181}Re reaction, the measured recoil range distribution has been fitted with Gaussian peaks using the ORIGIN software as shown in Figure 4.53(a), and the areas under the two peaks have been determined. The peak represented by the dark solid curve gives the ICF contribution while the dotted curve represents the contribution due to the complete fusion process. The relative contributions of these processes (CF and ICF) are obtained by

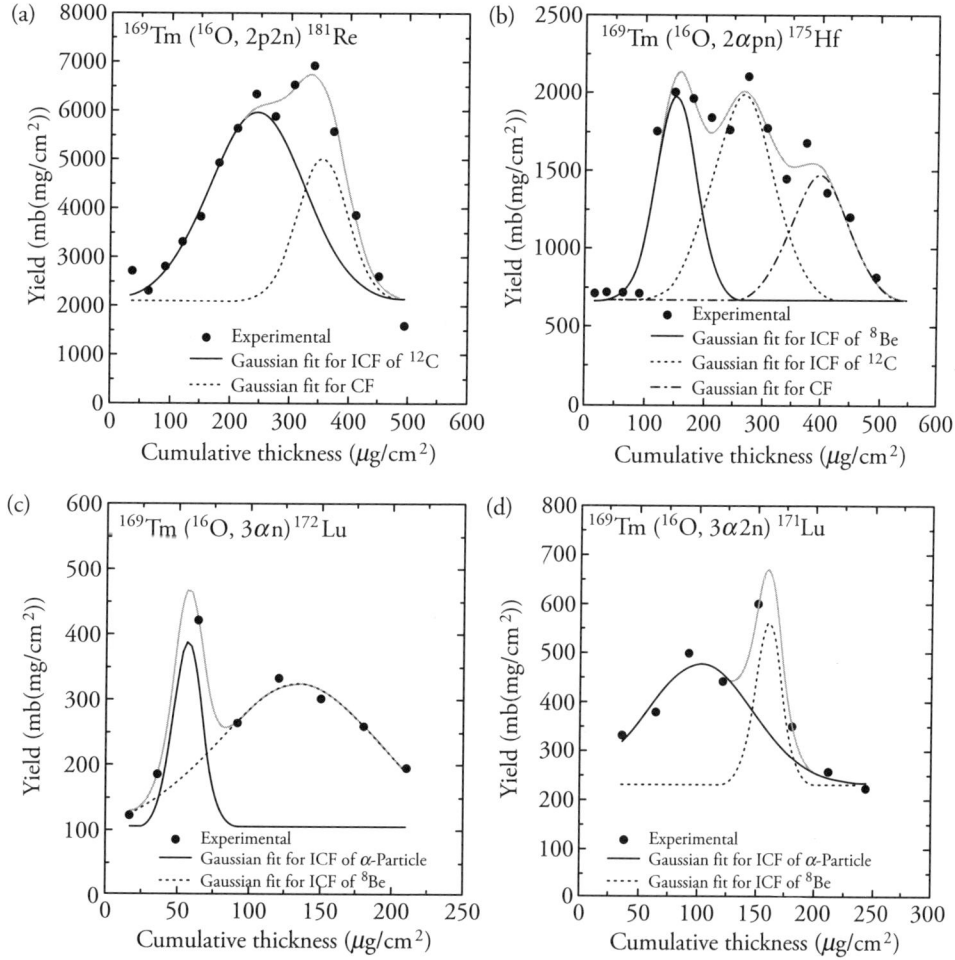

Figure 4.53 Measured recoil range distributions fitted with Gaussian peaks to determine the relative contributions of complete and incomplete fusion

dividing the area of the respective peak by the total area of the curve. The incomplete fusion (ICF) contribution in this case is found to be ≈ 65% and the CF contribution is obtained as about 35%, with an uncertainty of nearly 5%. For the reaction ^{169}Tm(^{16}O,2αpn)^{175}Hf, the experimentally measured distribution has been found to be fitted with three Gaussian peaks at cumulative thicknesses ≈150 mg/cm², ≈ 260 mg/cm² and ≈ 370 mg/cm² respectively, as shown in Figure 4.53(b). The relative contributions of CF, ICF for the fusion of fragment ^{12}C and the ICF contribution corresponding to the fusion of ^{8}Be are found to be nearly ≈ 25%, ≈ 46% and ≈ 29%, respectively, for these channels. Similarly, the relative contributions of ICF, as indicated in Figures 4.53 (c) and (d), of the α particle and ^{8}Be have been found to be ≈ 20% and ≈ 80% for the residues ^{172}Lu, while it was found to be ≈ 74% and ≈ 26% for the residues ^{171}Lu, respectively.

4.2.4 Recoil range distribution for the system ^{16}O+^{181}Ta

As has been mentioned, the degree of linear momentum transfer (ρ_{LMT}) from the projectile to the target nucleus may be used to differentiate the CF and ICF processes in heavy ion interactions. To study the energy dependence of incomplete fusion processes, the recoil range distributions for several residues produced in reactions in system ^{16}O+^{181}Ta have been measured. The experiments were carried out[34] to study the RRD of the residues resulting from reactions ^{181}Ta(^{16}O,xn), ^{181}Ta(^{16}O,pxn), $^{181\text{Ta}}$(^{16}O,αn), ^{181}Ta(^{16}O,α2n), ^{181}Ta(^{16}O,α3n) and ^{181}Ta(^{16}O,2α3n) at three distinctly different energies at 81, 90 and 96 MeV. As a representative case and to present the different linear momentum transfer components in various CF and ICF processes, the measured RRDs for ^{192}Hg(p4n), ^{191}Au^{g}(α2n) and ^{186}Ir^{g}(2α3n) residues are shown in Figures 4.54 (a), (b) and (c) at three beam energies.

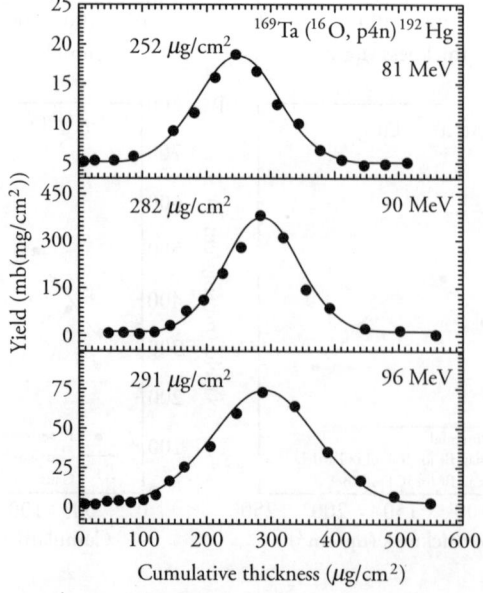

Figure 4.54(a) Experimentally measured recoil range distributions for ^{192}Hg(p4n) at projectile energies of ≈ 81, 90 and 96 MeV

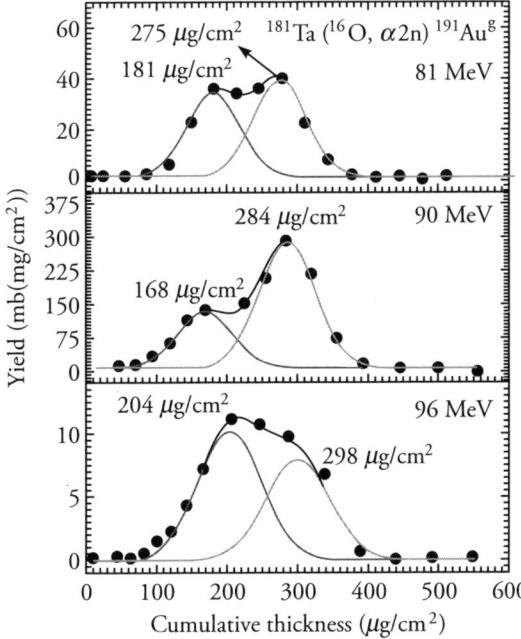

Figure 4.54(b) Experimentally measured recoil range distributions for $^{191}Au^g(\alpha 2n)$ at projectile energies ≈ 81, 90 and 96 MeV

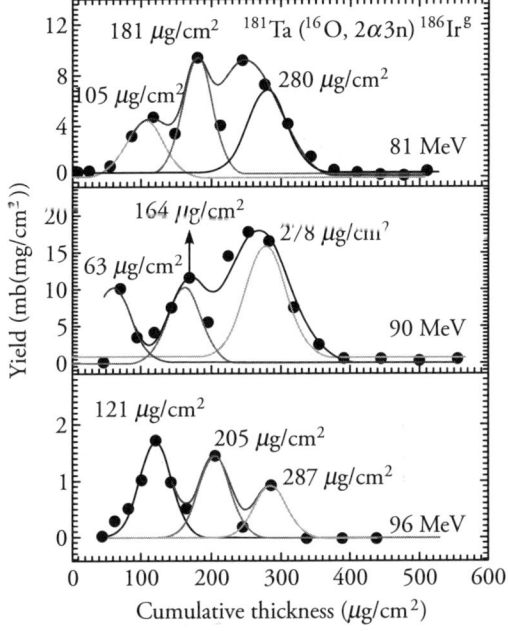

Figure 4.54(c) Experimentally measured recoil range distributions for $^{186}Ir^g(2\alpha 3n)$ at projectile energies ≈ 81, 90 and 96 MeV

As can be seen from Figure 4.54(a), for the p4n channel, the measured RRD represents only a single peak at all the three bombarding energies, indicating only a single linear momentum transfer component in the production of ^{192}Hg. Further, it may be observed in this figure that the recoil range peak shifts to a higher cumulative catcher thickness as the energy of the incident beam increases. This is due to the fact that the linear momentum transfer increases with incident beam energy. It may be remarked that the particle emission from the recoiling nuclei may change the energy/momentum of the recoiling residues depending on their direction of emission. This is reflected in the width (FWHM) of the measured RRDs. Straggling effects may also contribute to the observed width of the RRD peak.

The experimentally measured most probable ranges $R_{p(exp)}$ for all the identified CF residues along with the calculated (using the code SRIM) mean ranges $R_{p(the)}$, are given in Table 4.23. The most probable ranges listed in the table have been estimated theoretically assuming that in case of CF, the incident heavy ion fuses completely with the target nucleus transferring its total linear momentum to the fused system, which then recoils in the forward direction to conserve the input linear momentum. As such, for example, the production of ^{192}Hg(p4n) is associated with the entire LMT and may be represented as

$$^{16}O + {}^{181}Ta \rightarrow {}^{197}Tl^* \rightarrow {}^{192}Hg + p4n$$

In the same way, the recoil range distributions of the residues ^{194}Tl, ^{193}Tl, ^{192}Tl, ^{193}Hg^g, ^{193}Hg^m, ^{191}Hg^g, and ^{191}Hg^m are found to have a single peak associated with complete/full linear momentum transfer from the projectile to the composite system, indicating their population only through the CF process. On the other hand, in case of (αn), (α2n) and (α3n) reaction channels, where the residues ^{192}Au^g, ^{191}Au^g, and ^{190}Au^g are produced, each of the recoil range distribution has been found to have a two-peak structure. These observed RRDs were resolved into two Gaussian peaks – one corresponding to the fusion of the entire projectile and the other due to the fusion of ^{12}C (if ^{16}O breaks up into ^{12}C+α and ^{12}C fuses) with the ^{181}Ta. In Figure 4.54(b), the representative RRDs for the residues ^{191}Au^g are plotted at three different energies.

Table 4.23 Experimentally measured recoil ranges $R_{p(exp)}$ deduced from RRD data, and the theoretically calculated most probable mean ranges $R_{p(the)}$ for CF components at $\approx$ 81, 90 and 96 MeV, obtained using the range energy relation for the residues produced in the interaction of ^{16}O with ^{181}Ta

Residues	Energy (E) ≈ 81 MeV		Energy (E) ≈ 90 MeV		Energy (E) ≈ 96 MeV	
	$R_{p(exp)}$ ($\mu g/cm^2$)	$R_{p(the)}$ ($\mu g/cm^2$)	$R_{p(exp)}$ ($\mu g/cm^2$)	$R_{p(the)}$ ($\mu g/cm^2$)	$R_{p(exp)}$ ($\mu g/cm^2$)	$R_{p(the)}$ ($\mu g/cm^2$)
^{194}Tl(3n)	265 ± 76	267	275 ± 47	287	286 ± 48	298
^{193}Tl^g(4n)	260 ± 77	267	254 ± 39	287	286 ± 67	298
^{192}Tl(5n)	244 ± 58	267	255 ± 21	287	264 ± 75	298

$^{193}Hg^g(p3n)$	261 ± 82	267	257 ± 75	287	290 ± 52	298
$^{193}Hg^m(p3n)$	275 ± 75	267	270 ± 60	287	292 ± 51	298
$^{192}Hg(p4n)$	252 ± 61	267	282 ± 57	287	291 ± 80	298
$^{191}Hg^g(p5n)$	276 ± 47	267	256 ± 47	287	277 ± 50	298
$^{191}Hg^m(p3n)$	249 ± 53	267	230 ± 65	287	287 ± 69	298

It may be observed from this figure that for the residues $^{191}Au^g$, there are two linear momentum transfer components, one having the mean ranges at 275 ± 37, 284 ± 40 and 298 ± 45 $\mu g/cm^2$ at ≈ 81, 90 and 96 MeV beam energies (corresponding to the fusion of ^{16}O) and at 181 ± 37, 168 ± 40 and 204 ± 45 $\mu g/cm^2$ (corresponding to the fusion of ^{12}C) at the respective three incident energies. Further, it may also be observed from Figure 4.54(b) that the peak value of the ranges, that is, $R_{p(exp)}$ shifts toward a higher cumulative catcher thickness as the incident energy increases, as expected. Thus, it may be inferred that the residues $^{191}Au^g$ populated through the $^{181}Ta(^{16}O,\alpha 2n)$ channel have contributions both from the complete as well as incomplete fusion channels, i.e., via, the reactions

a. Complete fusion of ^{16}O as

$$^{16}O + {}^{181}Ta \Rightarrow {}^{197}Tl^* \Rightarrow {}^{191}Au^g + \alpha + 2n$$

b. Incomplete fusion of ^{16}O as

$$^{16}O(^{12}C + \alpha) + {}^{181}Ta \Rightarrow {}^{193}Au^* + \alpha \text{ (spectator)}$$

$$^{193}Au^* \Rightarrow {}^{191}Au^g + 2n$$

The experimentally measured recoil ranges for the reaction channels (αn), $(\alpha 2n)$ and $(\alpha 3n)$ via the complete fusion and incomplete fusion processes are shown in Table 4.24 and are found to agree with those calculated using code SRIM, based on the breakup fusion model. In these calculations, it is assumed that no energy is lost during the breakup of the incident ion and that the α particle essentially acts as spectator during the reaction, so that the linear momentum transfer of the residue is reduced to 3/4 of the CF value.

Table 4.24 Experimental ranges $R_{p(exp)}$, deduced from RRD data and theoretically calculated ranges $R_{p(the)}$, for various incomplete fusion components at ≈ 81, 90 and 96 MeV incident energy

Residues	$R_{p(exp)}$ $\mu g/cm^2$ (CF of ^{16}O)	$R_{p(the)}$ $\mu g/cm^2$ (CF of ^{16}O)	$R_{p(exp)}$ $\mu g/cm^2$ (ICF of ^{12}O)	$R_{p(the)}$ $\mu g/cm^2$ (ICF of ^{12}O)	$R_{p(exp)}$ $\mu g/cm^2$ (ICF of ^8Be)	$R_{p(the)}$ $\mu g/cm^2$ (ICF of ^8Be)
Energy (E) $\approx$ 81 MeV						
$^{192}Au^g(\alpha n)$	275 ± 48	267	145 ± 45	198	–	–
$^{191}Au^g(\alpha 2n)$	275 ± 37	267	181 ± 37	198	–	–
$^{190}Au^g(\alpha 3n)$	282 ± 45	267	181 ± 45	198	–	–

$^{186}Ir^g(2\alpha3n)$	280 ± 30	267	181 ± 28	198	105 ± 30	108
Energy (E) ≈ 90 MeV						
$^{192}Au^g(\alpha n)$	256 ± 43	287	168 ± 43	215	–	–
$^{191}Au^g(\alpha2n)$	284 ± 40	287	168 ± 40	215	–	–
$^{190}Au^g(\alpha3n)$	282 ± 32	287	196 ± 32	215	–	–
$^{186}Ir^g(2\alpha3n)$	278 ± 28	287	164 ± 28	215	63 ± 28	117
Energy (E) ≈ 96 MeV						
$^{192}Au^g(\alpha n)$	290 ± 55	298	200 ± 55	227	–	–
$^{191}Au^g(\alpha2n)$	298 ± 45	298	204 ± 45	227	–	–
$^{190}Au^g(\alpha3n)$	286 ± 55	298	213 ± 55	227	–	–
$^{186}Ir^g(2\alpha3n)$	287 ± 23	298	205 ± 23	227	121 ± 23	122

Similarly, it may also happen that ^{16}O may break up into four alpha particle fragments, out of which two α particles (^8Be) may fuse with the target nucleus, and the remaining two may escape without any interaction. One such case has been observed in the present work where $^{186}Ir^g$ ($T_{1/2}$ = 16.64 h) residues produced via the $2\alpha3n$ channel have been identified. The measured RRDs for residues $^{186}Ir^g$ are shown in Figure 4.54 (c). As can be seen from this figure, the RRDs for this case may be resolved clearly into three Gaussian peaks, indicating the presence of more than one linear momentum transfer component associated with the CF of ^{16}O and ICF of ^{12}C and ^8Be. From this figure, it may be resolved that the population of $^{186}Ir^g$ residues, at the energies of interest, may take place via the three linear momentum transfer components. The peaks at ranges 280 ± 30, 278 ± 28 and 287 ± 23 $\mu g/cm^2$ at ≈ 81, 90 and 96 MeV energies, respectively may be attributed to the fusion of ^{16}O with the target nucleus ^{181}Ta. The ranges at cumulative catcher thickness values 181 ± 28, 164 ± 28 and 205 ± 23 $\mu g/cm^2$ (fusion of ^{12}C) and 105 ± 30, 63 ± 28 and 121 ± 23 $\mu g/cm^2$ (fusion of ^8Be) at the respective energies have also been observed. As such, the residues $^{186}Ir^g$ produced through the $^{181}Ta(^{16}O, 2\alpha3n)$ reaction channel have contributions from both the CF and ICF processes, which may be represented as

a. Complete fusion of ^{16}O, that is,

$$^{16}O + {}^{181}Ta \Rightarrow {}^{197}Tl^* \Rightarrow {}^{186}Ir^{g*} + 2\alpha + 3n$$

b. Incomplete fusion of ^{16}O, that is,

$$^{16}O(^{12}C + \alpha) + {}^{181}Ta \Rightarrow {}^{193}Au^* + \alpha \text{ (spectator)} + 3n$$

c. Incomplete fusion of ^{16}O, that is,

$$^{16}O(^8Be + {}^8Be) + {}^{181}Ta \Rightarrow {}^{189}Ir^* + 2\alpha \text{(spectator)}$$
$$^{189}Ir^* \Rightarrow {}^{186}Ir^g + 3n$$

Here, it is assumed that in the case of the ICF process, the incident ^{16}O ion breaks into two fragments (e.g., ^{12}C and α or ^8Be and ^8Be) as soon as it comes in the vicinity of the nuclear field of the target nuclei. The fragments produced in this way are considered to travel with the same velocity as that of the incident ion. One of the fragments of ^{16}O (^{12}C or ^{8}Be or α) then fuses with the target nucleus forming a composite system which recoils in the forward direction to conserve the input linear momentum. It may be pointed out that the events due to the fusion of single α particles have not been observed in this case.

4.3 Measurement of Angular Distribution of Heavy Residues and their Analysis

As already mentioned in Chapter 3 of this monograph, the angular distribution of heavy residues may provide information of considerable value about the reaction mechanism involved. In the present work, the angular distributions of some heavy residues produced in the interaction of ^{16}O ion beam with target ^{169}Tm and ^{27}Al have been measured at projectile energy $\approx$ 81 and 85 MeV respectively as listed in Table 4.25. The target–catcher arrangement used for the angular distribution measurement has already been described in the Chapter 3. However, for the sake of completeness, some important details are given here.

Table 4.25 List of systems for which angular distributions of fusion residues have been measured. The energy of the incident ion and the Coulomb barrier for the system are also listed

S.No.	System	Energy of the incident ion (MeV)	Coulomb barrier (MeV)
1	^{16}O+^{27}Al	85.0	51.05
2	^{16}O+^{169}Tm	81.0	67.0

4.3.1 Angular distribution of residues emitted from the system ^{16}O+^{169}Tm

In this experiment,[35] the ^{169}Tm target sample was made by depositing spectroscopically pure material of thickness $\approx$ 50.0 µg/cm^2 on an aluminium backing. The sample was kept normal to the beam direction with the Al surface facing the beam. The incident energy of ^{16}O ions on the Al surface was $\approx$ 84.5 MeV, which after energy degradation of about 3.5 MeV in the backing aluminium foil, is incident with energy $\approx$ 81 MeV on the target material. The reaction residues which recoil out of the target at different angles were collected by a set of thick annular Al catchers of thickness 0.3 mm (sufficient to stop the most energetic recoiling products produced by full linear momentum transfer) with diameters of 0.81, 1.29, 1.95, 2.64, 3.27, 5.46 and 6.4 cm respectively. The annular catcher arrangement was placed 1.8 cm behind the target covering angular zones of 0–13° (most forward cone), 13–21°, 21–30°, 30–39°, 39–45°, 45–60° and 60–64° successively. The sample was irradiated for about 11 hrs with a beam current of 10 pnA. As soon as the irradiation was over, the activities induced in the annual catcher rings were counted. It may be pointed out that the post-irradiation analysis was similar to that which was used for the recoil range distribution measurements.

From the intensity of the activities induced in each catcher foil, the angular distributions were obtained for nine reaction residues viz., ^{182}Ir(3n), ^{181}Ir(4n), ^{182}Os(p2n), ^{181}Os^g(p3n), ^{181}Re(α), ^{175}Hf(2αpn), ^{171}Hf^g(2αp5n), ^{172}Lu^g(3αn) and ^{171}Lu^g(3α2n) populated in the ^{16}O+^{169}Tm system at projectile energy of ≈ 81 MeV. The measured angular distributions of the identified reaction residues are shown in Figure 4.55(a)–(h). It may be observed from these figures that the recoiling products ^{182}Ir(3n), ^{181}Ir(4n), ^{182}Os(p2n) and ^{181}Os^g(p3n) are emitted within the forward angles, peaking around 0–13° with respect to the incident beam direction. On the other hand, the reaction products ^{181}Re(α), ^{175}Hf(2αpn), and ^{171}Hfg(2αp5n) are found to have contributions both in the 0–13° and 45–60° angular zones. The angular distribution of CF products may be obtained theoretically using the code PACE4, which shows that CF products are emitted within a folding angle of ±18°, peaking at around 5°.

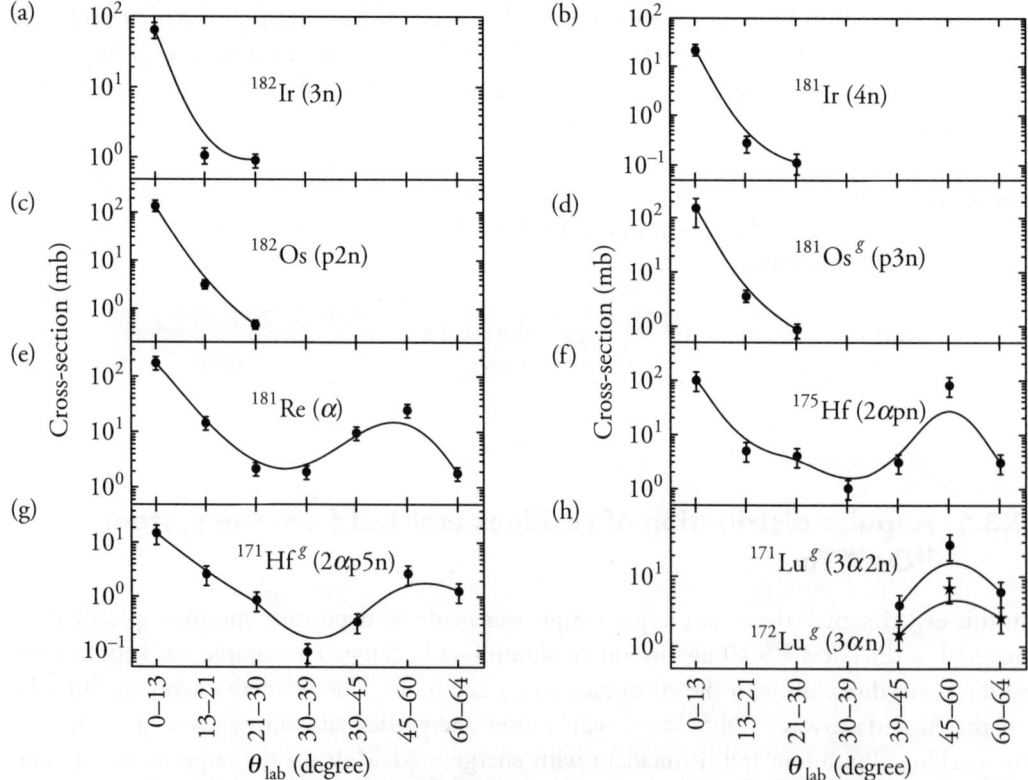

Figure 4.55 Experimentally measured angular distributions of the various reaction residues populated via CF and/or ICF mechanisms in the ^{16}O+^{169}Tm system at an energy of ≈ 81 MeV. Solid lines are drawn to guide the eyes. Different reaction products have been labeled by their corresponding emission channels

It may, therefore, be concluded that the residues emitted at folding angles larger than ±18° may correspond to the linear momentum transfer events (ICF products), where only a fraction of the projectile fuses with the target nucleus. This is because the folding angle is inversely proportional to the velocity of the reaction residue, which in turn is a measure of magnitude of the linear momentum transfer. In Figure 4.55(e) to (g), angular distributions for the residues ^{181}Re(α), ^{175}Hf(2αpn), and ^{171}Hfg(2αp5n) are found to spread from angular zone (0–13°) all the way up to angular zone (60–64°). The intensity distribution for these residues from 0° up to 30° may be assigned to the production of these residues through the complete fusion process, while the fraction of the observed activity falling in the angular zone beyond 30° may represent those events in which these residues are populated by ICF process. The measured angular distributions of the reaction residues ^{172g,171g}Lu are shown in Figure 4.55(h), where it is found that there is no data up to ≈ 39°. The angular distributions for ^{172g}Lu and ^{171g}Lu are found to be peaking in the angular range ≈ 39–64°, which clearly indicates that these residues are populated only by ICF process in peripheral collisions, where for the conservation of angular momentum, these residues are emitted at relatively large angles with respect to the beam direction.

4.3.2 Angular distributions of the residues emitted from ^{16}O+^{27}Al system

To measure the angular distribution of recoiling residues in the ^{16}O+^{27}Al system, another experiment[36] was carried out at ≈ 85 MeV beam energy. Here, an Al target supported by a Tm material of thickness 0.48 mg/cm² followed by a stack of annular concentric and thick Al-catcher foils was mounted in the irradiation chamber normal to the direction of the incident beam, The annular concentric aluminium catchers of thickness 0.3 mm with diameter 0.81, 1.29, 1.95, 2.64, 3.27, 5.46 and 6.4 cm were used to trap the recoiling residues emitted at different angles. The arrangement of the target and the catcher assembly used for the angular distribution measurements is shown in Figure 4.56. The annular catcher arrangement was placed 1.8 cm behind the target for trapping the residues ejected in seven different angular zones viz., 0–13° (most forward cone), 13–21°, 21–30°, 30–39°, 39–45°, 45–60° and 60–64°. The bombardment was done for about 11 hours using a beam of ^{16}O with current ≈ 7 pnA. As has been done in all offline experiments, the activities induced in each annular catcher were followed for several days.

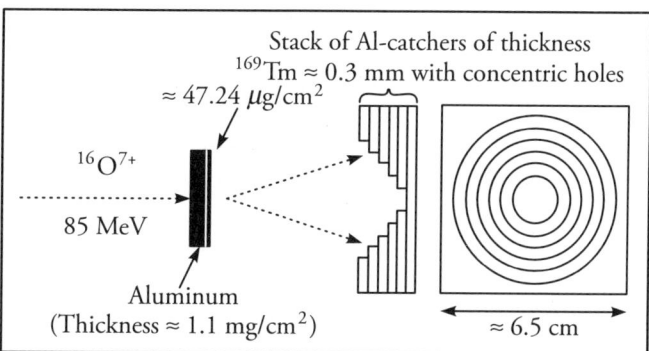

Figure 4.56 Typical arrangement of the target–catcher assembly used for the angular distribution measurements covering the annular range from (0–13°) to (60–64°)

A typical recorded gamma-ray spectrum indicating the region of interest for different annular Al-catcher rings covering the angular range from 0–13° to 45–60°, is presented in Figure 4.57. The residues were identified through their characteristic gamma lines and also by measuring their half-lives. Here, only the γ-ray of 146.5 keV corresponding to the reaction $^{27}Al(^{16}O,2\alpha n)^{34}Cl$ could be identified from its energy as well as the half-life of residues ^{34}Cl. Further, the intensities of the characteristic gamma line in different annular catcher foils were used to obtain production cross-sections in various angular zones.

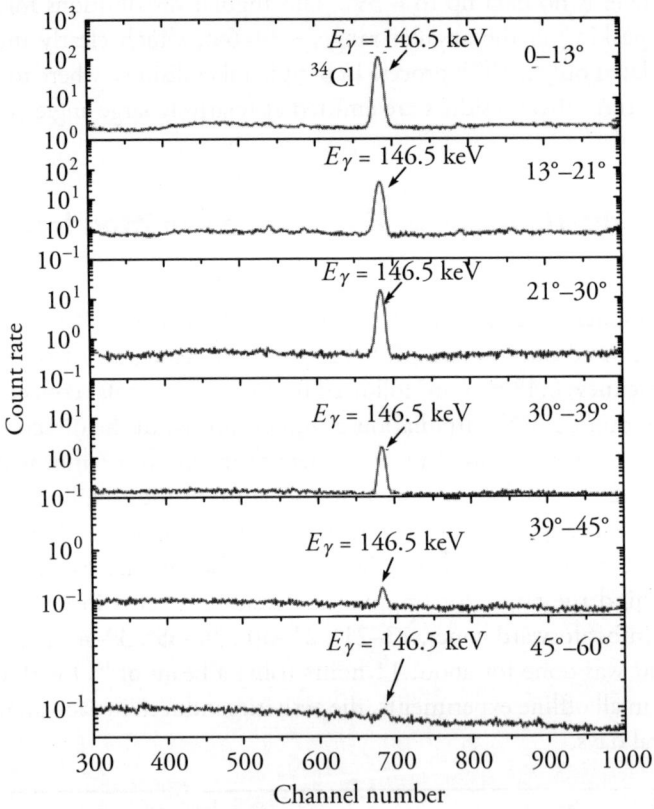

Figure 4.57 Typical γ-spectra of Al-catcher rings covering the angular zones from (0–13°) to (45–60°)

The measured cross-sections in annular concentric catcher rings were used to generate the angular distribution for the reaction $^{27}Al(^{16}O,2\alpha n)^{34}Cl$, which is shown in Figure 4.58. Depending on the momentum transfer from the incident beam particle to the composite system, the residues formed by CF and ICF processes got trapped in annular catcher rings at different angles. Two peaks are observed in the angular distribution shown in Figure 4.58 – one peak is around 0–13°, which may be assigned to the residues populated by full momentum transfer in a complete fusion process, while the other peak in the angular range 45–60°, may be

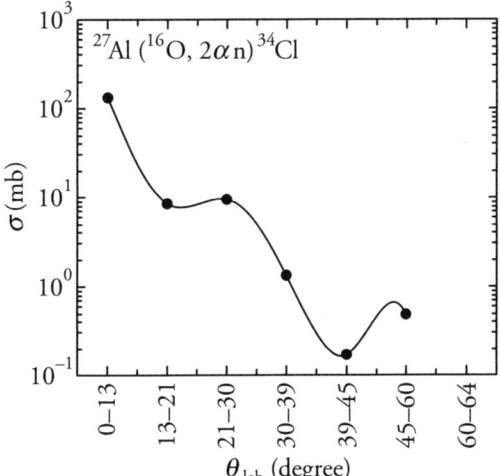

^{27}Al (^{16}O, 2αn) ^{34}Cl

Figure 4.58 Measured angular distribution of residues in the reaction ^{27}Al(^{16}O,2αn)^{34}Cl

assigned to the residues produced by ICF processes. It is thus clear that the angular distribution of heavy residues may be used to study the CF and ICF processes.

4.4 Measurement of Spin Distribution and Feeding Intensity Profiles

The measurement of fast projectile-like fragments associated with the ICF reaction dynamics has been a topic of interest in recent years, particularly at energies near and/or above the fusion barrier. At these energies, only complete fusion is expected to be the dominant process. The heavy ion induced CF and ICF reactions can be disentangled semi-classically based on the driving input angular momenta (ℓ) imparted to the system due to various interaction trajectories. According to the critical angular momentum or sharp cut-off model, all angular momentum values from $\ell = 0$ up to a limiting value $\ell = \ell_{crit}$ contribute to the complete fusion process and the fusion is considered to be zero for $\ell > \ell_{crit}$. In complete fusion, it is assumed that the attractive nuclear potential overcomes the repulsive Coulomb and centrifugal potentials, at smaller values of impact parameters; consequently, the target nucleus traps the entire projectile leading to the formation of an equilibrated compound nucleus that decay statistically. On the other hand, at relatively higher values of impact parameter, the driving angular momenta may exceed its critical limit (ℓ_{crit}) for fusion to occur; the attractive nuclear potential is now not strong enough to capture the entire projectile. As such, fusion cannot occur unless a part of the projectile is ejected to release the excess input angular momenta. In this way, the emission of a fraction (or part) of the projectile takes place to provide the sustainable input angular momenta to the system. Hence, the remaining nuclear system has angular momenta less than or equal to its own critical limit for fusion with the target nucleus. Therefore, the ICF is assumed to be associated with ℓ-values above the ℓ_{crit} for complete fusion. It may be remarked that Tserruya et al., observed that there is no sharp limit of input angular momenta for CF and ICF, but that both these reaction processes are expected to contribute significantly below and above their input angular momenta limits. To get the information regarding which of the driving input angular momenta is associated with the complete and incomplete reaction channel, a few particle gamma coincidence experiments have been carried out at energies ≈ 4–7 MeV/nucleon. The experiments were carried out at the Inter University Accelerator Centre (IUAC), New Delhi, India. Spin distribution of different reaction residues have been measured to get information regarding CF and ICF reaction dynamics. The projectile–target combinations have been chosen for the reason that in our earlier experiments, detailed analysis of excitation functions and recoil range distributions for these systems indicated considerable contribution from incomplete fusion channels. The two systems that were studied and their respective energies for the spin distribution measurements are listed in Table 4.26.

Table 4.26 List of systems for which spin distributions have been measured along with the energy of the incident ion and the Coulomb barrier for each system

S.No.	System	Energy of the incident ion (MeV)	Coulomb barrier (MeV)
1	$^{16}O+^{169}Tm$	87.0	67.0
2	$^{12}C+^{169}Tm$	67.0 and 78.0	51.0

4.4.1 Measurement of spin distribution and feeding intensity for the system $^{16}O+^{169}Tm$

To get the particle–gamma coincidence, the gamma detector array (GDA) along with the charged particle array (CPDA) setups have been used[37]. The gamma detector array (GDA) is a combination of 12 Compton-suppressed, high-resolution gamma-ray spectrometers at angles 45°, 99° and 153° with respect to the incident beam direction; there are 4 detectors at each of these angles. On the other hand, the CPDA is a set of 14 Phoswich detectors contained inside a 14 cm diameter chamber, which covers about 90% of the total solid angle. The CPDA which has 14 detectors has been divided into the following angular zones: (i) forward angle (F) 10–60°, (ii) sideways (S) 60–120° and (iii) backward angles (B) 120–170°. The protons and the alpha particles have been identified using the slow and fast components of the CPDA. A self-supporting, spectroscopically pure ^{169}Tm (100%) target of 0.93 mg/cm^2 thickness was bombarded with an oxygen beam of 90 MeV energy, using the Pelletron accelerator of the IUAC, New Delhi, India. In order to get rid of the scattered beam, the charged particle detectors were covered by Al-foils of appropriate thickness. In principle, at the forward angle, the detectors are expected to detect both (i) the alpha particles emitted through CF reaction, i.e., E_α (evaporation) ≈18 MeV and (ii) ICF α particles having the same velocity as that of the incident projectile, i.e., of energy 22.5 MeV. To cut down the evaporation alpha particles from the forward cone, additional Al-absorbers of appropriate thickness were kept on the forward angle (F) detectors so that only fast alpha particles may be detected. The in-beam prompt gamma-ray spectra have been recorded in coincidence with α and 2α particles in backward, forward and 90° angles. To identify the xn channels, the singles data was also recorded. In the first step of the experiment, efficiency determination as well as gain matching of HPGe detectors was performed by using the standard ^{152}Eu source of known strength, kept at the sample position. Different coincidence conditions have been projected on to the gamma spectra to generate particle ($Z = 1, 2$) gated spectra. To have improved data statistics, assuming the angular distribution of gamma rays to be isotropic, all gated spectra for a given gating condition have been summed up. The various reaction products likely to be populated via CF and/or ICF have been identified from their characteristic gamma lines from the gated and/or singles spectra. The pxn channels have been identified by subtracting the backward (B) α-gated spectra from the backward (B) proton, and the α particle ($Z = 1, 2$)-gated spectra to obtain the pure proton gated spectra. On the other hand, the αxn/2αxn (CN-α) channels populated via CF have been identified from the backward (B) α-gated spectra. Since the ICF alpha particles are expected to be emitted mostly in the forward cone (F), the αxn/2αxn (direct α) channels populated through ICF have been identified from the forward (F) α-gated spectra. To determine the relative production yield, the efficiency corrected intensity and area under the photo-peaks of the characteristic gamma lines have been used. It may be remarked that the γ-ray energies and their intensities used in the present work have been taken from the RADWARE level scheme[38]. To obtain the spin distribution of CF and ICF channels, the relative production yields have been plotted as a function of observed spin (J_{obs}^{exp}) corresponding to prompt gamma transition. In order to compare the spin distributions of various reaction channels (xn, αxn and 2αxn) in a panel, the relative production yields have been normalised with their own highest measured values (Y_{obs}^{max}) at lowest J_{obs}^{min}. To get the representation of data, the experimentally measured spin distributions obtained as discussed earlier are fitted by a function of the form

$$Y = Y_o/[1+\exp(J - J_o)/\Delta)$$

In this expression, Δ, is related to the width of the mean input angular momenta (J_o) and Y_o is the normalization constant. J_o is a sensitive parameter providing qualitative information about the driving input angular momenta associated with various reaction channels.

The measured spin distributions for xn, αxn and 2αxn channels that are populated via CF and/or ICF are shown in Figure 4.59(a) and (b). The errors of measurements are estimated to be < 10% and are not shown in these figures. It may be remarked that the normalised yield Y_{nor} at different J_{obs} for a particular reaction product have been obtained as the ratio of Y_{obs}^{max} at J_{obs}^{min} to Y_{obs} at different J_{obs} values, and hence, the fractional errors are significantly reduced in the values of Y_{nor}. Further, the inclusion of these errors does not affect the fitting of spin distributions and hence, the analysis of data. It may be observed from Figures 4.59(a) and (b), that there is a striking difference in the spin distributions for different reaction products expected to be populated via CF and ICF processes, which indicates the involvement of an entirely different reaction dynamics in the population of these reaction products. As can be

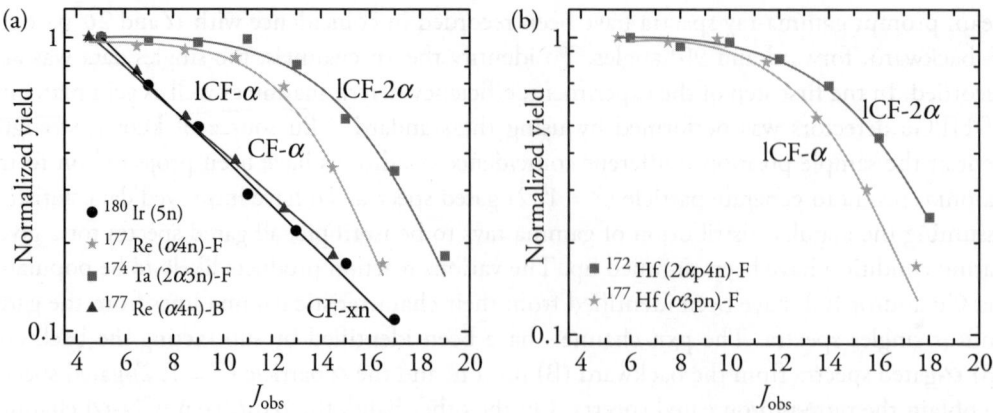

Figure 4.59 Experimentally measured spin distributions for different residues populated via xn (CF product) and αxn/2αxn (both the CF and/or the ICF products) in the ^{16}O+^{169}Tm system at $\approx$ 5.6 MeV/nucleon energy. The nomenclature used in the plots indicates the expected reaction dynamics involved – ICF-α and ICF-2α means that the involved reaction dynamics is ICF respectively with one α and 2α multiplicity; CF-α indicates complete fusion (CF) with one α multiplicity identified from the backward α-gated spectra. The nomenclature also shows that the exit channels are composed of the specific residual nucleus, α particle(s) (M_α = 1–2), neutron(s) and/or proton(s). F and B represent the reaction products identified respectively from forward and backward α-gated spectra. The lines and curves through data points are the result of the best fit procedure

seen from these figures, the intensity of the xn channels, which are predominantly populated via CF, falls off rapidly with J_{obs} or, in other words, the intensity increases steeply towards the band head indicating strong feeding during the de-excitation of CN. On the other hand, for αxn and 2αxn channels that are identified from the forward (F) α-gated spectra (associated with ICF), the intensity appears to be almost constant up to a certain value of observed spin, i.e., $J_{obs} \approx 10\,\hbar$ for direct α-emitting channels and $J_{obs} \approx 12\,\hbar$ for direct 2α-emitting channels. Hence, based on these trends, it may be inferred that the intensity does not increase towards band head after $\approx$10 $\hbar$ and $\approx$12 $\hbar$ for direct α/2α-emitting channels. This indicates the absence of feeding to the lowest members of the Yrast band and/or the population of low spin states are strongly hindered in the αxn, and/or 2αxn channels, which are associated with the ICF reaction dynamics. In case of αxn, and 2αxn channels, it may be remarked that intensity increases with J_{obs} up to $J \approx$10 –12 $\hbar$, respectively, which indicates significant feeding up to J $\approx$ 10 − 12 $\hbar$ from the entry point. In order to present the entirely different observed trends of spin populations for CF and ICF products, the same residues produced via forward-α (ICF) and backward-α (CF) emitting channels (Re isotopes) have been compared and plotted in Figure 4.59 (a). It may be observed from this figure that the spin distribution for ^{177}Re residues identified from backward α-gated spectra has been found to be distinctly different from that observed from forward α-gated spectra. This indicates the involvement of an entirely different reaction dynamics in the two cases. The spin distribution for ^{177}Re(α4n) identified from the backward α-gated spectra shows the same trend as has been observed for xn channels (CF). Moreover, Figure 4.59(b) presents the spin distribution of ^{172,177}Hf residues identified from the forward α-gated spectra, which are likely to reflect similar characteristics as that found for the direct α-emitting channels. From these figures, the mean input angular momenta (J_0) is found to be 10 $\hbar$ for xn channels while for direct αxn and 2αxn channels (ICF residues), the value of J_0 approaches $\approx 13\,\hbar$ and $\approx 16\,\hbar$ respectively. The smaller value of J_0 in case of CF channels compared to ICF channels indicate the involvement of less input angular momenta in CF reactions as compared to ICF reactions. These measurements clearly demonstrate how the driving input angular momenta (ℓ) increases with the direct alpha multiplicity; it may be presented as

$$\ell_{(ICF-\alpha xn)} \approx 1.3\,\ell_{(CF-xn/\alpha xn)}$$

$$\ell_{(ICF-2\alpha xn)} \approx 1.23\,\ell_{(ICF-\alpha xn)} \approx 1.6\,\ell_{(CF-xn/\alpha xn)}$$

These experiments clearly show that even at energy as low as 5.6 MeV/nucleon, a significant part of the reaction proceeds through incomplete fusion (ICF) and the multiplicity of direct α particles in the forward cone (ICF-α) increases with J_0, demonstrating the role played by the angular momenta. On the basis of these findings, it may further be inferred that the ICF takes place in the peripheral interactions at relatively larger values of impact parameters. The accuracy and the self-consistency of the measured spin distributions have been checked by estimating the relative production yield of each reaction residue from the spin distribution data. For this, the experimentally measured relative yield of each reaction product has been extrapolated up to $J = 0\,\hbar$ and the yield value at $J = 0\,\hbar$ $(Y^{J=0})$ has been normalised with the total yield (sum of all fusion evaporation channels) to get the relative yield value for each reaction residue. In the same way, the relative production yields of the individual reaction products, calculated employing the code PACE4 have also been normalised with the total yield of the fusion evaporation channels. The ratio of the experimentally obtained and calculated relative yields (Y_{EXP}/Y_{PACE4}) for all the fusion channels has been plotted in Figure 4.60.

As may be seen from Figure 4.60, the experimentally measured and theoretically calculated values agree reasonably well within the experimental uncertainties that strengthen the validity of the measured spin distributions. It may be observed from Figures 4.59(a) and (b) that the intensity of the Yrast line transitions decreases gradually with high spin for CF; on the other hand, in case of ICF, the intensity remains almost constant up to a certain limiting spin value and decreases rapidly for transitions of higher spin, indicating entirely different de-excitation patterns for CF and ICF residues from entry state to the Yrast line, This implies a rather smooth and broad feeding distribution for the Yrast states in case of CF process. On the other hand, for the ICF process, this distribution must have a narrow window meaning thereby a well-localised angular momentum region where a given projectile-like fragment is ejected in contrast to a large window for complete fusion reactions.

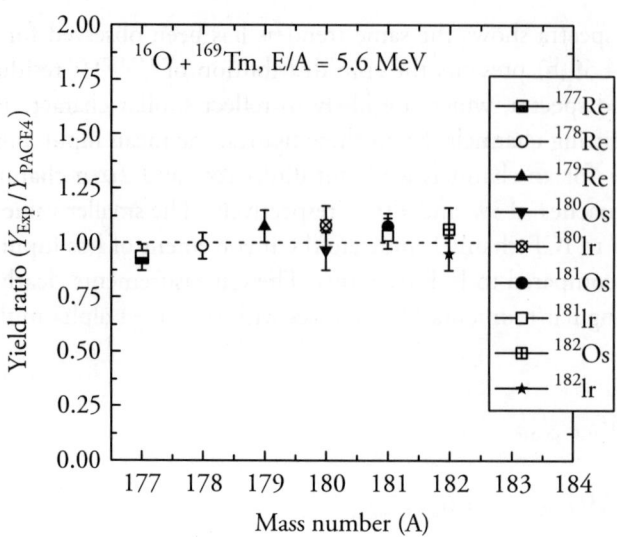

Figure 4.60 Ratio of experimentally measured and theoretically estimated relative production yields of residues populated only via CF in the $^{16}O+^{169}Tm$ system at ≈ 5.6 MeV/nucleon energy

Therefore, to understand the feeding pattern in various reaction channels associated with CF and ICF processes, the feeding intensity of gamma population have been obtained from the measured spin distribution of residues. The deduced feeding intensities for the presently studied reaction channels have been drawn as a function of J_{obs} and are given in Figures 4.61(a) and (b).

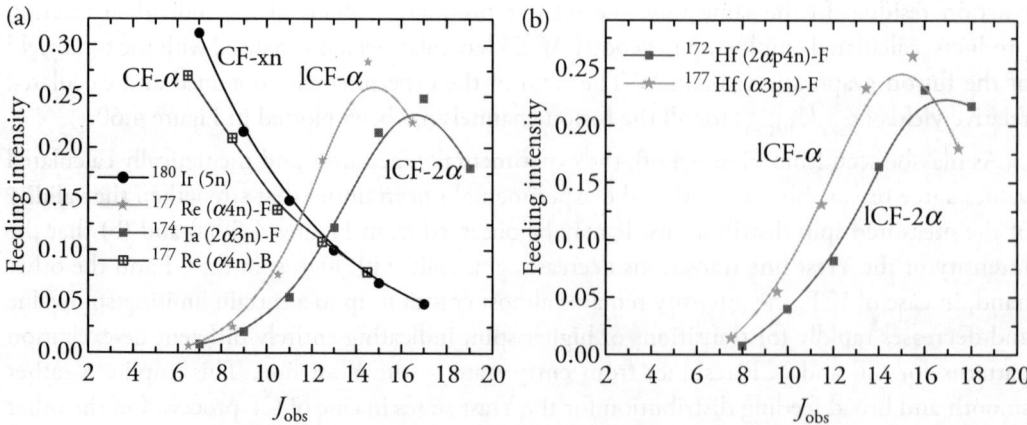

Figure 4.61 Deduced feeding intensities of gamma cascades of different ERs expected to be produced via xn, αxn and/or 2αxn channels in the $^{16}O+^{169}Tm$ system at ≈ 5.6 MeV/nucleon. The lines and curves through data points are drawn only to guide the eye

It may be observed from these figures that the feeding intensity for the forward-gated αxn and 2αxn channels is found to increase up to $J \approx 14\hbar$ and $J \approx 17\hbar$, respectively from higher (entry side) spin states and then decrease with the decrease of the spin. This indicates that high spin states are strongly fed even in ICF channels and that low spin states are relatively less fed. It, therefore, indicates the absence of feeding to the lowest members of the Yrast band or the fact that low spin states are less populated in αxn and 2αxn channels. It may be noted that these ICF channels are identified from the forward α-gated spectra. Such type of feeding intensity profile is expected to arise from the narrow ℓ- window, localised near and/or above the critical angular momentum for complete fusion that is associated with the incomplete fusion process. It may also be observed from Figures 4.61(a) and (b) that the feeding intensity profile shows a sharp exponential rise toward the low spin states for all the xn channel (CF), indicating here a strong feeding. In order to observe the feeding intensity patterns for backward α-channels along with the forward α-channels, these have been plotted in Figure 4.61(a) and (b). It may be observed from this figure that the feeding intensity of the αxn channel identified from the backward α-gated spectra indicates exponential rise towards lower spin states, as expected for the complete fusion dynamics where the band is fed over a broad spin range. It may be remarked that the feeding intensity is found to be less in the production of ^{177}Re isotopes identified from backward alpha-gated spectra, as compared to the xn channels. This may be due to the fact that neutron emission carries almost negligible value of angular momentum from the compound nucleus, while the alpha emission from the compound nucleus removes significant value of angular momentum as well as the excitation energy, which finally gives rise to a broad feeding range towards the band head.

4.4.2 Measured spin distribution and feeding intensity profile for the ^{12}C+^{169}Tm system

In order to study the role of high angular momentum involvement in the onset of ICF reactions, another experiment for the measurement of spin distribution was carried out[39] for the system ^{12}C+^{169}Tm at 5.64 MeV and 6.54 MeV. Here, spin distributions of various xn/pxn/αxn/2αxn channels were measured at these energies. An isotopically pure Tm (^{169}Tm, abundance = 100%) target of thickness 1.83 mg/cm^2 was bombarded by a ^{12}C^{5+} beam of current $\approx$ 30–35 nA, delivered from the Pelletron accelerator of IUAC, New Delhi, India. Similar to earlier experiments, the thickness of the sample was measured by an α-transmission method. The experimental setup and the technique were similar to the previous one; however, a short account of the experimental conditions and data reduction procedure are presented in the following.

In this work, the particle ($Z = 1,2$)–gamma coincidences are recorded employing the gamma detector array (GDA) along with the charged particle detector array (CPDA) to identify various reaction channels involved in the interaction. As already mentioned in the previous section, the gamma detectors array consists of 12 Compton-suppressed high-resolution high purity germanium (HPGe) gamma-ray spectrometers, which are arranged at 45°, 99° and 153° with four detectors at each of these angles. At the same time, the CPDA array has 14-Phoswich detectors arranged in two truncated hexagonal pyramid shapes. These detectors cover a solid angle of about 90% of the total. The array of CPDs was divided into three angular zones: (i) forward (F) 10–60°, (ii) sideways (S) 60–120° and (iii) backward (B) 120–170°. The purpose of dividing the array into angular zones is to employ the gating conditions for particles (Z = 1, 2) in coincidence with gamma rays at various angles. During the experiment, coincidences were demanded between particles and prompt gamma rays by putting three gating conditions corresponding to the given angular zones for each value of Z (1, 2). To record only the fast alpha particles which are expected to be associated with the ICF process in the forward cone, it is required to stop the slow alpha particles by adding an absorber onto the forward CPDs. Therefore, the energy spectrum of the slow α particles was generated using PACE4 code, as shown in Figure 4.62. It may be observed from this figure that the estimated most probable energy of the slow alpha particles (E_{CF}-α) is found to be nearly 18.5 MeV at 5.6 A MeV. The energy of the fast alpha component emitted as ICF α particles (E_{ICF}-α) is about 22.5 MeV at 67.5 MeV for the ^{12}C beam. Therefore, to stop the 18.5 MeV slow α particles, an Al absorber of thickness around 3 mg/cm^2 was put on the forward CPDs, so that only the fast α particles in the forward cone may be detected.

Data reduction is an important aspect in such experiments. In the present work, the off-line data analysis was carried out with the software INGAsort[40]. First, energy calibration as well gain matching of the HPGe detectors was carried out using a standard ^{152}Eu radioactive source. This was done before as well as after the experiment by putting the sources at the target position. The particle (Z = 1, 2)–gamma coincidence spectra are generated to identify the different reaction channels. Various gating conditions are projected onto the gamma spectra

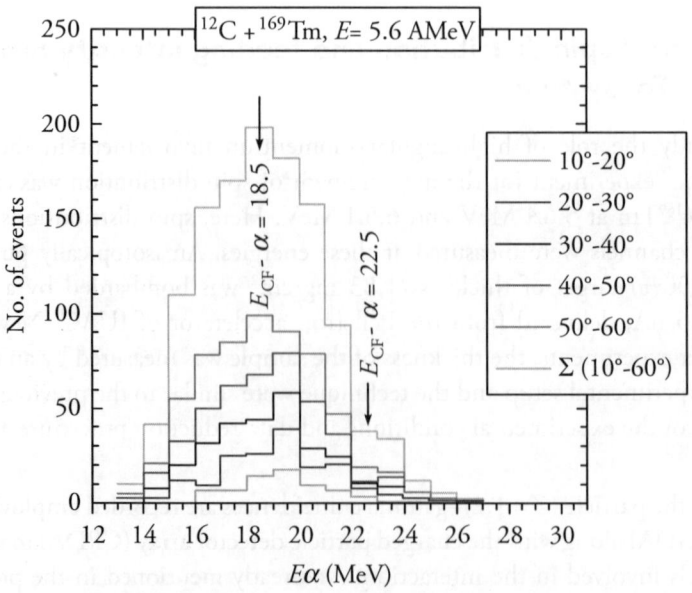

Figure 4.62 Fusion–evaporation (CF) α-energy profile for forward (F) zone (10–60°) at projectile energy $E = 5.6A$ MeV in the ^{12}C+^{169}Tm system predicated by PACE4. Different angular slices from 10° to 60° are also shown in the figure

to generate the particle ($Z = 1, 2$)-gated spectra for each angular zone. The reaction residues populated via CF and/or ICF are identified by observing their characteristic gamma lines in the particle gated and/or singles spectra. The fusion xn channels are identified by employing the singles spectra which were recorded by two coaxial gamma detectors. To identify the pxn channels populated through CF channel, the backward (B) alpha-gated spectra were separated from the backward (B) particle and ($Z = 1,2$)-gated spectra to get the pure backward (B) proton-gated spectra. The αxn channels populated via CF process (which consists of slow α-component) are identified from the backward (B) α-gated spectra. It may be mentioned that in the ICF process, the direct α particles are expected to be concentrated in the forward cone only. The slow alpha component is filtered out by putting the Al absorber on the forward cone charged particle detectors. At the same time, in order to remove any contamination from the slow alpha component in the forward (F) cone, the backward (B) alpha-gated spectra are subtracted from the forward (F) alpha-gated spectra. The ICF αxn/2αxn channels are identified from the forward (F) alpha-gated spectra, which was corrected for the slow alpha component. Then the relative production yield of the identified reaction residues are obtained from the intensity and area under peak of the characteristic prompt gamma ray of the particular reaction product. The prompt gamma ray energies and their intensities are obtained from the RADWARE level scheme directory.

In order to get the information of the decay patterns of the CF and ICF reaction products, the spin distributions of various residues are generated. The relative population yields are plotted as a function of experimentally observed spin J_{obs}^{exp}, and normalised with their highest yield values Y_{obs}^{max}, at the lowest observed spin to compare the different energy data in one single

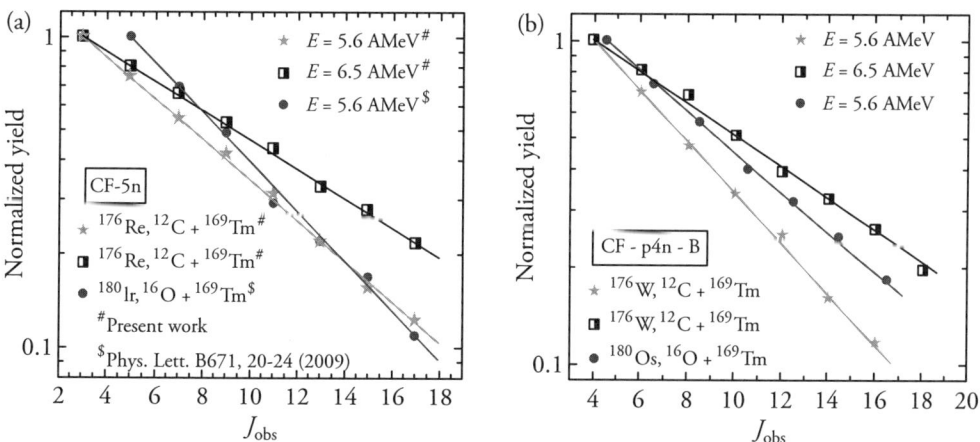

Figure 4.63 Experimentally measured spin distributions for (a) CF-5n channel (identified from singles spectra) and (b) CF-p4n channel (identified from backward proton-gated spectra) are plotted along with the spin distributions of the same channels from Singh et al.[37] Reaction products are labeled by self-explanatory notations and emission cascades. The nomenclature shows that the exit channels are composed by the one given residual nucleus, neutron(s), and/or proton(s). Lines through the data points are the result of the best-fit

panel. The measured spin distributions of the residues are fitted by a Fermi function. The measured spin distributions have also been compared with those for the system ^{16}O+^{169}Tm. In Figure 4.63, the spin distribution of 5n and p4n channels that is (a) ^{176}Re(5n) identified from the singles spectra and (b) ^{176}W(p4n) identified from the backward (B)-proton gated spectra are shown.

Further, in Figures 4.64 and 4.65, the measured spin distributions of alpha-emitting channels ^{173,174}Ta(α3n/α4n) are identified from the (a) backward (B) α-gated spectra (associated with CF) and (b) forward (F) α-gated spectra (associated with ICF). The measured spin distribution of ^{171}Lu(2α2n), which are identified from the forward (F) α-gated spectra is shown in Figure 4.66.

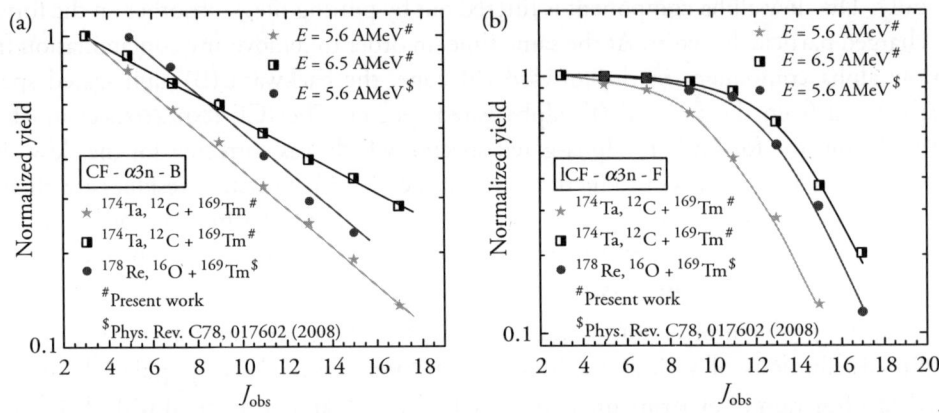

Figure 4.64 Experimentally measured spin distributions for (a) CF-α4n channel and (b) ICF-α4n channel along with the spin distributions of the same channels from Singh et al.[37] B and F represent the reaction products identified from backward (B) α-gated spectra and from forward (F) α-gated spectra, respectively. Lines and curves are the result of the best-fit procedure

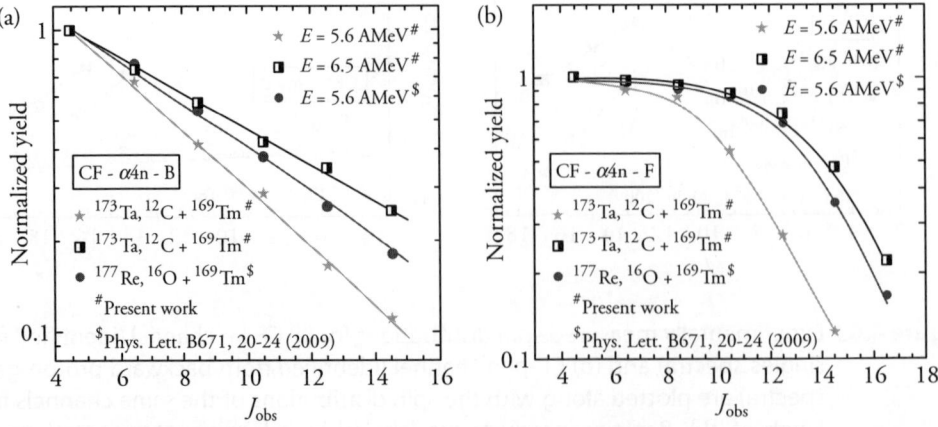

Figure 4.65 Experimentally measured spin distributions for (a) CF-α3n channel, and (b) ICF-α3n channel along with the spin distributions of the same channels from Singh et al.[37]

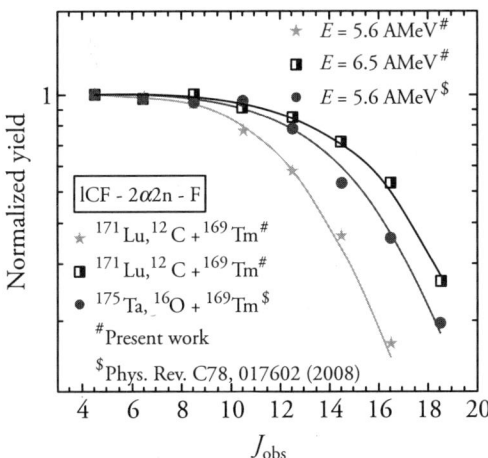

Figure 4.66 Experimentally measured spin distribution for ICF-2α2n channel identified from the forward α-gated spectra. ICF-2α2n means the involved reaction dynamics is ICF with a multiplicity of $\alpha = 2$

In order to have a comparison with similar data, the measured spin distributions of ^{180}Ir(5n), ^{180}Os(p4n), ^{177,178}Re(α4n/α4n) and ^{175}Ta(2α2n) populated via similar emission channels in the ^{16}O+^{169}Tm system at 5.6 A MeV are plotted in these figures. The curves and lines plotted through the data points represent the least square fits to the function given in the text. It may be observed from Figures 4.63–4.65 that the trends of the spin distribution for alpha-emitting channels (ICF products), identified from the forward (F) α-gated spectra, are found to be entirely different as compared to that for the CF products. This entirely different feature represents the involvement of an entirely different de-excitation pattern in CF and ICF residues. It may also be pointed out that the spin distributions of ^{176}Re(5n), ^{176}W(p4n)-B and $^{173,\,174}$Ta(α4n/α3n)-B produced in the ^{12}C+^{169}Tm reaction are found to show similar decay characteristics as that of ^{181}Ir(5n), ^{180}Os(p4n)-B and ^{177,178}Re(α4n/α3n)-B identified to be populated through the complete fusion in the ^{16}O+^{169}Tm system. It may be observed from Figures 4.63, 4.64(a) and 4.65(a), that there is a gradual monotonic increase in the intensity towards the band head, which indicates a broad population and/or strong feeding over a broad spin range during the de-excitation of these reaction products. On the other hand, for alpha-emitting channels ^{173,174}Ta(α4n/α3n)-F and ^{171}Lu(2α2n)-F identified from the forward (F) alpha-gated spectra, the intensity increases up to a certain value of J_{obs}, and then remains constant down to the band head (J_{obs}^{min}). This pattern of spin distribution is expected to arise from the narrow spin population only up to a certain value of J_{obs}. Apart from this, the constant behaviour of intensity for low spin states in the case of ICF α4n/α3n/2α2n-F channels indicate almost negligible feeding for low spin states. It may be pointed out that the increase in intensity up to a certain value of J_{obs} may be because of the feeding for some high spin states from the entry side. As a result, it may be inferred that forward going alpha particles tend to be from the ICF process. Moreover, the intensity of CF-xn/pxn/αxn channels as seen from Figures 4.63, 4.64(a) and 4.65(a) falls off sharply for projectile energy $\approx 5.6A$ MeV (^{12}C+^{169}Tm system) as compared to that for $\approx 6.5A$ MeV at the higher spin side or entry side, which indicates the involvement of rather less driving angular momenta at $\approx 5.6A$ MeV, as expected. As such, it may be concluded from the trends of the spin distributions that the CF products are strongly fed over a broad spin range, but the ICF products are likely to be associated with narrow spin population and/or less feeding probability for low spin states during the de-excitation process.

From the aforementioned description of the measured spin distributions, it may be observed that CF products are strongly fed over a broad spin range as compared to ICF products. Apart from this, it is important to have direct evidence of the feeding probability of the gamma population in CF and/or ICF channels. Here, the feeding intensity profiles for various reaction products populated via CF and ICF are generated from the best fitting procedure of the measured spin distributions. To get the feeding intensity profiles, the feeding probability of each observed gamma transition for different reaction products are plotted as a function of J_{obs} in Figures 4.67 and 4.68. It may be observed from Figure 4.67(a) that the feeding intensity for CF channels (xn/pxn-B) show a sharp exponential rise towards low spin states, which indicates a regular population with a strong feeding contribution for each gamma transition up to J_{obs}^{min}. In order to have a better comparison of direct α-emitting channels (ICF) and fusion–evaporation α-emitting channels, the feeding intensity profiles for CF and ICF products are plotted in a single panel. It may be observed from Figures 4.67(b) and 4.68(a) that the feeding intensity of αxn channels (^{173,174}Ta-CF-B) identified from the backward (B) alpha-gated spectra shows a similar trend as that which was observed for xn/pxn-B channels, where the band is fed over a broad spin range.

Further, as shown in Figures 4.67(a), 4.68(a) and 4.68(b), the feeding intensity for forward (F)αxn/2αxn channels is found to increase up to a certain value of J_{obs} and then decrease gradually towards the band head. At $\approx 5.6A$ MeV, the feeding intensity for the forward (F)

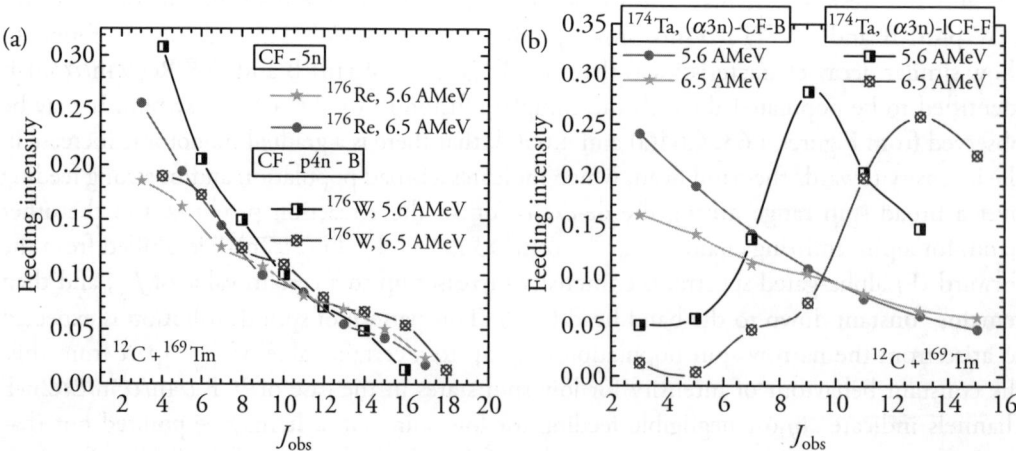

Figure 4.67 Deduced feeding intensities of γ cascades of different reaction products expected to be populated via (a) CF-5n, CF-p4n and (b) CF-α3n (identified from backward α-gated spectra) and CF-α3n (identified from forward α-gated spectra) channels. Lines and curves are drawn only to guide the eyes

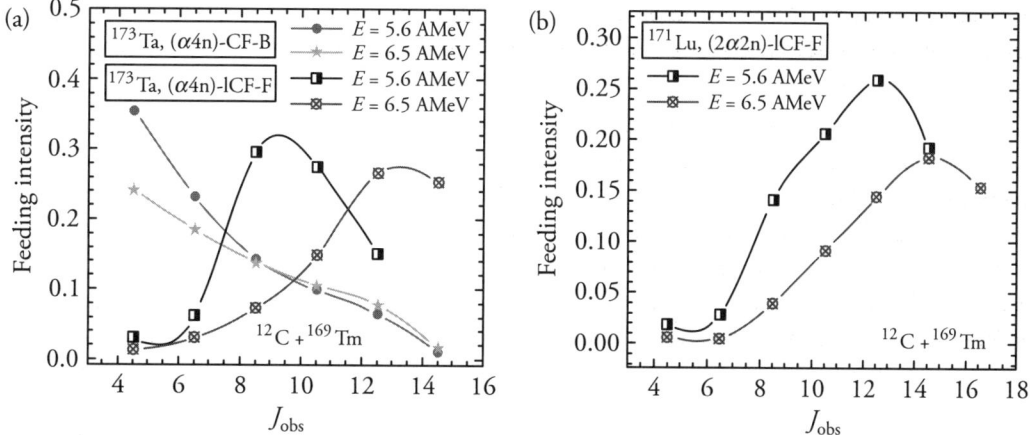

Figure 4.68 Deduced feeding intensities of γ cascades of different reaction products expected to be populated via (a) CF-α4n (identified from backward α-gated spectra), ICF-α4n (identified from backward α-gated spectra) and (b) ICF-2α2n (identified from forward α-gated spectra) channels. Lines and curves are drawn only to guide the eyes

αxn increases up to $\cong 9\,\hbar$ from the higher spin states (entry side). On the other hand, at $\approx 6.5A$ MeV, it increases only up to $14\,\hbar$. This clearly indicates that the high spin states are strongly fed even in the case of ICF channels. As the residual nucleus de-excites, the feeding intensity decreases gradually with available excitation energy and/or angular momenta, which indicates the absence of feeding to the lowest members of the Yrast band, or the fact that the low spin states are less likely to be populated in the ICF-αxn/2αxn-F channels. Such a feeding intensity profile is expected to arise from the narrow angular momentum window, localised near and/or above the critical angular momentum for CF. It is also observed that the feeding intensity is found to be less at $\approx 6.5A$ MeV, as compared to $\approx 5.6A$ MeV, due to the high angular momentum imparted into the system.

4.5 Measurement of Pre-equilibrium Component in Heavy Ion Reactions at < 10 MeV/n Energy

Both in complete fusion (CF) and incomplete fusion (ICF) of heavy ions, an excited composite nucleus is formed: in the former, by the amalgamation of the whole of the incident ion with the target; while in the latter case, by the fusion of a part of the incident ion with the target nucleus. The excited composite system undergoes relaxation to attain thermal equilibrium and become a compound nucleus (CN). In the first few steps of relaxation process, the composite system may emit light nuclear particles (LNPs) if its excitation energy is more than the particle emission threshold. These LNPs and/or light clusters emitted during the initial stages of the relaxation when the system has not yet reached the stage of thermal equilibrium are emitted predominantly in the forward direction and are called pre-equilibrium (PEQ) or pre-compound particles. Some of the important signatures of PEQ-emission are: (i) relatively large number of energetic LNPs as compared to the LNPs emitted by an equilibrated compound nucleus, (ii) slowly descending tails of the excitation functions (EFs) and (iii) forward packed angular distribution of LPNs.

Though, a large amount of data on pre-equilibrium emission has been obtained during the last couple of decades in reactions initiated by light ions, PEQ-data on heavy ion induced reactions is very limited. This is because distinction between LNPs emitted via PE-emission and via EQ-emission is difficult in HI reactions due to the large angular momenta carried by the incident heavy ion and also because of the presence of ICF along with CF in HI reactions. However, PEQ data on HI reactions has assumed greater significance in recent times because of their application in recently proposed accelerator-driven energy systems, in nuclear waste management, and in the optimization of production yields of medically important radio nuclides.

4.5.1 Measurement of pre-equilibrium component in the $^{16}O+^{169}Tm$ system

In HI reactions, LNPs are likely to be emitted from the product nuclides of both CF and ICF reaction channels. These LNPs may be further grouped into two classes – those emitted via PE process and those evaporated from the equilibrated compound nuclides. The PEQ-component of LNPs is forward peaked while LNPs emitted from the equilibrated compound nuclei have isotropic distribution. As such, excess number of particles of a given kind in the forward cone as compared to their number in the backward cone, may be assigned to the PE-process. This method has been used to study[41] the interplay of EQ and PEQ process for the system $^{16}O+^{169}Tm$ at $E/A \approx 5.6$ MeV. The forward-to-backward yield ratios $[R_{Y(F/B)}]$ for different LNPs were experimentally measured in an experiment carried out at the Inter University Accelerator Centre (IUAC), New Delhi, India. A spectroscopically pure (abundance 100%), self-supporting ^{169}Tm target of 0.93 mg/cm^2 thickness was bombarded by a ^{16}O beam of most probable charge state 7$^+$ and 5.6 MeV per nucleon energy. The beam current was kept steady at about 30 nA. The choice of the system was dictated by the well-known decay schemes of reaction products, particularly of the prompt gamma rays that were used to identify the reaction products. In the on-line experiment, both the gamma detection array (GDA) and charged particle detectors (CPDs) were used to identify the prompt gamma rays in coincidence with charged particles. The schematic diagram of the experimental setup is shown in Figure 4.69 and has been discussed in detail in Section 4.4.

As already mentioned, the array of 14 Phoswich charged particle detectors was divided into three angular zones: (i) forward (F) 10–60° (ii) sideways (S) 60–120° and (iii) backward (B) 120–170°. In order to cut off the scattered beam, the CPDs were covered by aluminium absorbers of appropriate thicknesses. The coincidences were demanded between particles of charge $Z = 1, 2$ and prompt gamma rays employing three gating conditions corresponding to given angular zones for each value of Z. Multi-parameter data was recorded in LIST mode.

The detectors used in the experiment were earlier calibrated over the entire energy range using standard sources of known strengths. In the first step of the analysis, the forward alpha-

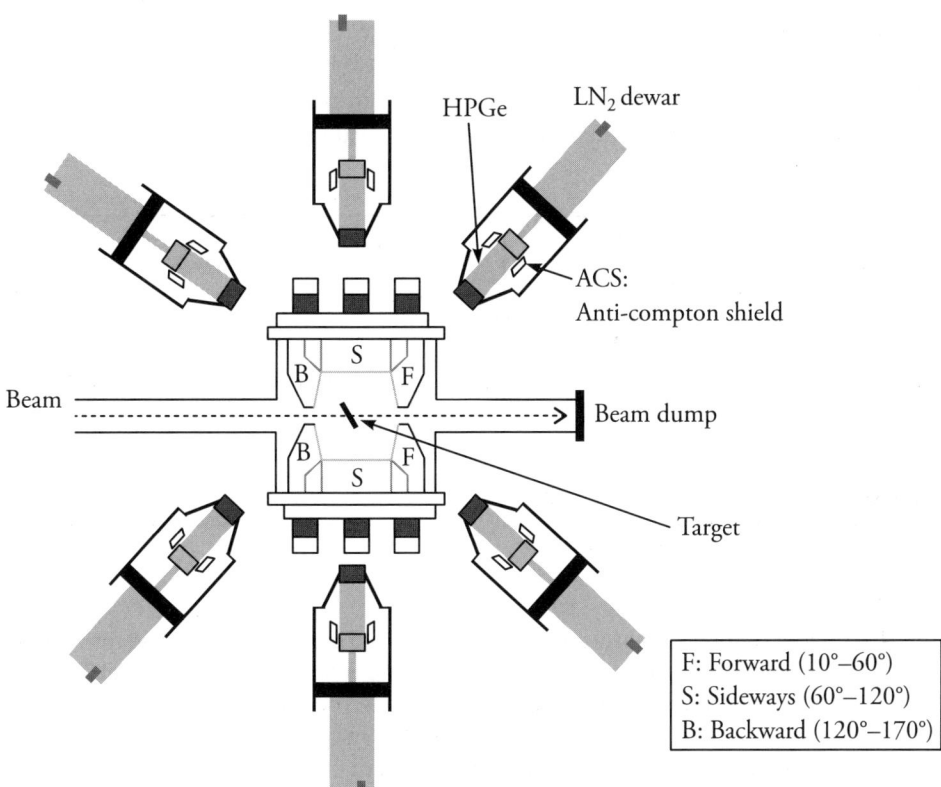

Figure 4.69 Schematic diagram of an experimental setup for the measurement of PEQ components in the system $^{16}O+^{169}Tm$

gated spectra were subtracted from the forward particle $(Z = 1, 2)$-gated spectra to obtain forward proton-gated spectra. In a similar way, the proton-gated spectra for the backward cone and for the sideways cone were also obtained. In order to improve the statistics, all gated spectra for a particular angular zone or gating condition were added up, assuming that the identified prompt gammas were isotropically distributed. The efficiency and dead-time corrected photo peak intensities and area under the peak were used to calculate the yield of reaction products.

Yield profiles of individual reaction products for each angular zone were plotted as a function of the spins of the levels of residual nuclide identified through the prompt gamma rays. Further, yields for different spin states of the given residue were normalised with respect to the yield at lowest spin and designated by Y^{nor}. For example, the normalised yield profile of levels (having different spins) of product nucleus ^{181}Os populated via p3n channel are shown in Figure 4.70(a) for the forward $\left(Y^{nor}(F)\right)$, backward $\left(Y^{nor}(B)\right)$ and sideways $\left(Y^{nor}(S)\right)$ zones. Figure 4.70(b) shows the ratios of the normalised forward to backward yield $(R_{Y(F/B)})$, forward to sideways yield $(R_{Y(F/s)})$ and sideways to backward yield $(R_{Y(S/B)})$. It is interesting to note in Figure 4.70(a) that the normalised backward and sideways yield profiles agree remarkably well with each other (within experimental uncertainties $\approx 10\%$; while there is significant enhancement in the normalised yield distribution of the forward angular zone, clearly demonstrating the presence of additional protons in the forward cone that may be assigned to pre-equilibrium emission. The same effect is also visible in Figure 4.70(b) where ratios $(R_{Y(F/B)})$ and $(R_{Y(F/S)})$ have values larger than one for all spins while the ratio $(R_{Y(S/B)})$ has a value ≈ 1.

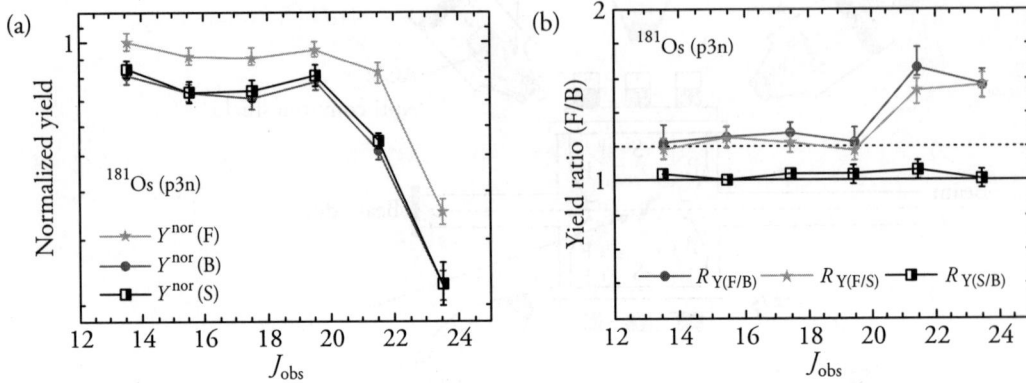

Figure 4.70 Experimentally measured (a) normalised yield profiles and (b) yield ratios for levels with different spins (J_{obs}) in ^{181}Os residue populated via p3n channel. Data points are joined with straight lines only to guide the eye. The horizontal dotted line at ratio =1.2 in Figure 4.70(b) indicates the maximum limit of forward–backward asymmetry expected from linear momentum effects

In HI reactions, the incident heavy ions carry large amount of linear momentum and therefore, the LNPs emitted from reaction products are likely to show forward peaking in the laboratory frame even if they are evaporated from an equilibrated compound nucleus. To estimate the forward–backward asymmetry in the evaporation spectra due to the linear momentum effect, the angular and energy distribution of protons emitted from a fully equilibrated compound nucleus have been generated theoretically using the statistical code PACE4. The theoretical spectra had a Gaussian shape with the peak at $\approx 10 - 12\,\mathrm{MeV}$. Further analysis of the theoretical proton spectra generated by statistical code showed that the maximum enhancement in proton yield in the forward cone as compared to the backward cone cannot be more than 20%. A horizontal dotted line at the value 1.2 has been drawn in all graphs showing yield ratios to indicate the upper limit of the forward–backward asymmetry due to input linear momentum effect. It may, however, be remarked that the experimentally observed forward–backward or forward–sideway ratios are much larger than 1.2 and cannot be explained by the linear momentum effects. The observed large ratios, more than 1.2, can be explained only by assuming that LNPs are emitted both by the compound nuclear process and also by the pre-equilibrium process.

Forward to backward yield ratios for other residues ^{183}Re (produce via 2p emission), ^{182}Re (produced via 2pn emission), ^{181}Re (produced via 2p2n emission), ^{182}W, ^{181}W, ^{179}W produced respectively through 3p, 3pn and 3p3n emission and ^{180}Ta (via 4pn channel) and ^{179}Ta (via 4p2n channel) are shown in Figure 4.71(a), (b) and (c). As may be observed in all these figures, there are significant contributions of pre-equilibrium emission in all cases. In general, it appears that pre-equilibrium emission is higher for higher spin states.

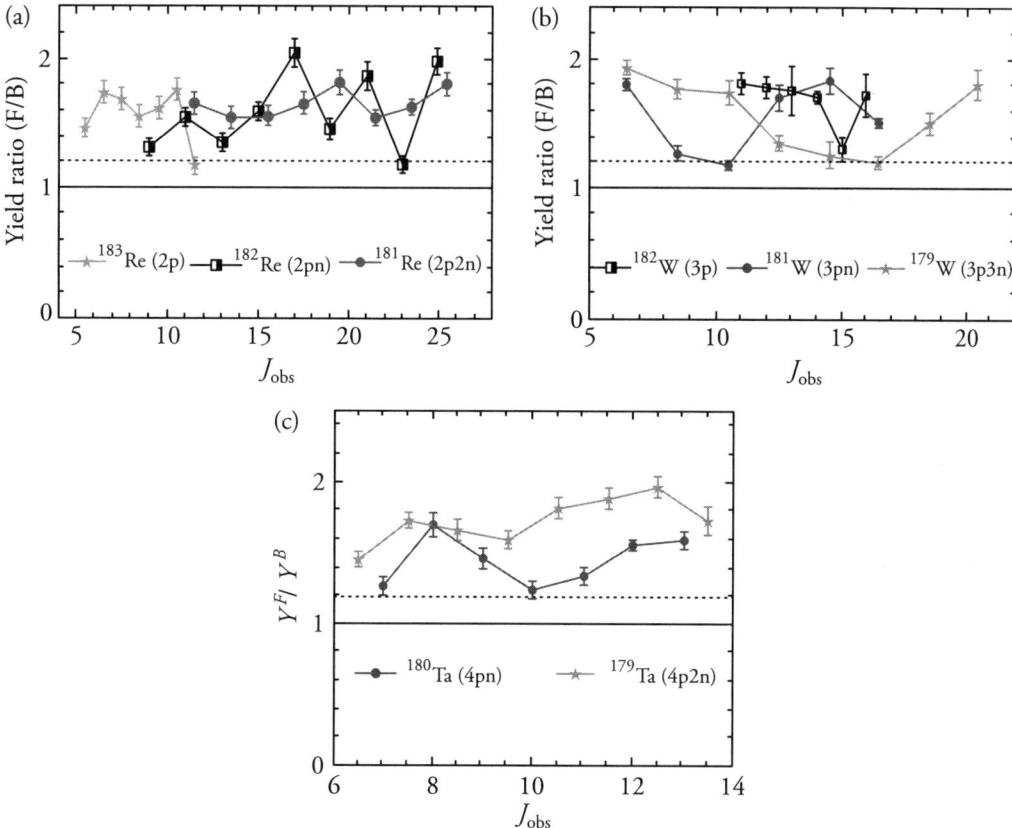

Figure 4.71 Experimentally measured yield ratios for different residues populated via 2p, 2pn, 2p2n, 3p, 3pn, 3p3n, 4pn and 4p2n channels

4.5.2 Measurement of pre-equilibrium components in $^{16}O+^{159}Tb$, $^{16}O+^{169}Tm$ and $^{16}O+^{181}Ta$ systems

The residues of the reactions $^{159}Tb(^{16}O,2n)^{173}Ta$, $^{159}Tb(^{16}O,np)^{173}Hf$, $^{159}Tb(^{16}O,3n)^{173}Ta$, $^{169}Tm(^{16}O,2n)^{183}Ir$ and $^{181}Ta(^{16}O,2n)^{195}Tl$ are all expected to be produced through the complete fusion of the incident oxygen ions with the target nuclei and therefore, should be free from contamination of incomplete fusion products. Further, all these residues are radioactive and decay with measurable half-lives emitting their characteristic gamma rays. As such, their yields may be easily determined in a given activation experiment. However, a given residue may be produced via the emission of nucleons from an equilibrated compound nucleus (CN component) or via the pre-equilibrium emission of nucleons in the first few steps of the de-excitation of the excited composite nucleus, followed by the evaporation of nucleons from an

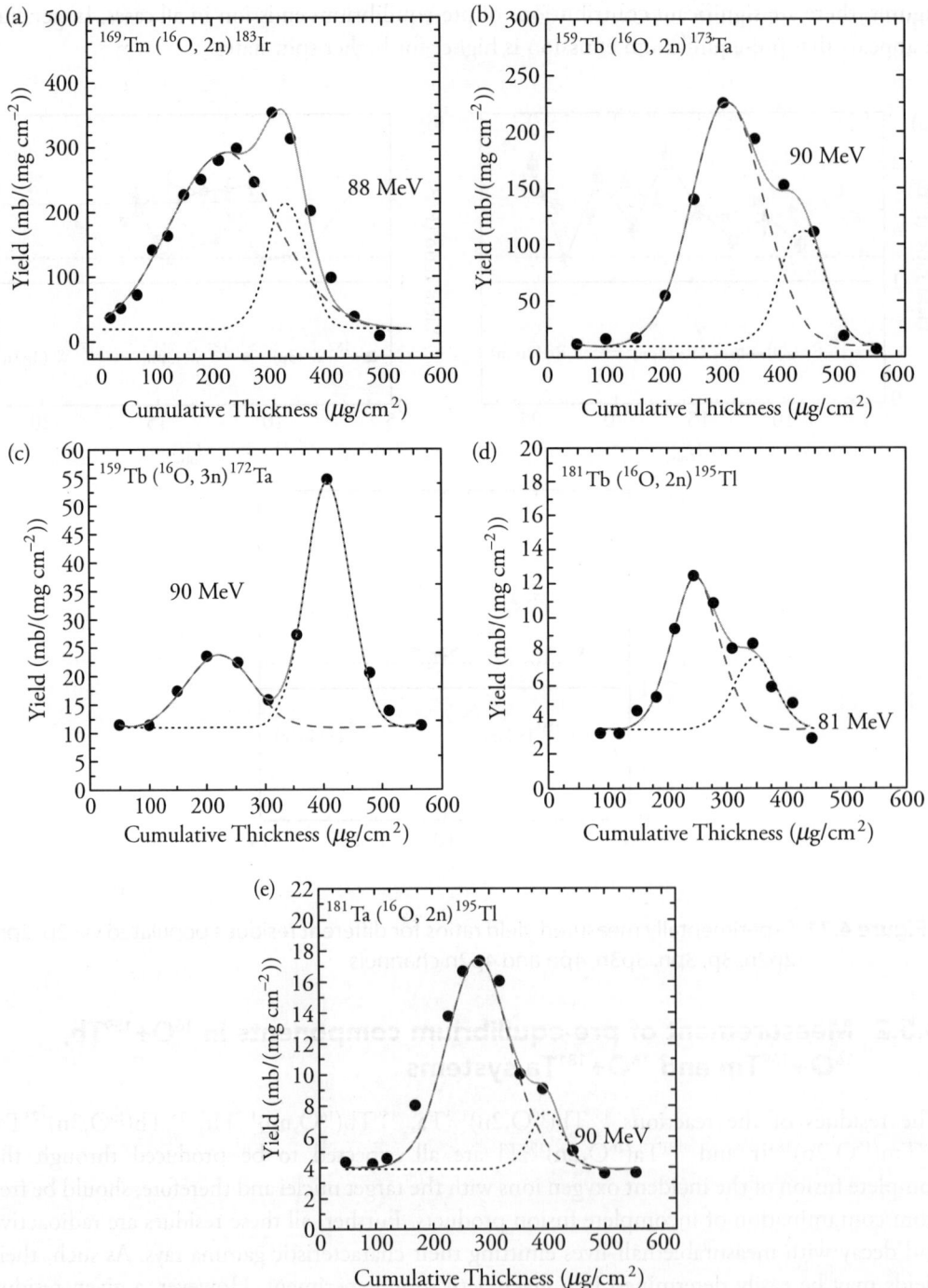

Figure 4.72 Experimentally measured recoil range distribution (RRDs) curves for different reactions. The reaction and the energy of the incident ^{16}O ion are listed in each panel

equilibrated compound nucleus. Thus, the yield of a particular residue may have two components: (i) that resulted from PE-emission and (ii) those resulting from CN evaporation. The two kinds of residues will have different values of average linear momentum. Those produced via PE-emission mode will have a smaller forward linear momentum (as a substantial part of the input linear momentum may be carried away by the pre-equilibrium nucleon) as compared to those produced via the CN evaporation route. This difference in the linear momentum may be explored by measuring the recoil range distributions of residues.

With the view to separate the residues produced via PE-emission and CN evaporation routes, recoil range distributions (RRDs) of residues in aluminium catcher foils were measured at different incident energies of oxygen ions[42]. Details of the experimental measurement of recoil ranges are already described in Section 4.2 (as well as in Chapter 3). RRD measurements were done for residues of reactions ^{169}Tm(^{16}O,2n)^{183}Ir at incident energy $\approx$ 88 MeV; ^{159}Tb(^{16}O, 2n)^{173}Ta, ^{159}Tb(^{16}O,pn)^{173}Hf, ^{159}Tb(^{16}O, 3n)^{172}Ta at incident energy $\approx$ 90 MeV and ^{181}Ta(^{16}O, 2n)^{195}Tl at incident energies $\approx$ 81 MeV, $\approx$ 90 MeV and $\approx$ 96 MeV. Typical measured RRD curves for the residues of reactions ^{169}Tm(^{16}O, 2n)^{183}Ir at incident energy $\approx$ 88 MeV, ^{159}Tb(^{16}O,2n)^{173}Ta, ^{159}Tb(^{16}O,3n)^{172}Ta at energy $\approx$ 90 MeV, ^{181}Ta(^{16}O,2n)^{195}Tl at incident energy 81 MeV and at incident energy 96 MeV are shown in Figures 4.72(a)–(e).

A common feature of all these experimental RRD curves is that they have structures that may be resolved into two components peaking at different values of cumulative thickness. The peak at higher cumulative thickness may be assigned to the CN evaporation residues while the peak at the lower cumulative thickness may be attributed to the residues populated via PE-emission route. Relative contribution of PE and CN components may be estimated from the areas lying under the two peaks. Range for the CN component where an equilibrated compound nucleus is formed and the total incident linear momentum is transferred to the compound nucleus, may be calculated theoretically using energy range relationships. The experimentally measured range for the CN component in each of the measured case is found to be consistent with the value calculated theoretically. The relative strengths (in %) of PE and CN components deduced from the analysis of RRD data is tabulated in Table 4.27.

Table 4.27 Relative strengths of PE and CN components deduced from the analysis of RRD data for different reactions at different incident energies

Reaction	Incident energy (MeV)	Relative strength of PE-component (%)	Relative strength of CN-component (%)
^{169}Tm(^{16}O,2n)	$\approx$ 88	$\approx$ 80	$\approx$ 20
^{159}Tb(^{16}O, 2n)	$\approx$ 90	$\approx$ 70	$\approx$ 30
^{159}Tb(^{16}O, pn)	$\approx$ 90	$\approx$ 73	$\approx$ 27
^{159}Tb(^{16}O, 3n)	$\approx$ 90	$\approx$ 35	$\approx$ 65
^{181}Ta(^{16}O,2n)	$\approx$ 81	$\approx$ 68	$\approx$ 32
^{181}Ta(^{16}O,2n)	$\approx$ 90	$\approx$ 79	$\approx$ 21
^{181}Ta(^{16}O,2n)	$\approx$ 96	$\approx$ 88	$\approx$ 12

In order to confirm the findings on PE and CN components obtained from the RRD measurements, in an auxiliary experiment, the excitation functions (EFs) for the aforementioned reactions were measured using the stacked foil activation technique, described in Section 4.2. The measured EFs for some typical cases are presented in Figure 4.73(a)–(d).

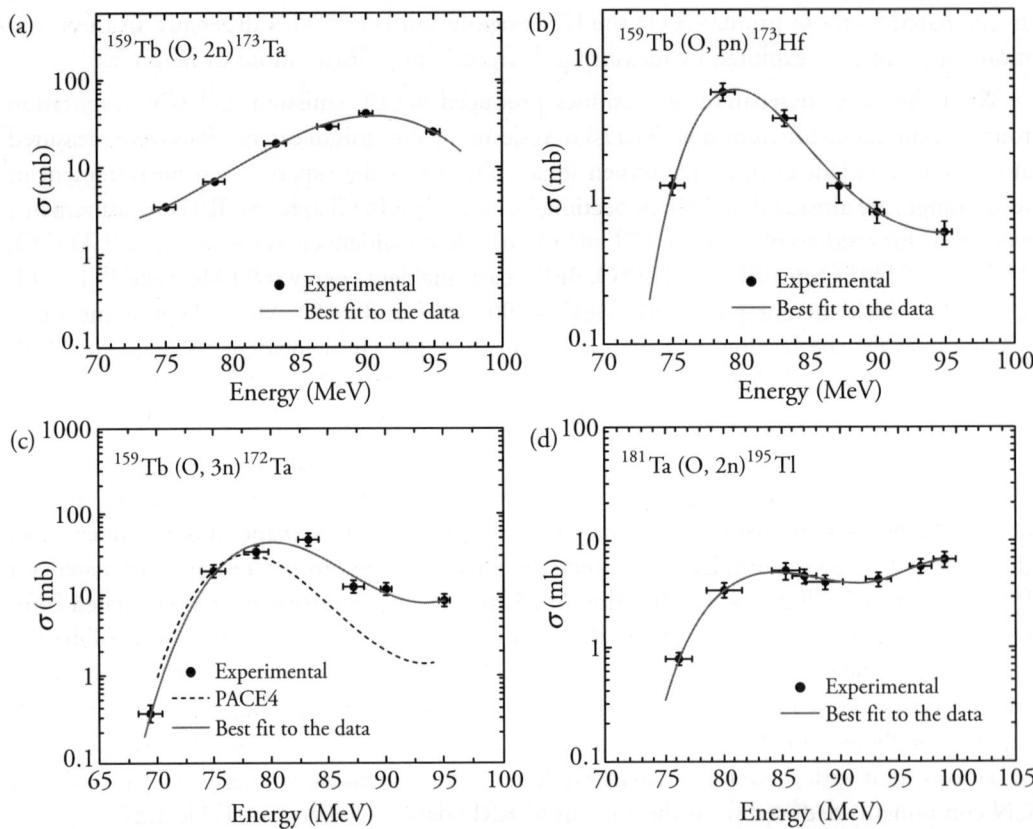

Figure 4.73 Measured excitation functions (EFs) for reactions specified in each panel

Excitation functions, calculated theoretically using the statistical code PACE4 for the reaction ^{159}Tb(^{16}O,3n) is also shown in Figure 4.73(c). However, for other reactions, the theoretical values were negligibly small, and therefore, not shown in figures. Enhancement of experimental excitation function(s) over the corresponding theoretical EFs has been assigned to the PE component. Analysis of the EF for the reaction ^{159}Tb(^{16}O,3n) gives a PE contribution of $\approx 35\%$ and a CN contribution of $\approx 65\%$ at incident energy ≈ 88 MeV. These values of relative strengths of PE and CN contributions are consistent with the corresponding values obtained from RRD data at the specified incident energy. These findings on PE and CN contributions are further supported by the measurement of spin distributions of reaction residues that show that two different reaction mechanisms populate residues emitted in the forward cone (predominantly PE residues) and in the backward cone (predominantly CN residues).

Results and Conclusions

5.1 Incomplete Fusion below 10 MeV/A Energy and its Dependence on Entrance Channel Parameters

The study of incomplete fusion (ICF) reactions induced by heavy ions (HI) has been a topic of resurgent interest because of its presence in considerable strength at energies as low as from Coulomb barrier to well above it[1-9]. At these energies, it is expected that only the complete fusion (CF) process will contribute to the total fusion cross-section[4-6]. In case of CF, at projectile energies higher than the Coulomb barrier (V_b), having input angular momentum $\ell < \ell_{crit}$ (where ℓ_{crit} is the critical angular momentum of the system), the entire projectile is likely to fuse with the target nucleus. As a result, an excited composite system is formed that undergoes equilibration to become the compound nucleus (CN), which has a predetermined mass, charge and angular momentum. The CN so formed may decay by the emission of neutron, proton, alpha particles and light clusters followed by the emission of characteristic gamma radiations. At relatively higher projectile energies, for which the $\ell > \ell_{crit}$, the fusion of the complete projectile is hindered from sustaining the excess input angular momentum associated with the system. As a result, the projectile breaks up into fragments; one of the fragments may fuse with the target nucleus forming the so-called incompletely fused composite system, with the remnant moving in the forward direction with the same velocity as that of the incident projectile. Projectile-like fragments in massive transfer reactions were reported for the first time by Britt and Quinton[10] and were also observed later by Wolfgang and Kauffmann[11]. Since then considerable efforts have been made by different research groups to explain the reaction dynamics of such partial fusion processes. Experiments involving particle–gamma coincidences carried out by Inamura et al.[12-13] and Zolnowski et al.[14] contributed a great deal to the understanding of ICF reaction dynamics. It was suggested by Geoffery et al.[15] that ICF reactions are produced in non-central interactions. Trautmann et al.[16] and Inamura et al.[17] also emphasized the non-central nature of such reactions. The localization of the ℓ window in ICF reactions were reported in the spin distribution measurements[18–20] presented in Chapter 4 of this monograph.

In order to explain the dynamics of ICF reactions, several theoretical models have been put forward; some of these are the (i) breakup fusion model,[21] (ii) the exciton model,[22] (iii) the SUM RULE model,[23–24] (iv) the hot spot model,[25] (v) the promptly emitted particles (PEP) model[26] etc. A brief description of some of these models is already presented in Chapter 2 of this monograph. It may, however, be remarked that the aforementioned models could explain the ICF reaction data only to a limited extent and that too at energies larger than 10 MeV/ nucleon. Moreover, at present, there is no theoretical model that can explain the low energy incomplete fusion reaction data satisfactorily. In view of the presence of the ICF component in sufficient strength at energies as low as $\approx$ 4–7 MeV/nucleon and also due to the unavailability of any reliable theoretical model for low energy ICF reactions, the study of incomplete fusion in HI reactions at low energies is still interesting and an open area of investigation. At present, efforts are being made to generate large amount of experimental data on different aspects of ICF process with the view to understand the ICF reaction dynamics and to develop a viable theoretical model that can explain the experimental data on low energy ICF.

In the present report, an attempt has been made to study the effect of the entrance channel parameters on the strength of ICF component. Some of the important entrance channel parameters chosen for this study are the following: (i) the kinetic energy of the projectile, (ii) the projectile structure, (iii) the alpha Q value of the projectile, i.e., the binding energy of alpha particle in the projectile, (iv) mass asymmetry and (v) the Coulomb energy of the interacting partners. The conclusions drawn here are based on the analysis of the experimental data on excitation functions, recoil range distributions, angular distributions, spin distribution and the feeding profiles of reaction residues described in Chapter 4. However, some experimental data from literature has also been used for the study of systematic trends.

As mentioned in Chapter 4, for a large number of target–projectile pairs, ICF strengths have been deduced from complimentary experiments on excitation functions, recoil range distributions and angular distributions[27–28] etc. It is worth mentioning that in most of the cases, the results obtained from different measurements agree well with each other within experimental uncertainties. A good agreement in the results of complimentary experiments indicates the reliability of the data and the analysis technique. Further, it may also be mentioned that the ICF contributions deduced from the measured recoil range distributions and from the angular distribution measurements are totally independent of any theoretical model for the reaction. Recoil range distribution arises because of the full linear momentum transfer in the case of complete fusion (CF) and partial linear momentum transfer in the case of ICF. Similarly, the angular distribution of reaction residues depends on the mass of the fragment fused with the target. However, the ICF strength deduced from the analysis of the excitation function depends on the statistical model. A good agreement in the values of the relative strength of the ICF components deduced from model independent measurements and model dependent measurements shows that the parameters used for the statistical model calculations are properly chosen. For example, as discussed in Chapter 4, the αxn channels may be populated via both the CF as well as the ICF processes. In the measurement of recoil range and angular distributions, the CF contributions in the production of αxn channels are satisfactorily reproduced via PACE4 predictions done with the same set of parameters used to calculate the cross-sections for the xn and pxn channels. The reproduction of the CF and ICF contributions in the alpha-emitting channels clearly validates the choice of parameters used for statistical calculations and instils confidence in the methodology adopted for analysis.

To compare the strength of the complete and incomplete fusion components, a quantity referred to as the ICF strength function (F_{ICF}) is often used. The F_{ICF} is a measure of the strength of the ICF relative to the total fusion cross-section. It is defined as

$$F_{ICF}(\%) = \left(\frac{\sum \sigma_{ICF}}{\sigma_{Total\ Fusion}} \right) \times 100; \quad \sigma_{Total\ Fusion} = \sum \sigma_{CF} + \sum \sigma_{ICF}$$

Further, to compare the ICF components for the same system obtained by different experiments and also for different systems, one defines the normalised (or reduced) beam energy E_{Norm} as

$$E_{Norm} = \frac{\text{Incident energy of the projectile } E_{beam}}{\text{Coulomb barrier (CB) between the interacting ions}}$$

The use of normalised beam energy washes out the effect of different Coulomb barriers for different systems. Figure 5.1 shows a typical case for the system $^{16}O+^{169}Tm$, where F_{ICF} has been deduced from the analysis of the forward recoil range distributions (FRRDs), the angular

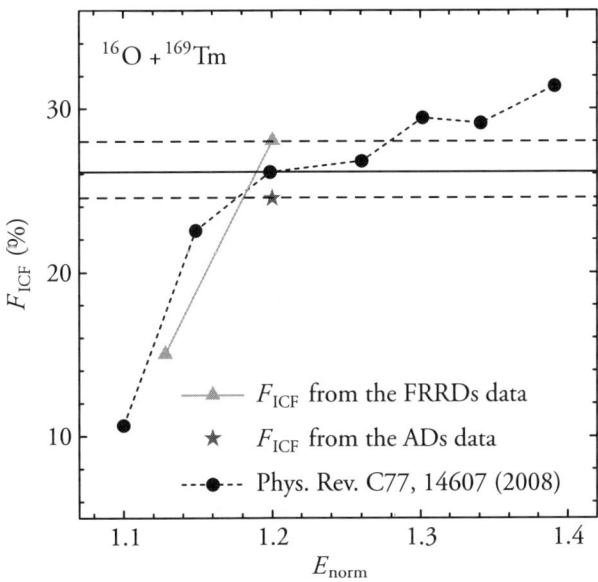

Figure 5.1   The incomplete fusion fraction deduced from the analysis of excitation functions, forward recoil range distribution (FRRDs) and angular distributions (ADs) data, as a function of normalised beam energy. The line drawn through the data points is just to guide the eyes

distribution of reaction residues (ADs) and the excitation functions at the same normalised energy F_{Norm} = 1.2. As may be observed from the figure, the percent F_{ICF} values from three independent measurements agree[27] within 10% (indicated by dotted horizontal lines).

The aforementioned agreement between the three independent measurements as described in Chapter 4, strengthens confidence in the analysis procedure.

In the following sections, the dependence of ICF fractions on various entrance channel parameters is presented.

5.1.1 Dependence of ICF on projectile structure and incident energy

In order to study the dependence of ICF reaction dynamics on the projectile structure, the experimental F_{ICF} (%) values for systems having the same target (^{159}Tb) but different projectiles (^{12}C, ^{13}C, ^{16}O and ^{19}F) are plotted as a function of normalised beam energy E_{Norm} in Figure 5.2. As already mentioned, the normalised beam energy is used to wash out the effect of the Coulomb barrier that has different values for different systems.

Figure 5.2 demonstrates three facts: (i) At the same normalised energy, different systems have different values for the incomplete fusion strength function F_{ICF}. Since the target nucleus is the same (^{159}Tb) for each system, this clearly demonstrates that the ICF reaction dynamics strongly depends on the projectile structure. (ii) For each system, F_{ICF} increases with the increase in the normalised energy but with different rates. This again demonstrates strong dependence of ICF on the nature of the incident ion as well as on the incident energy. (ii) The threshold value of the normalised energy at which ICF appears for the first time is different for different systems. The ICF sets in at the lowest normalised energy for projectile ^{19}F followed by ^{12}C, ^{16}O and ^{13}C.

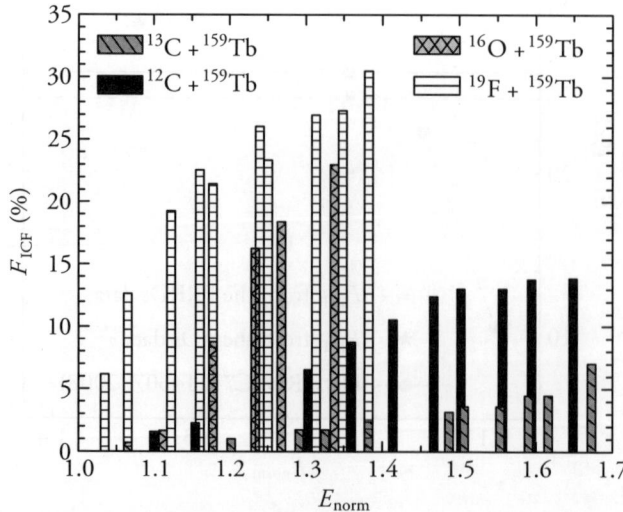

Figure 5.2 Comparison of experimental incomplete fusion strength functions for different projectiles on the same target. The normalised energy is plotted on the *x*-axis, to wash out the Coulomb effect

Dependence of F_{ICF} on the structure of the incident nucleus becomes still more evident in Figure 5.3, where experimental values for the incomplete fusion strength function F_{ICF} are plotted as a function of the normalised energy (E_{Norm}) for the two systems (^{12}C+ ^{159}Tb) and (^{13}C+^{159}Tb). Since the target nucleus for the two systems is same, the large difference in ICF strength functions for the two systems is entirely due to the difference in the structures of the incident projectiles ^{12}C and ^{13}C. The F_{ICF} for ^{12}C projectile, that has alpha cluster structure, has a larger value at each value of normalised energy as compared to the corresponding F_{ICF} value for (non-alpha cluster structure) ^{13}C projectile. For example, as may be seen from this figure, at the same normalised energy $(E_{Norm}=1.6)$ the value of ICF fraction F_{ICF}, for ^{13}C projectile (non-α cluster nucleus) is $\approx 4\%$, while for ^{12}C projectile (α cluster nucleus) it is around 14%.

This striking difference in the values of ICF fractions at the same normalised energy, as well as, over the entire range of energy, for ^{12}C and ^{13}C induced reactions on the same target ^{159}Tb, supports strong projectile structure dependence of ICF process. It may further be remarked that ^{13}C has only one neutron in excess to that ^{12}C, but since ^{12}C has an alpha clustered structure, it has a much larger component of ICF. Moreover, in the case of a ^{12}C projectile, the ICF shows up at a much lower value of the normalised energy as compared to the case of the ^{13}C projectile.

As may be observed in Figures 5.2 and 5.3, the ICF strength function, F_{ICF}, for all the presently studied systems increases with increase in the incident energy of the projectile. This is expected, as the increase in the incident energy of the projectile imparts larger angular momentum to the system, which leads to the flattening of the fusion pocket in the effective potential energy curve. In order to restore the fusion pocket and to provide the sustainable angular momentum, the incident heavy ion may break into constituent clusters leading to the incomplete fusion. Since the ^{12}C nucleus has a well-known alpha cluster structure, it may break into 3 alpha particles or into $(\alpha + {}_{4}^{8}\text{Be})$. Depending upon the conditions imposed by the input

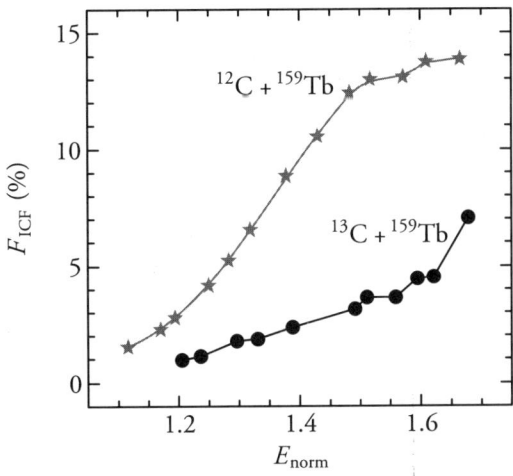

Figure 5.3 Comparison of F_{ICF} values for the two systems ($^{12}_{6}$C+^{159}Tb) and (^{13}C+^{159}Tb)

angular momentum, one of the fragments may fuse with the target nucleus giving rise to the formation of an incompletely fused composite system, also called the partially fused system.

In the case of the projectiles ^{12}C and ^{13}C, it was observed that the ICF strength function (F_{ICF}) has significantly larger values for ^{12}C as compared to ^{13}C at all values of normalised energies. This may lead to the conclusion that ICF strength strongly depends on whether the projectile has cluster structure or not. However, ^{19}F is a non-α cluster projectile and the values of the F_{ICF} for this projectile at corresponding energies are larger than for well-know alpha clustered projectiles like ^{16}O and ^{12}C (see Figure 5.2). Moreover, the ICF threshold energy, the minimum value of the normalised energy at which ICF shows up for the first time, is much lower in the case of projectile ^{19}F than for other cluster structured or non-cluster structured projectiles shown in Figure 5.2. These facts, as observed in Figures 5.2 and 5.3, points to the fact that some entrance channel parameter, other than the cluster nature of the projectile and the normalised energy, also plays a significant role in ICF dynamics.

5.1.2 Dependence of ICF on mass asymmetry

Another entrance channel parameter that may influence the ICF strength may be the mass asymmetry, i.e., the difference in the masses of the target and the projectile. To parameterize mass asymmetry, one defines mass asymmetry parameter μ, as follows;

$$\mu = \frac{A_T}{\left(A_T + A_P\right)}$$

where A_T and A_P are respectively the atomic mass numbers of the target and the projectile nuclei.

Morgenstern et al.[30] indicated that the probability of incomplete fusion depends strongly on the mass asymmetry of the interacting partners. The experimental data of the present set of experiments has been analysed in line with Mongenstern's mass asymmetry systematics[30]. The values of F_{ICF} for different systems have been compared[29] as a function of entrance channel mass asymmetry parameter at constant normalised beam energy (E_{Norm}) in Figure 5.4.

It may be noted that for some projectiles, the data is limited to a few target nuclei only. As may be observed in Figure 5.4, the ICF strength functions increase almost linearly with the increase in the mass asymmetry, separately for each projectile. The magnitudes of F_{ICF} for systems that have nearly the same mass asymmetry but different projectiles are quite different. This is not in accordance with the mass asymmetry systematics proposed by Morgenstern et al.[30] Morgenstern in his analysis had concluded that the ICF strength depends only on the degree of entrance channel mass asymmetry. However, the present analysis does not support Morgenstern's conclusion.

Figure 5.4 also shows that the rates of increase in the magnitude of F_{ICF} with the asymmetry parameter μ have different values for different projectiles. This further supports the conclusion of the present analysis that incident energy, the projectile structure as well as the entrance channel mass asymmetry contribute significantly to the ICF reaction dynamics at these low

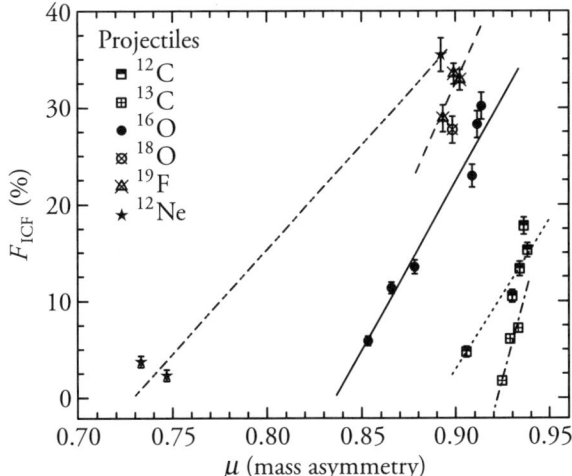

Figure 5.4 Comparison of F_{ICF} values for various systems as a function of the mass asymmetry parameter of interacting partners at a constant normalised energy. The lines are drawn to get the linear fit to the data separately for each projectile

energies of interest. Further, it may also be concluded that present results indicate a projectile dependent mass asymmetry systematics.[29]

5.1.3 Dependence of ICF on α Q value of the projectile

In the process of incomplete fusion (ICF), it is assumed that the projectile nucleus breaks up into fragments in the nuclear field of the target nucleus and a fragment of the incident nucleus fuses with the target to form a partially fused composite system. If so, the projectile with less binding energy is likely to show larger component of ICF. However, in Figure 5.4, the projectile ^{19}F with a larger magnitude of binding energy as compared to ^{12}C, ^{13}C and ^{16}O, show higher magnitudes of ICF fractions at the same normalised energy (E_{Lab}/V_{CB}) than that for carbon and oxygen projectiles. Therefore, the binding energy of the projectile does not seem to play any significant role in ICF.

Let us now turn to the projectile alpha Q value (Q_α), defined as the energy required to remove an alpha particle from the projectile. Alpha Q values for projectiles ^{19}F, ^{18}O, ^{16}O, ^{13}C and ^{12}C are respectively -4.01 MeV, -6.22 MeV, -7.16 MeV, -10.64 MeV and -7.16 MeV. The experimental values of the incomplete fusion strength function F_{ICF} (%) for the aforementioned projectiles and the same target ^{159}Tb at the fixed value of the normalised energy are plotted in Figure 5.5. It is evident from this figure, that the ICF strength function has an inverse correlation with the Q_α value; smaller the magnitude (negative) of the alpha Q value of the projectile, larger is the magnitude of the incomplete fusion strength function. ^{19}F has the smallest alpha Q value (-4.01 MeV) and largest ($\approx 29\%$) magnitude of F_{ICF}, while ^{13}C with the largest magnitude of Q_α (-10.64 MeV) has the least magnitude of F_{ICF} ($\approx 2\%$). The other projectiles show the same trend.

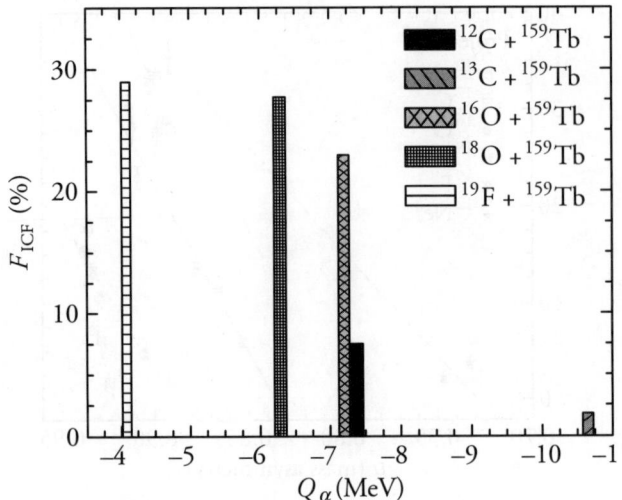

Figure 5.5   A comparison of incomplete fusion fraction F_{ICF} in terms of the α Q value of the projectile at a constant E_{norm} value for different projectiles on the same target, ^{159}Tb

To further generalise the Q_α dependence of F_{ICF}, data for different projectiles on two different targets ^{181}Ta and ^{169}Tm is plotted respectively in Figure 5.6. As may be observed in these plots, the incomplete fusion strength function decreases with the increase in the (negative) Q_α magnitude.

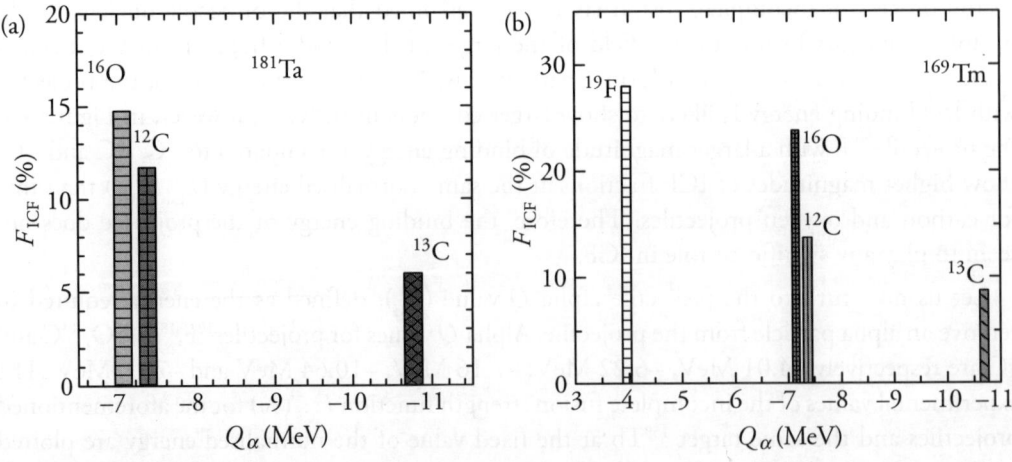

Figure 5.6   Incomplete fusion strength function F_{ICF}(%) for different projectiles on targets, ^{181}Ta and ^{169}Tm, as a function of the α Q value of the projectile at a fixed value of the normalised energy (E_{Norm})

The decrease in the incomplete fusion strength function with increase in the (negative) magnitude of the alpha Q value of the projectile may be understood in terms of the breakup probability of the projectile. A projectile that has a large (negative) magnitude of Q_α has small probability to break into alpha fragments and hence, has small probability for incomplete fusion.

5.1.4 Dependence of ICF on the Coulomb factor ($Z_p.Z_T$)

Incomplete fusion of heavy ions may be considered as a two-step process – the projectile ion breaking into fragments is the first step and one or more fragments of the incident ion fusing with the target is the second. If so, the Coulomb interaction between the two ions may play a role in fragmentation of the projectile. Since the strength of the Coulomb interaction between the projectile of charge Z_p and the target of charge Z_T is proportional to ($Z_p.Z_T$), it is interesting to investigate the ($Z_p.Z_T$), called Coulomb factor, dependence of the incomplete fusion strength function F_{ICF}.

Incomplete fusion strength function F_{ICF} (in %) for different combinations[31] of projectile–target nuclei at a fixed value of the normalised energy are plotted as a function of Coulomb factor ($Z_p.Z_T$) in Figure 5.7. As may be seen in the figure, incomplete fusion strength function increases, almost linearly, with the increase of the Coulomb factor. This suggests that for two different projectile–target combinations with Coulomb factors $\left(Z_{P_1}Z_{T_1}\right)$ and $\left(Z_{P_2}Z_{T_2}\right)$, the incomplete fusion strength functions should have almost the same value if $\left(Z_{P_1}Z_{T_1}\right) \approx \left(Z_{P_2}Z_{T_2}\right)$. However, in a recent research,[29] it was found that this does not hold. As such, it may be concluded that the Coulomb factor alone cannot account of the incomplete fusion of heavy ions at energies below 10 MeV/A.

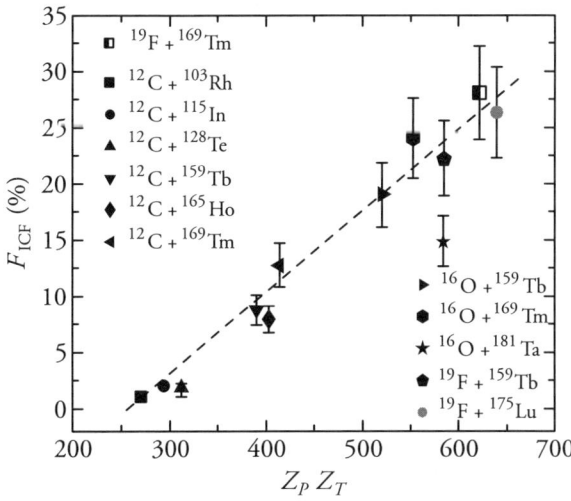

Figure 5.7 The ICF strength function F_{ICF} of various systems as a function of $Z_p.Z_T$. The dashed line is drawn to guide the eye, indicating a linear fitting

5.1.5 Angular momentum (ℓ) distribution and mean input angular momentum for complete and incomplete fusion reactions

As is evident from the total potential energy curve (Coulomb + Nuclear + Centrifugal) for heavy ions, the orbital angular momentum ℓ of the system plays an important role in the fusion of the ions. The attractive pocket in the total potential gets shallower with the increase of the angular momentum and vanishes for a particular value of the angular momentum, called the critical angular momentum and denoted by ℓ_{crit}. Critical angular momentum has different values for different projectile–target systems. For example, in the case of projectile ^{19}F impinging on target ^{169}Tm, the value of ℓ_{crit} is $\approx 64\hbar$. On the other hand, the value of the maximum angular momentum ℓ_{max} (defined as the maximum value of ℓ for which the colliding nuclei penetrate into the region where the total nucleus–nucleus potential is attractive or the distance of closest approach is smaller than the sum of the half density radii) and the cross-section σ_ℓ for each value of the angular momentum may be calculated using the semi-classical approach for a given projectile–target pair if the projectile energy is known. Fusion angular momentum distributions for the system (^{19}F+^{169}Tm) over a broad energy range between 80 MeV and 105 MeV have been calculated using the computer code CCFULL[33] and are shown in Figure 5.8. Since the value of the critical angular momentum ℓ_{crit} for the system (^{19}F+^{169}Tm) is around $64\,\hbar$, and the value of the maximum angular momentum ℓ_{max} for the same system up to the incident energy of 105 MeV is less than $64\,\hbar$, all ℓ bins below ℓ_{crit} contribute to the complete fusion of the ions. However, at still higher energies, where ℓ_{max} values exceed ℓ_{crit}, and incomplete fusion may occur, a window of ℓ bins, around ℓ_{crit} may feed to the incomplete fusion. It is interesting to note that in the present work, significant ICF fractions have been observed, indicating that partial waves with ℓ below ℓ_{crit} may contribute to the ICF processes.

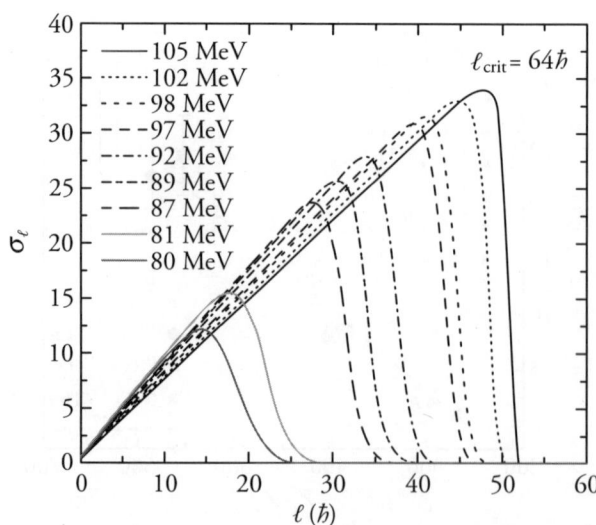

Figure 5.8   The fusion ℓ distribution calculated using the code CCFULL to get the population of ℓ bins at different values of energies

Similar observations have been found for all the other systems studied in the present set of experiments at energies below 10 MeV/A. The present results, therefore, suggests a diffused boundary for the ℓ values, contrary to the sharp cut-off model that may penetrate even close to the barrier.

The aforementioned conclusions are supported by our experiments on the measurements of spin distributions. In these experiments, described in detail in Chapter 4, the observed value of the mean input angular momentum (observed angular momentum value for which the yield becomes half the maximum value at the lowest spin) have been deduced for xn and αxn and 2αxn modes of decay. Typical mean input angular momentum values for the system (^{16}O+^{169}Tm) at E/A = 5.6 MeV for the three modes of decay are plotted in Figure 5.9. As can be seen from this figure, the mean input angular momentum increases with alpha multiplicity; moreover, the mean input angular momenta <ℓ> is larger for ICF reactions (αxn and 2αxn) than for CF reactions (xn). For example, as may be observed in Figure 5.9, the mean angular momentum for CF (corresponding to the xn exit channel) is ≈ 10ℏ, while for the ICF channels αxn and 2αxn it has values ≈ 13ℏ and ≈ 16ℏ, respectively. Larger values for the mean angular momentum for ICF channels imply the peripheral nature of the interaction, i.e., interaction takes place at increasingly larger distances from the centre with increasing alpha multiplicity.

Summing up, it may be concluded that the incomplete fusion dynamics depends strongly on the incident projectile energy and other entrance channel parameters like projectile structure, mass asymmetry, alpha Q value for the projectile, Coulomb factor and the input angular momentum in a complex way. As has been discussed earlier, the incomplete fusion strength function strongly depends on the alpha Q value of the projectile and is not so sensitive to the binding energy of the projectile. Further, the target–projectile mass asymmetry does play an important role in ICF dynamics. It has been observed that the ICF strength function for a given projectile increases almost linearly with the mass asymmetry. However, the rate of increase and the magnitudes of F_{ICF} have different values for different projectiles. The present

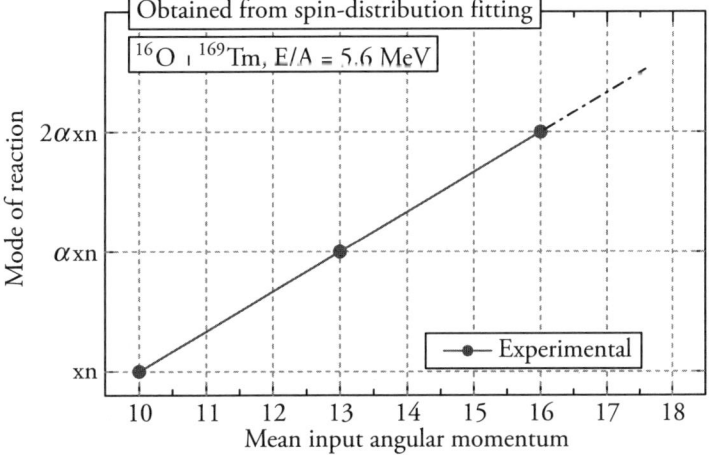

Figure 5.9 Experimentally obtained mean values for angular momentum as a function of the complete and incomplete fusion modes for the system ^{16}O+^{169}Tm at 5.6 MeV/A

analysis also shows a strong dependence of incomplete fusion strength function on the Coulomb factor as discussed earlier. Measurement of the spin distributions in residual nuclei reveals that, while in the case of complete fusion, all values of angular momentum below the critical angular momentum contribute to the fusion, angular momenta lying in a narrow window in the angular momentum space feed to the ICF process. It is expected that the results and the analysis presented here may help in developing a theoretical model for low energy incomplete fusion reactions. Further, more data at low energies (< 10 MeV/nucleon) covering the low to high mass region of the target nuclei with a variety of projectiles is required to disentangle the effects of various entrance channel parameters.

5.2   Pre-equilibrium Emission in Heavy Ion Reactions at Energies < 10 MeV/A

The compound reaction mechanism proposed by Bohr[34], some 70 years back, is still the most frequently used theoretical approach to describe all types of binary nuclear reactions, particularly at low and intermediate energies. The compound nuclear mechanism assumes that a nuclear reaction proceeds in two steps: in step I, the projectile nucleus fuses with the target nucleus forming an excited composite system that remains intact till a thermodynamic equilibration is established. Bohr called the composite nucleus in thermal equilibrium the compound nucleus (CN). In the Bohr mechanism, it is inherently assumed that the excited composite nucleus does not decay during the time it undergoes equilibration. In step II of the reaction, the compound nucleus decays through the possible exit channels. An essential assumption of the Bohr mechanism is the independence hypothesis which makes step I and step II independent of each other.

Sufficient experimental evidences accumulated during the last thirty years or so have demonstrated that the excited composite system formed initially on the amalgamation of the projectile with the target does not stay intact but decay by the emission of light ions and clusters during equilibration and before becoming the compound nucleus. Emission of nucleons and/or light clusters during the process of equilibration and before the establishment of thermodynamic equilibrium is termed as pre-equilibrium (PE) or pre-compound (PCN) emission. Several semi-classical and at least one totally quantum mechanical theories, some of them discussed in Chapter 2, have been proposed to describe pre-equilibrium emission in statistical nuclear reactions. Extending the statistical framework to include heavy ion interactions at low and moderate energies, efforts have been made to investigate pre-equilibrium emission in such reactions. Since the time scales for (PE) and (CN) emissions (respectively $\approx 10^{-21}$ s and 10^{-16} s) are beyond the present limits of experimental observations, PE products are mostly identified through some other characteristic, like predominant emission in the forward direction and larger number of emitted particles/clusters with higher energies than what is expected from CN emission etc. In the case of HI reactions, the problem of identifying PE products becomes more involved because of the presence of ICF residues, which the present set of experiments have proved to be present in considerable strength at energies as low as below 10 MeV/A.

In light of the aforementioned discussion, PE contributions for those reaction channels that are fed only through complete fusion, like (HI, xn), (HI, xp) and (HI, xpyn) etc., (where x and y may vary from 1 to 5) have been investigated for extracting the PE component of the emitted nucleons. In a set of experiments, the excitation functions for the production of radioactive residues populated via complete fusion of the projectile with the target, have been experimentally measured. The stacked foil activation technique, described in chapter 3, has been used for these measurements. In this technique, different targets of the stack are irradiated with incident ions of different energies, making it possible to generate the excitation functions for several reactions over a given energy range in a single irradiation. The measured excitation functions are then compared with their counterparts, theoretically calculated using computer codes based either only on CN mechanism or on a mixture of CN and PE mechanisms. Subtracting the contribution of pure CN emission, calculated theoretically, from the experimentally measured excitation function, one may extract the energy dependent PE contribution. However, PE contribution may also be extracted by adjusting the free parameters of the code based on the mixture of PE+CN so as to reproduce the measured excitation function. In a recent publication,[35] as detailed in Chapter 4, the PE contribution for some exit channels fed by complete fusion of heavy ions have been deduced from the measured excitation functions of radioactive residues. In a complementary experiment, forward recoil ranges (RRD) of the same radioactive residues were measured in an appropriate absorbing material (aluminium or carbon) that showed two distinct peaks, one corresponding to the CN emission and the other corresponding to the residues from which PE emission has taken place. Relative contributions of PE components were deduced from the areas lying under the two peaks, as detailed in Chapter 4. It is satisfying to note that the relative contributions of CN and PE components obtained from the analysis of excitation functions and recoil range distributions agree with each other within experimental errors.

The percent PE component for the reaction ^{181}Ta (^{16}O, 2n) obtained from the analysis of recoil range distributions as a function of energy is shown in Figure 5.10(a). As expected, the PE component increases with the incident energy. Figure 5.10(b) shows the percentage of PE components in the system ^{16}O+^{159}Tb respectively for the emission of one neutron and one proton (pn), two neutrons (2n) and three neutrons (3n) from the composite system before the establishment of thermodynamic equilibrium at the incident energy of 90 MeV. As may be seen from this figure, the PE component for the emission of two neutrons and one neutron plus one proton is roughly the same ($\approx$ 70%) but is almost half (35%) for the emission of three neutrons. This is not unexpected, as the probability of PE emission decreases with the number of particles since each emitted particle decreases the excitation energy of the composite system.

In another experiment, spin distributions have been measured for some of the residues for which RRD and excitation functions were measured earlier. Details of spin distribution measurements are provided in the reference [18] and are mentioned in Chapter 4. It was observed in these experiments that the spin distribution for residues emitted in the forward direction was distinctly different from the spin distribution in residues emitted in the backward cone as shown in Figure 5.11. This different pattern in forward and backward spin distributions have been assigned to the residues of compound nuclear emission (predominantly in the backward

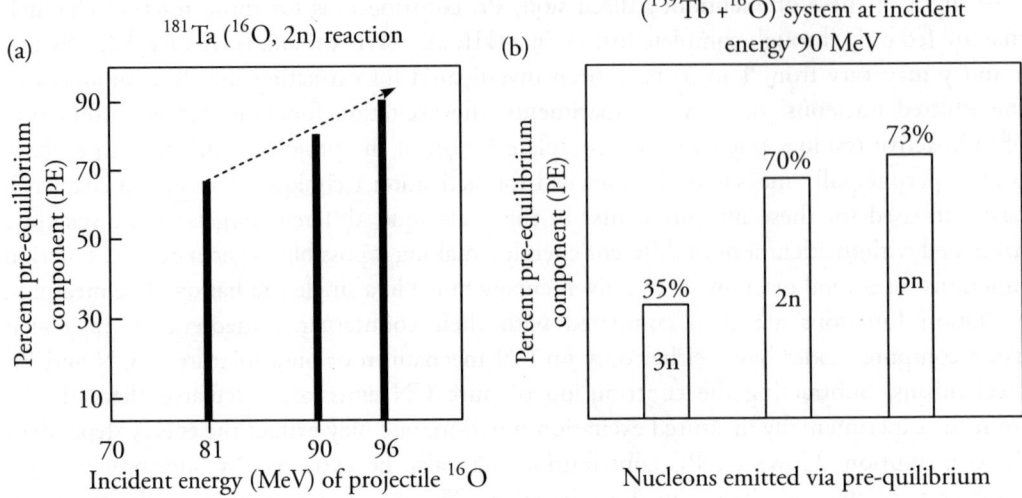

Figure 5.10 (a) Energy dependence of the percent pre-equilibrium emission for reaction ^{181}Ta(^{16}O,2n) (b)Percent pre-equilibrium component for the emission of (pn), (2n) and (3n) from the system ^{159}Tb+^{16}O at 90 MeV

direction) and the PE emission in the forward direction. As indicated in the figure, the mean angular momenta associated with the compound (CN) and pre-equilibrium (PE) emissions have different values, as expected. However, these investigations need further confirmation.

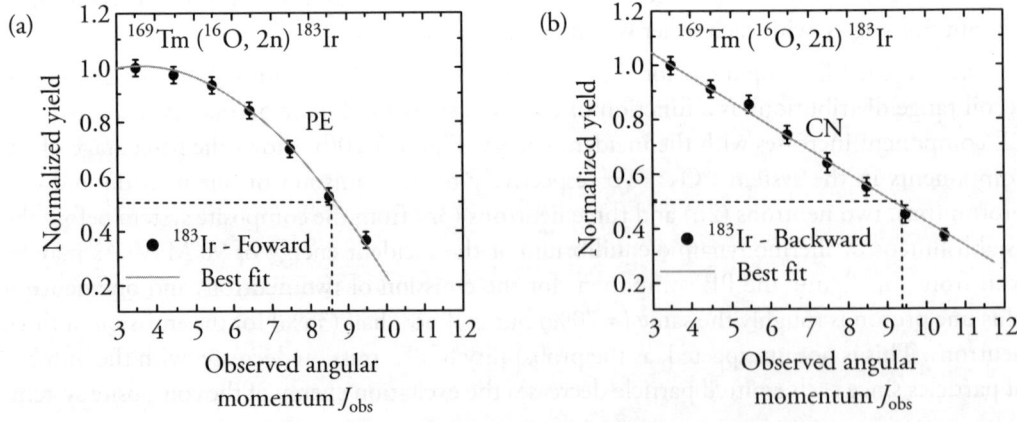

Figure 5.11 Experimentally measured spin distributions for the reaction ^{169}Tm (^{16}O, 2n)^{183}Ir

Abhishek et al.[36] used the difference in the intensity of particles emitted in the forward and backward directions to extract information on pre-equilibrium emission. Details and the results of these experiments were discussed in Chapter 4. As shown there, the intensity of emitted particles in the forward and the backward cones are quite different from each other, while the intensity of particles in the sideways cone is essentially equal to that in the backward cone. This difference in the intensity by more than 10% may be attributed to the pre-equilibrium emission of particles predominantly in the forward direction. The fact that the measured intensities in the sideways and backward cones agree with each other indicates that particle emission from the equilibrated compound nucleus is isotropic.

It may, however, be remarked that there are only few measurements on PE emission in heavy ion reactions. As such, it is not possible to develop a systematic for the process. More experimental data covering a large number of projectile–target pairs spreading over the periodic table and covering a large range of energies are required for any systematic study of the process.

5.3 Applications of Heavy Ion Reactions at Energy < 10 MeV/A

Some applications of heavy ion incomplete fusion reactions at low energies are discussed in the following sections.

5.3.1 Study of high spin states populated via incomplete fusion

Gamma spectroscopy of heavy residues populated via complete or incomplete fusion reactions has been a topic of considerable interest because of the large spins associated with such reactions. In earlier studies, focus was on populating a narrow spin distribution in the compound nucleus[37] formed as the result of CF of the interacting ions, corresponding to collisions near grazing. However, gamma spectroscopy of residues populated via incomplete fusion (ICF) at low energies has acquired considerable importance in recent times because of the fact that ICF results in slightly neutron rich nuclei, which are otherwise difficult to populate. Further, the maximum spin reached by the ICF is also larger than that reached by CF. The one possible way to populate slightly neutron or proton rich nuclei is by using radioactive ion beams, which at present are not available in desired intensity. One method of producing radioactive ion beams is the fragmentation of nuclei at relativistic energies. However, the products of fragmentation at relativistic energies are highly energetic and are not likely to initiate fusion–evaporation reactions. Incomplete fusion at low energies may itself be considered as a method of producing low energy radioactive ion beams with energies appropriate for fusion–evaporation reactions. For example, let us consider the breakup of the low energy stable beam of $^{9}_{4}$Be as given here.

$$^{9}_{4}\text{Be} \rightarrow {}^{4}_{2}\text{He}\left(\alpha \text{ particle}\right) + {}^{5}_{2}\text{He}$$

This breakup has appreciable cross-section around the Coulomb barrier and the fragment $^{5}_{2}$He is emitted with energy suitable for fusion with the target. As such, incomplete fusion of ^{9}Be with a stable target may result in a slightly neutron rich nucleus, just away from the neutron stability line. It may be argued that the same residue may also be populated when ^{9}Be undergo complete fusion with the target and an alpha particle is evaporated from the compound nucleus formed by complete fusion. However, the probability of evaporation of an alpha particle from the compound nucleus is considerably lower. Further, experiments show that the yield of emitted alpha particles is much enhanced over what is expected of evaporation from the compound nucleus, indicating that they are emitted via ICF/pre-equilibrium process.

Similarly, the following breakup of $^{7}_{3}$Li into an alpha particle and $^{3}_{1}$H and in case of incomplete fusion of $^{3}_{1}$H with the target heavy ion may produce a slightly neutron rich residual nucleus;

$$^{7}_{3}\text{Li} \rightarrow {}^{4}_{2}\text{He}\left(\text{alpha particle}\right) + {}^{3}_{1}\text{H}$$

Apart from the production of slightly neutron rich residues, which is interesting in itself, other differences in the CF and ICF dynamics may also be exploited for in-beam gamma-spectroscopy of neutron rich residues. The evaporation products from an equilibrated compound nucleus are emitted isotropically in the centre of mass while the spectator fragment of the incomplete fusion moves essentially with the beam velocity in a direction peaked around the grazing angle. As such, in the case of ICF, there is a correlation between the angle at which the charged particles are emitted and the nature of the heavy residual nucleus. As discussed by Dracoulis,[38] more forward focused charged particles are associated with heavy residual nuclei from which fewer neutrons are evaporated. This difference in the angle of emission of charged particles may be used to tag the residual nucleus and assign the associated gamma rays.

Gorgen et al.[39] investigated the level structure of ^{200}Hg nucleus produced via incomplete fusion in the ^{198}Pt+^{9}Be reaction. Assuming that the projectile ^{9}Be breaks up into ^{4}He+^{5}He in the nuclear field of target ^{198}Pt and that ^{5}He fuses with the target while ^{4}He moves on as spectator, the incomplete fusion reaction ^{198}Pt(^{5}He,3n) will produce the residual nucleus ^{200}Hg. The same nucleus ^{200}Hg may also be produced in the conventional complete fusion evaporation reaction ^{198}Pt (^{4}He, 2n)^{200}Hg. However, considerably lower spin values are populated in the conventional complete fusion evaporation method because of the small mass of incident alpha particle and the central nature of collision. On the other hand, in the case of the incomplete fusion, higher spin states got populated primarily on account of the peripheral nature of incomplete fusion. Thus, using incomplete fusion at low energies, Gorgen et al. extended the level scheme of ^{200}Hg up to higher spins and also identified some new band crossings.

Another interesting case investigated using ICF at low energies is of isotopes ^{211,212}Po. Both these nuclei have long-lived isomeric states that have high spins, respectively, $J^{\pi} = \left(\dfrac{25}{2}\right)^{+}$ and $(18)^{+}$ that decay by alpha emission. Till some time back, no states higher than the isomeric states were known for these nuclides. Production of these neutron rich nuclides using conventional reactions like ^{208}Pb (^{4}He, n) ^{211}Po is not possible as they do not have favourable thresholds. The other alternative reactions are ^{208}Pb (^{9}Be, αn)^{212}Po and ^{208}Pb (^{9}Be, α2n)^{211}Po. In the framework of statistical compound reaction mechanism, which assumes that projectile ^{9}Be undergoes complete fusion with target ^{208}Pb and that alpha particle and neutrons are evaporated from the equilibrated compound nucleus, the cross-sections for these reactions turn out to be negligible. Experiments have, however, indicated enhanced production of the two isotopes, most likely through ICF channels;

$$_4^9\mathrm{Be} \to {}_2^4\mathrm{He}\,(\alpha - \mathrm{particle}) + {}_2^5\mathrm{He}$$

and $({}_2^5\mathrm{He} + {}_{82}^{208}\mathrm{Pb}) \to {}_{84}^{212}\mathrm{Po} + {}_0^1\mathrm{n}$ and $({}_2^5\mathrm{He} + {}_{82}^{208}\mathrm{Pb}) \to {}_{84}^{211}\mathrm{Po} + 2\,{}_0^1\mathrm{n}$

Employing in-beam particle gamma coincidence technique using charged particle and gamma detection arrays, high spin states above the isomeric state $\left(\dfrac{25}{2}\right)^+$ in nucleus $^{211}\mathrm{Po}$ have been studied by Lane et al.[40] using the correlation between multiplicities of emitted neutrons and the angle of emission of spectator alpha particle. Incomplete fusion followed by nucleon emission has also been employed to study the level structure of other heavier nuclei, which are difficult to produce through conventional complete fusion route.

5.3.2 Incomplete fusion and synthesis of super heavy elements

Nuclei with atomic number $Z > 92$ are unstable and are, therefore, not found in any measurable quantity in nature. These heavy nuclei, $Z > 92$ (also called super heavy elements, SHE), may be synthesised using complete and incomplete fusion of two heavy ions under stringent controlled conditions as they are highly unstable with extremely short half-lives. Theoretical calculations, however, predict an 'island of stability' around extended magic numbers $Z = 108, 110, 114$ with N around 184. In this region of enhanced stability, nuclei with half-lives ranging from seconds to days have been predicted. Theoretical calculations have also predicted strange shapes, like a bubble configuration with a hole at the centre, for these nuclei. With the view to further explore the nature of nuclear forces as exhibited in these bloated super heavy nuclei, attempts have been made to synthesise super heavy nuclei using accelerated heavy ion beams to reach the extended island of stability. Synthesis of the following super heavy elements has been confirmed by the International Union on Pure and Applied Chemistry (IUPAC),

$Z = 104$ (Rutherfordiaum); $Z = 113$ (Nihonium, Nh); $Z = 115$ (Moscovium Mc); $Z = 117$ (Tennessine, Ts) and $Z = 118$ (Oganesson, Og).

Kosuke Morita on August 12, 2012 announced the synthesis of a super heavy element with $Z = 113$ through the fusion of $^{70}\mathrm{Zn}$ and $^{209}\mathrm{Bi}$ at Japan's RIKEN Nishina Centre for Accelerator Based Sciences. They also estimated the production cross-section for this element to be of the order of 22 femtobarn ($\approx 22 \times 10^{-15}$ b). Earlier, an element with $Z = 117$ was synthesised at the Joint Institute for Nuclear Research, Dubna, Russia and also at GSI Helmholtz Centre for Heavy Ion Research at Darmstadt, Germany.

Since the aim of the synthesis of super heavy elements is to reach the largest possible atomic number, complete fusion rather than incomplete fusion of heavy ions is advantageous. The line of action for the synthesis of SHE is to chose the appropriate projectile–target pair of heavy ions and control projectile energy so that the desired SHE nucleus is produced with maximum cross-section as the reaction residue of the complete fusion–evaporation sequence. Nasirov[41] elaborated on the sequence of events in heavy ion collisions that may follow as shown in Figure 5.12. As may be observed in this figure, inelastic collision, incomplete fusion and quasi-fission, all add up to deplete the production cross-section for the desired SHE residue via the complete fusion channel. However, in principle, it appears that it may be advantageous if the desired SHE nucleus is populated via incomplete fusion channel. This will require projectile and targets of higher masses and relatively larger input kinetic energy to overcome the Coulomb barrier.

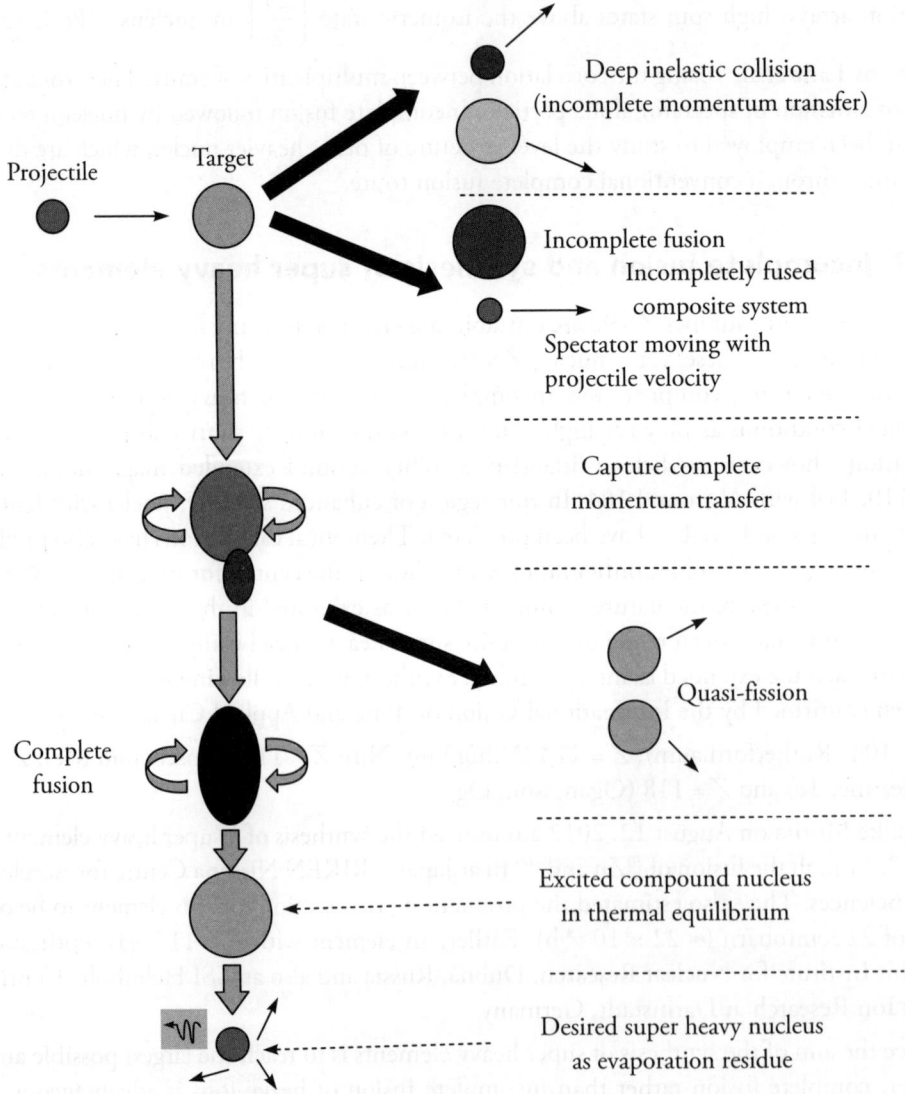

Figure 5.12 Schematic diagram of the event sequence for the production of SHE nucleus via complete fusion

5.2.3 Incomplete fusion and production of isotopes of special interest

Experiments carried out in the course of the present study have indicated that a given residue may be produced both via complete fusion as well as via incomplete fusion routes and that the cross-section for the incomplete fusion route is significantly larger. It, therefore, appears to be advantageous to produce these nuclides via incomplete fusion process to accumulate larger amounts of these isotopes. As such, one can choose proper irradiation conditions for the production of medically important isotopes as well. Some of the heavy radioisotopes which are generally used for various applications in the medical field are ^{123}I, ^{153}Sm, ^{213}Bi, ^{169}Er, ^{186}Re, ^{166}Ho, ^{192}Ir, ^{177}Lu etc. In the same way, various radioactive isotopes produced via the complete as well as incomplete fusion processes are of interest in the field of agricultural sciences and astrophysics.

Appendix

A consolidated list of some important research papers related to the heavy ion interactions at energies < 10 MeV/n, published by our group, is presented here.

Gupta, Sunita, B. P. Singh, M. M. Musthafa, H. D. Bhardwaj, and R. Prasad. 2000. "Complete and incomplete fusion of ^{12}C with ^{165}Ho below 7 MeV/nucleon: Measurements and analysis of excitation functions." *Physical Review C* 61 (6): 064613.

Gupta, Unnati, Pushpendra P. Singh, Devendra P. Singh, Manoj Kumar Sharma, Abhishek Yadav, Rakesh Kumar, S. Gupta, H. D. Bhardwaj, B. P. Singh, and R. Prasad. 2009. "Disentangling full and partial linear momentum transfer events in the ^{16}O+^{169}Tm system at $E_{proj} \approx 5.4$ MeV/nucleon." *Physical Review C* 80 (2): 024613.

Gupta, Unnati, Pushpendra P. Singh, Devendra P. Singh, Manoj Kumar Sharma, Abhishek Yadav, Rakesh Kumar, B. P. Singh, and R. Prasad. 2008. "Observation of large incomplete fusion in ^{16}O+^{103}Rh system at $\approx$ 3–5 MeV/nucleon." *Nuclear Physics A* 811 (1–2): 77–92.

Sharma, Manoj Kumar, B. P. Singh, Sunita Gupta, M. M. Musthafa, H. D. Bhardwaj, R. Prasad, and A. K. Sinha. 2003. "Complete and Incomplete Fusion: Measurement and Analysis of Excitation Functions in ^{12}C+^{128}Te System at Energies near and above the Coulomb Barrier." *Journal of the Physical Society of Japan* 72 (8): 1917–1925.

Sharma, Manoj Kumar, Abhishek Yadav, Vijay Raj Sharma, Devendra P. Singh, Pushendra P. Singh, Indu Bala, Rakesh Kumar, B. P. Singh, and R. Prasad. 2015. "Experimental study of cross sections in the ^{12}C+^{27}Al system at $\approx$ 3–7 MeV/nucleon relevant to the incomplete fusion process." *Physical Review C* 91 (2): 024608.

Sharma, Manoj Kumar, B. K. Sharma, B. P. Singh, H. D. Bhardwaj, Rakesh Kumar, K. S. Golda, and R. Prasad. 2004. "Complete and incomplete fusion reactions in the ^{16}O+^{169}Tm system: Excitation functions and recoil range distributions." *Physical Review C* 70 (4): 044606.

Sharma, Manoj Kumar, B. P. Singh, Rakesh Kumar, K. S. Golda, H. D. Bhardwaj, and R. Prasad. 2006. "A study of the reactions occurring in ^{16}O+^{159}Tb system: Measurement of excitation functions and recoil range distributions." *Nuclear Physics A* 776 (3–4): 83–104.

Sharma, Manoj Kumar, Devendra P. Singh, Pushpendra P. Singh, B. P. Singh, H. D. Bhardwaj, and R. Prasad. 2007. "Reaction mechanism in the ^{16}O+^{27}Al system: Measurements and analysis of excitation functions and angular distributions." *Physical Review C* 75 (6): 064608.

Sharma, Manoj Kumar, Pushendra P. Singh, Devendra P. Singh, Abhishek Yadav, Vijay Raj Sharma, Indu Bala, Rakesh Kumar, B. P. Singh, and R. Prasad. 2015. "Systematic study of pre-equilibrium emission at low energies in ^{12}C and ^{16}O-induced reactions." *Physical Review C* 91 (1): 014603.

Sharma, Vijay R., Abhishek Yadav, Pushpendra P. Singh, Devendra P. Singh, Sunita Gupta, M. K. Sharma, Indu Bala et al. 2014. "Influence of a one-neutron-excess projectile on low-energy incomplete fusion." *Physical Review C* 89 (2): 024608.

Sharma, Vijay R., Abhishek Yadav, Pushpendra P. Singh, Indu Bala, Devendra P. Singh, Sunita Gupta, M. K. Sharma et al. 2015. "Spin distribution measurements in ^{16}O+^{159}Tb system: incomplete fusion reactions." *Journal of Physics G: Nuclear and Particle Physics* 42 (5): 055113.

Sharma, Vijay R., Pushpendra P. Singh, Mohd Shuaib, Abhishek Yadav, Indu Bala, Manoj K. Sharma, S. Gupta et al. 2016. "Incomplete fusion in ^{16}O+^{159}Tb." *Nuclear Physics A* 946: 182–193.

Shuaib, Mohd, Vijay R. Sharma, Abhishek Yadav, Manoj Kumar Sharma, Pushpendra P. Singh, Devendra P. Singh, R. Kumar R. P. Singh, S. Muralithar, B. P. Singh, and R. Prasad 2017. "Influence of incomplete fusion on complete fusion at energies above the Coulomb barrier." *Journal of Physics G: Nuclear and Particle Physics* 44 (10): 105108.

Shuaib, Mohd, Vijay R. Sharma, Abhishek Yadav, Pushpendra P. Singh, Manoj Kumar Sharma, Devendra P. Singh, R. Kumar R. P. Singh, S. Muralithar, B. P. Singh, and R. Prasad 2016. "Incomplete fusion studies in the ^{19}F+^{159}Tb system at low energies and its correlation with various systematics." *Physical Review C* 94 (1): 014613.

Singh, Devendra P., Pushpendra P. Singh, Abhishek Yadav, Manoj Kumar Sharma, B. P. Singh, K. S. Golda, Rakesh Kumar, A. K. Sinha, and R. Prasad. 2010. "Energy dependence of incomplete fusion processes in the ^{16}O+^{181}Ta system: Measurement and analysis of forward-recoil–range distributions at $E_{lab} \approx 7$ MeV/nucleon." *Physical Review C* 81 (5): 054607.

Singh, Devendra P., Pushpendra P. Singh, Abhishek Yadav, Manoj Kumar Sharma, B. P. Singh, K. S. Golda, Rakesh Kumar, A. K. Sinha, and R. Prasad. 2009. "Investigation of the role of break-up processes on the fusion of ^{16}O induced reactions." *Physical Review C* 80 (1): 014601.

Singh, Devendra P., Vijay R. Sharma, Abhishek Yadav, Pushpendra P. Singh, M. K. Sharma, R. Kumar, B. P. Singh, and R. Prasad. 2014. "Experimental study of incomplete fusion reactions in the ^{16}O+^{130}Te system below 6 MeV/nucleon." *Physical Review C* 89 (2): 024612.

Singh, Pushpendra P., Abhishek Yadav, Devendra P. Singh, Unnati Gupta, Manoj K. Sharma, R. Kumar, D. Singh R. P. Singh, S. Muralithar, M. A. Ansari, B. P. Singh, R. Prasad and R. K. Bhowmik 2009. "Role of high ℓ values in the onset of incomplete fusion." *Physical Review C* 80 (6): 064603.

Singh, Pushpendra P., B. P. Singh, M. K. Sharma, Unnati Gupta, Rakesh Kumar, D. Singh, R. P. Singh S. Murlithar, M. A. Ansari, R. Prasad and R. K. Bhowmik 2009. "Probing of incomplete fusion dynamics by spin-distribution measurement." *Physics Letters B* 671 (1): 20–24.

Singh, Pushpendra P., B. P. Singh, Manoj Kumar Sharma, Devendra P. Singh, R. Prasad, Rakesh Kumar, and K. S. Golda. 2008. "Influence of incomplete fusion on complete fusion: Observation of a large incomplete fusion fraction at E ≈ 5–7 MeV/nucleon." *Physical Review C* 77 (1): 014607.

Singh, Pushpendra P., B. P. Singh, Manoj Kumar Sharma, R. Kumar, K. S. Golda, D. Singh, R. P. Singh S. Muralithar, M. A. Ansari, R. Prasad and R. K. Bhowmik 2008. "Spin-distribution measurement: A sensitive probe for incomplete fusion dynamics." *Physical Review C* 78 (1): 017602.

Singh, Pushpendra P., Manoj Kumar Sharma, Devendra P. Singh, Rakesh Kumar, K. S. Golda, B. P. Singh, and R. Prasad. 2007. "Observation of complete-and incomplete-fusion components in ^{159}Tb, ^{169}Tm (^{16}O, x) reactions: Measurement and analysis of forward recoil ranges at E/A ≈ 5–6 MeV." *The European Physical Journal A* 34 (1): 29–39.

Yadav, Abhishek, Pushpendra P. Singh, Mohd Shuaib, Vijay R. Sharma, Indu Bala, Sunita Gupta, D. P. Singh M.K. Sharma, R. Kumar, S. Murlithar, R. P. Singh, B. P. Singh, and R. Prasad 2017. "Systematic study of low-energy incomplete fusion: Role of entrance channel parameters." *Physical Review C* 96 (4): 044614.

Yadav, Abhishek, Pushpendra P. Singh, Vijay R. Sharma, Unnati Gupta, Devendra P. Singh, Manoj K. Sharma, B. P. Singh, and R. Prasad. 2011. "Signature of pre-equilibrium-emission in forward-to-backward yield ratio measurement." *International Journal of Modern Physics E* 20 (10): 2133–2142.

Yadav, Abhishek, Vijay R. Sharma, Pushpendra P. Singh, Devendra P. Singh, Manoj K Sharma, Unnati Gupta, R. Kumar, B. P. Singh, R. Prasad, and R. K. Bhowmik. 2012. "Large influence of incomplete fusion in ^{12}C+^{159}Tb at E_{lab} ≈ 4–7 MeV/nucleon." *Physical Review C* 85 (3): 034614.

Yadav, Abhishek, Vijay R. Sharma, Pushpendra P. Singh, Devendra P. Singh, R. Kumar, M. K. Sharma, B. P. Singh, R. Prasad, and R. K. Bhowmik. 2012. "Effect of entrance-channel parameters on incomplete fusion reactions." *Physical Review C* 85 (6): 064617.

Yadav, Abhishek, Vijay R. Sharma, Pushpendra P. Singh, R. Kumar, Devendra P. Singh, M. K. Sharma, B. P. Singh, and R. Prasad. 2012. "Effect of α-Q value on incomplete fusion." *Physical Review C* 86 (1): 014603.

References

Chapter 1

[1] Geiger, Hans, and Ernest Marsden. 1909. "On a diffuse reflection of the α-particles." *Proceedings of the Royal Society of London. Series A, Containing Papers of a Mathematical and Physical Character* 82 (557): 495–500.

[2] Rutherford, Ernest. 1911. "LXXIX. The scattering of α and β particles by matter and the structure of the atom." *The London, Edinburgh, and Dublin Philosophical Magazine and Journal of Science* 21 (125): 669–688.

[3] Hofstadter, Robert, H. R. Fechter, and J. A. McIntyre. 1953. "High-energy electron scattering and nuclear structure determinations." *Physical Review* 92 (4): 978.

Hofstadter, Robert, and Robert W. McAllister. 1955. "Electron scattering from the proton." *Physical Review* 98 (1): 217.

Hofstadter, Robert. 1956. "Electron scattering and nuclear structure." *Reviews of Modern Physics* 28 (3): 214.

Hofstadter, Robert, F. Bumiller, and M. R. Yearian. 1958. "Electromagnetic structure of the proton and neutron." *Reviews of Modern Physics* 30 (2): 482.

Hofstadter, Robert, and Robert Herman. 1961. "Electric and Magnetic Structure of the Proton and Neutron." *Physical Review Letters* 6 (6): 293.

[4] Joliot, F., and I. Curie. 1934. "Un nouveau type de radioactivité, in "CR Acad." *Sci. Paris* 198: 254–256.

[5] Fermi, Enrico, Edoardo Amaldi, Oscar D'Agostino, Franco Rasetti, and Emilio Segrè. 1934. "Artificial radioactivity produced by neutron bombardment." *Proceedings of the Royal Society of London. Series A, Containing Papers of a Mathematical and Physical Character* 146 (857): 483–500.

[6]   Hahn, Otto, and Fritz Straßmann. 1939. "Über den Nachweis und das Verhalten der bei der Bestrahlung des Urans mittels Neutronen entstehenden Erdalkalimetalle." *Naturwissenschaften* 27 (1): 11–15.

[7]   Ghoshal, Samarendro Nath. 1950. "An experimental verification of the theory of compound nucleus." *Physical Review* 80 (6): 939.

[8]   Wiley, J. R. 1974. Ph.D dissertation, Purdue University, Ind. USA.

[9]   Khurana, C. S. 1960. "Construction of 150kV Cockcroft-Walton Particle Accelerator and Study of Fast Neutron Reactions." Ph.D. dissertation, Aligarh Muslim University, India.

[10] Prasad, R., D. C. Sarkar, and C. S. Khurana. 1966. "Measurement of (n, p) and (n, α) reaction cross sections at 14.8 MeV." *Nuclear Physics* 85 (2): 476–480.

Prasad, R., and D.C. Sarkar. 1966. "Measurement of (n,2n) reactions cross-sections at 14.8 MeV." *Nuclear Physics* A 88:349-352

Prasad, R., and D. C. Sarkar. 1967. "Isomeric Cross-section ratios for (n,2n) reactions at 14.8 MeV." *Nuclear Physics* A 94: 476–480.

Prasad, R., and D. C. Sarkar. 1971. "Measured (n, p) reaction cross-sections and their predicted values at 14.8 MeV." *Il Nuovo Cimento A* 3 (4): 467–478.

[11] Singh, B. P., M. G. V. Sankaracharyulu, M. Afzal Ansari, H. D. Bhardwaj, and R. Prasad. 1993. "Equilibrium and preequilibrium emission in some *a*-induced reactions on enriched isotopes Te^{128,130} at moderate excitation energies." *Physical Review C* 47 (5): 2055.

[12] Cavinato, M., E. Fabrici, E. Gadioli, E. Gadioli Erba, P. Vergani, M. Crippa, G. Colombo, I. Redaelli, and M. Ripamonti. 1995. "Study of the reactions occurring in the fusion of C^{12} and O^{16} with heavy nuclei at incident energies below 10 MeV/nucleon." *Physical Review C* 52 (5): 2577.

[13] Abolmasov Sergey N, Alexandr A. Bizyukov, Yoshinobu Kawai, Andrei Y. Kashaba, Vasyl I. Maslov, and Konstantin N. Sereda. 2002. "Low-Energy Penning Ionization Gauge Type Ion Source Assisted by RF Magnetron Discharge." *Japanese Journal of Applied Physics* 41 (1): 5415.

[14] Alvarez, L.W. 1940. "High-energy carbon nuclei." Physical Review 58: 192.

[15] Walker, D., and J. H. Fremlin. 1953. "Acceleration of heavy ions to high energies." *Nature* 171 (4344): 189–191.

[16] Beringer, Robert, Robert L. Gluckstern, and Myron S. Malkin, Edward L. Hubbard, Lloyd Smith, and Chester N. Van Atta. 1954. "Linear Accelerators for Heavy ions". UCRL-2796.

[17] Lefort M. 1966. "Le Programme Scientifique du CEVIL d 'OSRAY'." *Industries Atomiques* 11: 41

[18] Basile, Robert, and Jean-Marie Lagrange. 1962. "Étude d'une source d'ions multichargés pour un cyclotron." *Journal de Physique Appliquée* 23 (S6): 111–116.

[19] Twin, P. J., B. M. Nyako, A. Hetal Nelson, J. Simpson, M. A_ Bentley, H. W. Cranmer-Gordon, P. D. Forsyth et al. 1986. "Observation of a Discrete-Line Superdeformed Band up to 60 $\hbar$ in Dy 152." *Physical review letters* 57 (7): 811.

[20] Negele, John W., and G. Rinker. 1977. "Density-dependent Hartree-Fock description of nuclei in the rare earth and nickel regions." *Physical Review C* 15 (4): 1499.

[21] Schröder, W. U., and J. R. Huizenga. 1984. "Damped nuclear reactions." In *Treatise on Heavy-Ion Science*, pp. 113–726. US: Springer.

[22] Birkelund, J. R., L. E. Tubbs, J. R. Huizenga, J. N. De, and D. Sperber. 1979. "Heavy-ion fusion: Comparison of experimental data with classical trajectory models." *Physics Reports* 56 (3): 107–166.

[23] Akyiiz, O., and A. Winther. 1981. "Course on nuclear structure and heavy ion reactions." North-Holland, Amsterdam: 491

[24] Cohen, S., F. Plasil, and W. J. Swiatecki. 1974. "Equilibrium configurations of rotating charged or gravitating liquid masses with surface tension. II." *Annals of Physics* 82 (2): 557–596.

[25] Dasgupta, M., D. J. Hinde, N. Rowley, and A. M. Stefanini. 1998. "Measuring barriers to fusion." *Annual Review of Nuclear and Particle Science* 48 (1): 401–461.

[26] Möller, P., and J. R. Nix. 1976. "Macroscopic potential-energy surfaces for symmetric fission and heavy-ion reactions." *Nuclear Physics A* 272 (2): 502–532.

Moller, P. and J. R. Nix. 1977. "Potential-energy surfaces for asymmetric heavy ion reactions." *Nuclear Physics A* 281 (2): 354–372.

Moller P., and J.R. Nix. 1981. "Nuclear mass formula with a Yukawa-plus-exponetial microscopic model and a folded-Yukava single-particle potential." *Nuclear Physics A* 361: 117–146.

[27] Dasgupta, Mahananda, D. J. Hinde, A. Mukherjee, and J. O. Newton. 2007. "New challenges in understanding heavy ion fusion." *Nuclear Physics A* 787 (1–4): 144–149.

[28] Diaz-Torres, Alexis, D. J. Hinde, Jeffrey Allan Tostevin, Mahananda Dasgupta, and L. R. Gasques. 2007. "Relating breakup and incomplete fusion of weakly bound nuclei through a classical trajectory model with stochastic breakup." *Physical review letters* 98 (15): 152701.

[29] Gasques, L. R., Mahananda Dasgupta, D. J. Hinde, T. Peatey, Alexis Diaz-Torres, and J. O. Newton. 2006. "Isomer ratio measurements as a probe of the dynamics of breakup and incomplete fusion." *Physical Review C* 74 (6): 064615.

[30] Gomes, P. R. S., I. Padron, M. D. Rodríguez, G. V. Marti, R. M. Anjos, J. Lubian, R. Veiga et al. 2004. "Fusion, reaction and break-up cross sections of weakly bound projectiles on ^{64}Zn." *Physics Letters B* 601 (1): 20–26.

[31] Dasgupta, Mahananda, P. R. S. Gomes, D. J. Hinde, S. B. Moraes, R. M. Anjos, A. C. Berriman, Rachel D. Butt et al. 2004. "Effect of breakup on the fusion of Li^6, Li^7, and Be^9 with heavy nuclei." *Physical Review C* 70 (2): 024606.

[32] Canto, L. F., R. Donangelo, Lia M. de Matos, M. S. Hussein, and P. Lotti. 1998. "Complete and incomplete fusion in heavy-ion collisions." *Physical Review C* 58 (2): 1107.

[33] Arena, N., Seb Cavallaro, S. Femino, P. Figuera, S. Pirrone, G. Politi, F. Porto, S. Romano, and S. Sambataro. 1994. "Complete and incomplete fusion in the Si^{28}+^{12}C, ^{13}C reactions around 5 MeV/nucleon." *Physical Review C* 50 (2): 880.

[34] Gadioli, E., C. Birattari, M. Cavinato, E. Fabrici, E. Gadioli Erba, V. Allori, F. Cerutti et al. 1998. "Angular distributions and forward recoil range distributions of residues created in the interaction of ^{12}C and ^{16}O ions with ^{103}Rh." *Nuclear Physics A* 641 (3): 271–296.

[35] Vergani, P., E. Gadioli, E. Vaciago, E. Fabrici, E. Gadioli Erba, M. Galmarini, G. Ciavola, and C. Marchetta. 1993. "Complete and incomplete fusion and emission of preequilibrium nucleons in the interaction of C^{12} with Au^{197} below 10 MeV/nucleon." *Physical Review C* 48 (4): 1815.

[36] Trautmann, W., Ole Hansen, H. Tricoire, W. Hering, R. Ritzka, and W. Trombik. 1984. "Dynamics of Incomplete Fusion Reactions from γ-Ray Circular-Polarization Measurements." *Physical Review Letters* 53 (17): 1630.

[37] Corradi, L., B. R. Behera, E. Fioretto, A. Gadea, A. Latina, A. M. Stefanini, Suzana Szilner et al. 2005. "Excitation functions for Fr 208–211 produced in the O^{18} + Au^{197} fusion reaction." *Physical Review C* 71 (1): 014609.

[38] Parker, D. J., J. Asher, T. W. Conlon, and I. Naqib. 1984. "Complete and incomplete fusion in C^{12} + V^{51} at $E(C^{12})$ = 36–100 MeV from analysis of recoil range and light particle measurements." *Physical Review C* 30 (1): 143.

[39] Cavinato, M., E. Fabrici, E. Gadioli, E. Gadioli Erba, P. Vergani, M. Crippa, G. Colombo, I. Redaelli, and M. Ripamonti. 1995. "Study of the reactions occurring in the fusion of C^{12} and O^{16} with heavy nuclei at incident energies below 10 MeV/nucleon." *Physical Review C* 52 (5): 2577.

[40] Tomar, B. S., A. Goswami, A. V. R. Reddy, S. K. Das, P. P. Burte, S. B. Manohar, and Bency John. 1994. "Investigations of complete and incomplete fusion in C^{12} + ^{93}Nb and O^{16} + ^{89}Y by recoil range measurements." *Physical Review C* 49 (2): 941.

[41] Crippa, M., E. Gadioli, P. Vergani, G. Ciavola, C. Marchetta, and M. Bonardi. 1994. "Excitation functions for production of heavy residues in the interaction of ^{12}C with ^{181}Ta." *Zeitschrift für Physik A Hadrons and Nuclei* 350 (2): 121–129.

[42] Britt, Harold C., and Arthur R. Quinton. 1961. "Alpha Particles and Protons Emitted in the Bombardment of Au^{197} and Bi^{209} by C^{12}, N^{14}, and O^{16} Projectiles." *Physical Review* 124 (3): 877.

[43] Galin, J., B. Gatty, D. Guerreau, C. Rousset, U. C. Schlotthauer-Voos, and X. Tarrago. 1974. "Study of charged particles emitted from Te^{117} compound nuclei. II. Comparison between Ar^{40} + Se^{77} and N^{14} + Rh^{103} reactions and determination of critical angular momenta." *Physical Review C* 9 (3): 1126.

[44] Pochodzalla, J., R. Butsch, B. Heck, and G. Rosner. 1986. "Influence of limiting angular momenta on complete and incomplete fusion-evaporation residue cross sections." *Physics Letters B* 181 (1–2): 33–37.

[45] Wilschut, H. W., G. J. Balster, P. B. Goldhoorn, R. H. Siemssen, and Z. Sujkowski. 1984. "Angular momentum dependence of incomplete fusion reactions." *Physics Letters B* 138 (1–3): 43–46.

[46] Tsang, M. B., D. R. Klesch, C. B. Chitwood, D. J. Fields, C. K. Gelbke, W. G. Lynch, H. Utsunomiya, K. Kwiatkowski, V. E. Viola, and M. Fatyga. 1984. "Limitations on linear momentum transfer in ^{14}N induced reactions on ^{238}U at EA = 15, 20, 25 and 30MeV." *Physics Letters B* 134 (3–4): 169-173.

[47] Oeschler, H., M. Kollatz, W. Bohne, K. Grabisch, H. Lehr, H. Freiesleben, and K. D. Hildenbrand. 1983. "Complete fusion in peripheral collisions observed in the reaction ^{16}O on ^{146}Nd at 10 MeV/u." *Physics Letters B* 127 (3–4): 177–180.

[48] Piattelli, P., D. Santonocito, Y. Blumenfeld, T. Suomijärvi, C. Agodi, N. Alamanos, R. Alba et al. 1998. "Impact parameter dependence of linear momentum transfer and the role of two-body dissipation mechanisms in heavy ion collisions around the Fermi energy." *Physics Letters B* 442 (1): 48–52.

[49] Arnell, S. E., S. Mattsson, H. A. Roth, Ö. Skeppstedt, S. A. Hjorth, A. Johnson, A. Kerek, J. Nyberg, and L. Westerberg. 1983. "Charged particle production in peripheral reactions induced by 118 MeV ^{12}C ions on ^{118}Sn." *Physics Letters B* 129 (1–2): 23–26.

[50] Barker, J. H., J. R. Beene, M. L. Halbert, D. C. Hensley, M. Jääskeläinen, D. G. Sarantites, and R. Woodward. 1980. "Direct evidence for a narrow window at high angular momentum in incomplete-fusion reactions." *Physical Review Letters* 45 (6): 424.

[51] Parker, D. J., J. J. Hogan, and J. Asher. 1989. "Complete and incomplete fusion of 6 MeV/nucleon light heavy ions on V^{51}." *Physical Review C* 39 (6): 2256.

[52] Wu, J. R., and I. Y. Lee. 1980. "Projectile fragmentation accompanied by incomplete fusion." *Physical Review Letters* 45 (1): 8.

[53] Hojman, D., M. A. Cardona, A. Arazi, O. A. Capurro, J. O. Fernandez-Niello, G. V. Marti, A. J. Pacheco et al. 2006. "Reaction-dependent spin population and evidence of breakup in O^{18}." *Physical Review C* 73 (4): 044604.

[54] Inamura, T., T. Kojima, T. Nomura, T. Sugitate, and H. Utsunomiya. 1979. "Multiplicity of g-rays following fast alpha;-particle emission in the 95 MeV ^{14}N + ^{159}Tb reaction." *Physics Letters B* 84 (1): 71–74.

Inamura, T., A. C. Kahler, D. R. Zolnowski, U. Garg, T. T. Sugihara, and M. Wakai. 1985. "Gamma-ray multiplicity distribution associated with massive transfer." *Physical Review C* 32 (5): 1539.

[55] Trautmann, W., Ole Hansen, H. Tricoire, W. Hering, R. Ritzka, and W. Trombik. 1984. "Dynamics of Incomplete Fusion Reactions from γ-Ray Circular-Polarization Measurements." *Physical Review Letters* 53 (17): 1630.

[56] Tserruya, I., V. Steiner, Z. Fraenkel, P. Jacobs, D. G. Kovar, W. Henning, M. F. Vineyard, and B. G. Glagola. 1988. "Incomplete Fusion Reactions Induced by C^{12} at 5.5-10 MeV/nucleon." *Physical review letters* 60 (1): 14.

[57] Gerschel, Claudie. 1982. "Incomplete fusion reactions." *Nuclear Physics A* 387 (1): 297–312.

[58] Inamura, T., M. Ishihara, T. Fukuda, T. Shimoda, and H. Hiruta. 1977. "Gamma-rays from an incomplete fusion reaction induced by 95 MeV ^{14}N " *Physics Letters B* 68 (1): 51–54.

[59] Zolnowski, D. R., H. Yamada, S. E. Cala, A. C. Kahler, and T. T. Sugihara. 1978. "Evidence for" Massive Transfer" in Heavy-Ion Reactions on Rare-Earth Targets." *Physical Review Letters* 41 (2): 92.

[60] Geoffroy, K. Ao, D. G. Sarantites, M. L. Halbert, D. C. Hensley, R. A. Dayras, and J. H. Barker. 1979. "Angular momentum transfer in incomplete-fusion reactions." *Physical Review Letters* 43 (18): 1303.

[61] Udagawa, T., and T. T. Tamura. 1980. "Breakup-fusion description of massive transfer reactions with emission of fast light particles." *Physical Review Letters* 45 (16): 1311.

[62] Wu, J. R., and I. Y. Lee. 1980. "Projectile fragmentation accompanied by incomplete fusion." *Physical Review Letters* 45 (1): 8.

[63] Wilczyński, J., K. Siwek-Wilczyńska, J. Van Driel, S. Gonggrijp, D. C. J. M. Hageman, R. V. F. Janssens, J. Łukasiak, and R. H. Siemssen. 1980. "Incomplete-Fusion Reactions in the N^{14} + Tb^{159} System and a" Sum-Rule Model" for Fusion and Incomplete-Fusion Reactions." *Physical Review Letters* 45 (8): 606.

Wilczynski, J., K. Siwek-Wilczynska, J. Van Driel, S. Gonggrijp, D. C. J. M. Hageman, R. V. F. Janssens, J. Łukasiak, Rh H. Siemssen, and S. Y. Van der Werf. 1982. "Binary l-matched reactions in N^{14} + Tb^{159} collisions." *Nuclear Physics A* 373 (1): 109–140.

[64] Bondrof, J. P., J. N. Fai, G. Karvinen, A. O. T. Jakobsson, B. J. Randrup. 1980. "Effect of prrojectile break-up threshold energy on incomplete fusion at energy = 4 MeV/nucleon." *Nuclear Physics A* 333: 285–301.

[65] Gross, D. H. E., and J. Wilczynski. 1977. "Does radial friction cause emission of fast *a*-particles?." *Physics Letters B* 67 (1): 1–4.

[66] Tricoire, H. 1983. "Emission of direct particles in heavy ions collisions." *Zeitschrift für Physik A Hadrons and Nuclei* 312 (3): 221–232.

[67] Weiner, R., and M. Weström. 1977. "Diffusion of heat in nuclear matter and preequilibrium phenomena." *Nuclear Physics A* 286 (2): 282–296.

[68] Awes, Terry C., G. Poggi, C. K. Gelbke, B. B. Back, B. G. Glagola, H. Breuer, and V. E. Viola Jr. 1981. "Precompound emission of light particles in the reaction O^{16} + U^{238} at 20 MeV/nucleon." *Physical Review C* 24 (1): 89.

[69] Blann, M. 1981. "Precompound decay in heavy-ion reactions." *Physical Review C* 23 (1): 205.

[70] Otsuka, Takaharu, and Kichinosuke Harada. 1983. "Pre-equilibrium description of fast light particle emission in heavy-ion reactions." *Physics Letters B* 121 (2–3): 106–110.

[71] Harvey, Bernard G., and H. Homeyer. 1985. "LBL-16882, Lawrence Berkeley Laboratory; Harvey, BG." *Nucl. Phys. A* 444: 498.

[72] Simbel, M. H., and A. Y. Abul-Magd. 1980. "Influence of angular momentum dependence of charge transfer in deeply inelastic collisions on fragment kinetic energies." *Zeitschrift für Physik A Hadrons and Nuclei* 294 (3): 277–280.

[73] Zagrebaev, V. I. 1990. "Semiclassical theory of direct and deep inelastic heavy ion collisions." *Annals of Physics* 197 (1): 33–93.

[74] Morgenstern, H., W. Bohne, K. Grabisch, H. Lehr, and W. Stöffler. 1983. "Formation and decay of the compound nucleus studied in the reaction ^{20}Ne + ^{27}Al." *Zeitschrift für Physik A Hadrons and Nuclei* 313 (1): 39–49.

Morgenstern, H., W. Bohen, W. Galster, K. Grabisch, and A. Kyanowski. 1984. "Influence of the Mass Asymmetry on the Onset of Incomplete and the Limit to Complete Fusion." *Physical Review Letters* 52 (13): 1104.

[75] Morgenstern, H., W. Bohne, W. Galster, and K. Grabisch. 1986. "Study of the reaction mechanism of incomplete fusion with the reactions ^{40}Ar + ^{11}B, ^{12,13}C at 7 MeV/amu." *Zeitschrift für Physik A Hadrons and Nuclei* 324 (4): 443–451.

[76] Singh, Pushpendra P., Manoj Kumar Sharma, Devendra P. Singh, Rakesh Kumar, K. S. Golda, B. P. Singh, and R. Prasad. 2007. "Observation of complete-and incomplete-fusion components in ^{159}Tb, ^{169}Tm (^{16}O, x) reactions: Measurement and analysis of forward recoil ranges at E/A≈ 5-6 MeV." *The European Physical Journal A* 34 (1): 29–39.

Singh, Pushpendra P., B. P. Singh, Manoj Kumar Sharma, Devendra P. Singh, R. Prasad, Rakesh Kumar, and K. S. Golda. 2008. "Influence of incomplete fusion on complete fusion: Observation of a large incomplete fusion fraction at E≈ 5-7 MeV/nucleon." *Physical Review C* 77 (1): 014607.

[77] Chakrabarty, S., B. S. Tomar, A. Goswami, G. K. Gubbi, S. B. Manohar, Anil Sharma, B. Bindukumar, and S. Mukherjee. 2000. "Complete and incomplete fusion reactions in the ^{12}C + ^{169}Tm." *Nuclear Physics A* 678 (4): 355–366.

[78] Gupta, Unnati, Pushpendra P. Singh, Devendra P. Singh, Manoj Kumar Sharma, Abhishek Yadav, Rakesh Kumar, B. P. Singh, and R. Prasad. 2008. "Observation of large incomplete fusion in ^{16}O + ^{103}Rh system at ≈ 3–5 MeV/nucleon." *Nuclear Physics A* 811 (1–2): 77–92.

[79] Jungclaus, Andrea, B. Binder, A. Dietrich, T. Härtlein, H. Bauer, Ch Gund, D. Pansegrau et al. 2002. "Backbending region study in ^{160,162}Dy using incomplete fusion reactions." *Physical Review C* 66 (1): 014312.

Jungclaus, Andrea, B. Binder, A. Dietrich, T. Härtlein, H. Bauer, Ch Gund, D. Pansegrau et al. 2003. "Excited bands and signature dependent electromagnetic decay properties in neutron-rich ^{159,161,163}Dy." *Physical Review C* 67 (3): 034302.

[80] Dracoulis, G. D., A. P. Byrne, T. Kibédi, T. R. McGoram, and S. M. Mullins. 1997. "Incomplete fusion as a spectroscopic tool." *Journal of Physics G: Nuclear and Particle Physics* 23 (10): 1191.

[81] Lane, G. J., G. D. Dracoulis, A. P. Byrne, A. R. Poletti, and T. R. McGoram. 1999. "Population of high-spin states in ^{234}U by an incomplete-fusion reaction." *Physical Review C* 60 (6): 067301.

[82] Mullins, S. M., A. P. Byrne, G. D. Dracoulis, T. R. McGoram, and W. A. Seale. 1998. "High-spin intrinsic and rotational states in the stable nucleus ^{177}Hf: Evidence for reaction-dependent spin population." *Physical Review C* 58 (2): 831.

Mullins, S. M., G. D. Dracoulis, A. P. Byrne, T. R. McGoram, S. Bayer, R. A. Bark, R. T. Newman, W. A. Seale, and F. G. Kondev. 2000. "Rotational and intrinsic states above the K π = 25/2⁻, $T_{1/2}$ = 25 day isomer in ^{179}Hf." *Physical Review C* 61 (4): 044315.

[83] Yadav, Abhishek, Vijay R. Sharma, Pushpendra P. Singh, R.Kumar, Unnati, M. K. Sharma, B. P. Singh and R. Prasad. 2012. "Effect of Alpha-Q value on incomplete fusion." *Physical Review C* 86: 014603.

[84] Sharma, Manoj Kumar, Pushendra P. Singh, Devendra P. Singh, Abhishek Yadav, Vijay Raj Sharma, Indu Bala, Rakesh Kumar, B. P. Singh, and R. Prasad. 2015. "Systematic study of pre-equilibrium emission at low energies in C^{12}- and O^{16}-induced reactions." *Physical Review C* 91 (1): 014603.

Sharma, Manoj Kumar, Pushpendra P. Singh, Vijay Raj Sharma, Mohd Shuaib, Devendra P. Singh, Abhishek Yadav, R. Kumar, B. P. Singh, and R. Prasad. 2016. "Precompound emission in low-energy heavy-ion interactions from recoil range and spin distributions of heavy residues: A new experimental method." *Physical Review C* 94 (4): 044617.

[85] Krappe, H. J., J. R. Nix, and A. J. Sierk. 1979. "Unified nuclear potential for heavy-ion elastic scattering, fusion, fission, and ground-state masses and deformations." *Physical Review C* 20 (3): 992.

Chapter 2

[1] Bohr, Niels. 1936. "Conservation laws in quantum theory." *Nature* 138 (25): 213.

[2] Bethe, H. A., and G. Placzek. 1937. "Resonance effects in nuclear processes." *Physical Review* 51: 450–484.

[3]   Weisskopf, V. F., and D. H. Ewing. 1940. "Erratum: On the Yield of Nuclear Reactions with Heavy Elements." Phys. Rev. 57: 472.

[4]   Wolfenstein, L. 1951. "Conservation of angular momentum in the statistical theory of nuclear reactions." *Physical Review* 82 (5): 690.

[5]   Hauser, Walter, and Herman Feshbach. 1952. "The inelastic scattering of neutrons." *Physical Review* 87 (2): 366.

[6]   Moldauer, P. A. 1976. "Evaluation of the fluctuation enhancement factor." *Physical Review C* 14 (2): 764.

[7]   Lynn, J. E. 1963. "Average Neutron Transmission in the keV Region." *Proc. Phys. Soc.* 82: 903.

[8]   Hodgson, P. E. 1971. Nuclear reactions and nuclear structure, Oxford: Clarendon Press.

[9]   Hilaire, S. 2000. "Statistical nuclear reactions." Paper presented on a workshop on Nuclear Reaction Data and Nuclear Reactor: Physics Design and Safety, Trieste, Italy: 105–120.

Hilaire, S. 2000. "Level Densities." Paper presented at a workshop on Nuclear Reaction Data and Nuclear Reactor: Physics Design and safety, Trieste, Italy: 12–136.

[10] Hill, David Lawrence, and John Archibald Wheeler. 1953. "Nuclear constitution and the interpretation of fission phenomena." *Physical Review* 89 (5): 1102.

[11] Weisskopf, Victor F. 1961. "Problems of nuclear Structure." *Physics Today* 14: 18.

[12] Griffin, James J. 1966. "Statistical model of intermediate structure." *Physical review letters* 17 (9): 478.

[13] Bertini, H. W., G. D. Harp, and F. E. Bertrand. 1974. "Comparisons of predictions from two intranuclear-cascade models with measured secondary proton spectra at several angles from 62- and 39-MeV protons on various elements." *Physical Review C* 10 (6): 2472.

[14] Chen, K., Z. Fraenkel, G. Friedlander, J. R. Grover, J. M. Miller and Y. Shimamoto. 1968. 'VEGAS: A monte carlo simulation of intranuclear cascade.' *Physical Review* 166: 949

[15] Cline, C. K., and M. Blann. 1971. "The pre-equilibrium statistical model: description of the nuclear equilibration process and parameterization of the model." *Nuclear Physics A* 172 (2): 225–259.

[16] Ericson, Torleif. 1960. "The statistical model and nuclear level densities." *Advances in Physics* 9 (36): 425–511.

[17] Böhning, M. 1970. "Density of Particle-hole States in the Equidistant-spacing Model." *Nuclear Physics A* 152 (3): 529–546.

[18] Williams, Frederick C. 1971. "Particle-hole state density in the uniform spacing model." *Nuclear Physics A* 166 (2): 231–240.

[19] Běták, E., and J. Dobeš. 1976. "The finite depth of the nuclear potential well in the exciton model of preequilibrium decay."*Zeitschrift für Physik A Hadrons and Nuclei* 279 (3): 319–324.

[20] Akkermans, J. M. 1979. "Angular distribution in exciton and hybrid models for preequilibrium reactions." *Physics Letters B* 82 (1): 20–22.

[21] Blann, Marshall. 1972. "Importance of the nuclear density distribution on pre-equilibrium decay." *Physical Review Letters* 28 (12): 757.

Blann, M. 1973. "A priori pre-equilibrium decay models." *Nuclear Physics A* 213 (3): 570–588.

Blann, M. 1996. "New precompound decay model." *Physical Review C* 54 (3): 1341.

[22] Feshbach, Herman, Arthur Kerman, and Steven Koonin. 1980. "The statistical theory of multistep compound and direct reactions." *Annals of Physics* 125 (2): 429–476.

[23] Weiner, R., and M. Weström. 1975. "Pre-equilibrium and heat conduction in nuclear matter." *Physical Review Letters* 34 (24): 1523.

Weiner, R., and M. Weström. 1977. "Diffusion of heat in nuclear matter and preequilibrium phenomena." *Nuclear Physics A* 286 (2): 282–296.

[24] Bondorf, Jacob P., J. N. De, G. Fai, A. O. T. Karvinen, B. Jakobsson, and J. Randrup. 1980. "Promptly emitted particles in nuclear collisions." *Nuclear Physics A* 333 (2): 285–301.

[25] Wilczynski, J., K. Siwek-Wilczynska, J. Van Driel, S. Gonggrijp, D. C. J. M. Hageman, R. V. F. Janssens, J. Łukasiak, Rh H. Siemssen, and S. Y. Van der Werf. 1982. "Binary l-matched reactions in ^{14}N + ^{159}Tb collisions." *Nuclear Physics A* 373 (1): 109–140.

[26] Bondorf, J. P., F. Dickmann, D. H. E. Gross, P. J. Siemens. 1971. 'Statistical aspects of heavy-ion reactions.' *J. Phys. Colloques.* 32-C6: 145–149.

[27] Myers, William D. 1973. "Geometric properties of leptodermous distributions with applications to nuclei." *Nuclear Physics A* 204 (3): 465–484.

[28] Britt, Harold C., and Arthur R. Quinton. 1961. "Alpha Particles and Protons Emitted in the Bombardment of Au^{197} and Bi^{209} by C^{12}, N^{14}, and O^{16} Projectiles." *Physical Review* 124 (3): 877.

[29] Inamura, T., M. Ishihara, T. Fukuda and T. Shimoda. 1977. "Gamma-rays for incomplete fusion reactions indiced by 95 MeV 14N." *Physics Letters B* 68: 51–54.

Inamura, T., T. Kojima, T. Nomura, T. Sugitate, T. Utsunomiya. 1979. "Multiplicity of gamma-rays following fast alpha-particle emission in the 95 MeV ^{14}N + ^{159}Tb reaction." *Physics Letters B* 84: 71–74.

[30] Zolnowski, D. R., H. Yamada, S. E. Cala, A. C. Kahler, and T. T. Sugihara. 1978. "Evidence for 'Massive Transfer' in Heavy-Ion Reactions on Rare-Earth Targets." *Physical Review Letters* 41 (2): 92.

[31] Kerman, A. K., and K. W. McVoy. 1979. "Fluctuations in two-step reactions through doorways." *Annals of Physics* 122 (1): 197–216.

[32] Udagawa, T., and T. T. Tamura. 1980. "Breakup-fusion description of massive transfer reactions with emission of fast light particles." *Physical Review Letters* 45 (16): 1311.

[33] Rossne, H. R., D. Hilscher, D. J. Hinde, B. Gebauer, M. Lehmann, M. Wilpert, and E. Mordhorst. 1989. "Analysis of pre- and post-scission neutrons emitted in the reaction ^{169}Tm(^{36}Ar,f) at E_{lab} = 205 MeV." *Physical review C* 40: 2629–2640.

[34] Perey, C. M., and F. G. Perey. 1976. "Optical Model Parameters." *At Data Nucl Data Tables* 17: 10–45.

[35] Huizenga, J. R., and George Igo. 1962. "Theoretical reaction cross sections for alpha particles with an optical model." *Nuclear Physics* 29: 462–473.

[36] Bass, R. 1977. "Nucleus-nucleus potential deduced from experimental fusion cross sections." *Physical Review Letters* 39 (5): 265.

[37] Sierk, Arnold J. 1986. "Macroscopic model of rotating nuclei."*Physical Review C* 33 (6): 2039.

[38] Cohen, S., F. Plasil, and W. J. Swiatecki. 1974. "Equilibrium configurations of rotating charged or gravitating liquid masses with surface tension. II." *Annals of Physics* 82 (2): 557–596.

[39] Bohr, Niels, and John Archibald Wheeler. 1939. "The mechanism of nuclear fission." *Physical Review* 56 (5): 426.

[40] Pühlhofer, F. 1977. "On the interpretation of evaporation residue mass distributions in heavy-ion induced fusion reactions."*Nuclear Physics A* 280 (1): 267–284.

[41] Becchetti Jr, F. D., and G. W. Greenlees. 1969. "Nucleon-nucleus optical-model parameters, A> 40, E< 50 MeV." *Physical Review* 182 (4): 1190.

[42] Satchler, G. R. 1965. "Analysis of the scattering of 28 MeV alpha particles." *Nuclear Physics* 70 (1): 177–195.

[43] Ramamurthy, V.S. 1970. "Excitation energy dependence of shell effects on nuclear level densities and fission fragment anisotropies." *Physical Review Letter* 25: 386–390.

[44] Dilg, W., Wl Schantl, H. Vonach, and M. Uhl. 1973. "Level density parameters for the back-shifted fermi gas model in the mass range 40< A< 250." *Nuclear Physics A* 217 (2): 269–298.

[45] Myers W.D., and W. J. Swiatecki. 1967. "Anomolies in nuclear masses." *Ark. Phys.* 36: 343.

[46] Kalbach C. 1994. "Multiple pre-equlibrium emission code PRECO." Report CEN-DPh-N/ BE/94/3.

[47] Kalbach, Cr. 1977. "The Griffin model, complex particles and direct nuclear reactions." *Zeitschrift für Physik A Hadrons and Nuclei* 283 (4): 401–411.

[48] Belgya, T., O. Bersillon, R. Capote, T. Fukahori, G. Zhigang, S. Goriely, M. Herman, A. V. Ignatyuk, S. Kailas, A. Kooning, P. Oblozinsky, V. Plujko, and P. Young. 2006. "Hand book for calcuations of nuclear reaction data." Reference Input Parameter Library RIPL-2, IAEA-TELDOC-1506 Vienna, Austria.

[49] Raynal, Jacques. 1994. "Notes on ECIS94." *CEA Saclay report CEA-N-2772.*

[50] Thomas, T. D. 1964. "Angular momentum effects in neutron evaporation." *Nuclear Physics* 53: 577–592.

[51] Dityuk, A. I., A. Yu Konobeyev, V. P. Lunev, and Yu N. Shubin. 1998. "New advanced version of computer code ALICE-IPPE." *Report INDC (CCP)-410,* Vienna: IAEA.

[52] Iwamoto, A., and Koji Harada. 1982. "Mechanism of cluster emission in nucleon-induced preequilibrium reactions." *Physical Review C* 26 (5): 1821.

[53] Sharma, Manoj Kumar, B. K. Sharma, B. P. Singh, H. D. Bhardwaj, Rakesh Kumar, K. S. Golda, and R. Prasad. 2004. "Complete and incomplete fusion reactions in the O^{16} + Tm^{169} system: Excitation functions and recoil range distributions." *Physical Review C* 70 (4): 044606.

Sharma, Manoj Kumar, Devendra P. Singh, Pushpendra P. Singh, B. P. Singh, H. D. Bhardwaj, and R. Prasad. 2007. "Reaction mechanism in the O^{16} + Al^{27} system: Measurements and analysis of excitation functions and angular distributions." *Physical Review C* 75 (6): 064608.

[54] Singh, Pushpendra P., Abhishek Yadav, Devendra P. Singh, Unnati Gupta, Manoj K. Sharma, R. Kumar, D. Singh et al. 2009. "Role of high ℓ values in the onset of incomplete fusion." *Physical Review C* 80 (6): 064603.

Singh, D., R. Ali, M. Afzal Ansari, K. Surendra Babu, Pushpendra P. Singh, M. K. Sharma, B. P. Singh et al. 2010. "Incomplete fusion dynamics by spin distribution measurements." *Physical Review C* 81 (2): 027602.

[55] Singh, Devendra P., Pushpendra P. Singh, Abhishek Yadav, Manoj Kumar Sharma, B. P. Singh, K. S. Golda, Rakesh Kumar, A. K. Sinha, and R. Prasad. 2010. "Energy dependence of incomplete fusion processes in the O^{16} + Ta^{181} system: Measurement and analysis of forward-recoil–range distributions at E lab ≈ 7 MeV/nucleon." *Physical Review C* 81 (5): 054607.

[56] Yadav, Abhishek, Vijay R. Sharma, Pushpendra P. Singh, Devendra P. Singh, Manoj K. Shamra, R. Kumar, B. P. Singh, R. Prasad, and R. K. Bhowmik. 2012. "Effect of alpha-Q-value on reaction dynamics at≈ 4-7 AMeV." In *EPJ Web of Conferences*, 21: 08005. EDP Sciences.

Yadav, Abhishek, Vijay R. Sharma, Pushpendra P. Singh, Devendra P. Singh, R. Kumar, Unnati, M. K. Sharma, B. P. Singh, R. Prasad, and R. K. Bhowmik. 2012. "Effect of entrance-channel parameters on incomplete fusion reactions." *Physical Review* C 85: 064617.

[57] Sharma, Vijay R., Abhishek Yadav, Devendra P. Singh, Pushpendra P. Singh, Indu Bala, R. Kumar, M. K. Sharma et al. 2014. "Incomplete fusion reactions at low energies in ^{13}C + ^{169}Tm system." *In EPJ Web of Conferences*, 66: 03079. EDP Sciences.

[58] Shuaib, Mohd, Vijay R. Sharma, Abhishek Yadav, Pushpendra P. Singh, Manoj Kumar Sharma, Devendra P. Singh, R. Kumar et al. 2016. "Incomplete fusion studies in the ^{19}F + ^{159}Tb system at low energies and its correlation with various systematics." *Physical Review C* 94 (1): 014613.

Chapter 3

[1] Koch, R. C. 1960. *Activation Analysis Handbook.* New York and London: Academic Press.

[2] Mughabghab, S. F., M. Divadeenam, and N. E. Holden. 1981. *Neutron Cross Sections*, Vol. 1 Part A, 89. New York: Academic Press

[3] Singh, Bhanu Prakash. 1991. "Equilibrium and pre-equilibrium emission in alpha induced reactions at moderate excitation energies." Ph.D. dissertation, Aligarh Muslim University, Aligarh, India.

[4] Mehta, G. K., and A. P. Patro. 1988. "15 UD pelletron of the nuclear science centre—status report." *Nuclear Instruments and Methods in Physics Research Section A: Accelerators, Spectrometers, Detectors and Associated Equipment* 268 (2–3): 334–336. (http://www.iuac.res.in/refac/np/inga/citation.html) Accessed on 27-12-2017.

[5] Kumar, BP Ajith, J. Kannaiyan, P. Sugathan, and R. K. Bhowmik. 1994. "The NSC 16 MV tandem accelerator control system." *Nuclear Instruments and Methods in Physics Research Section A: Accelerators, Spectrometers, Detectors and Associated Equipment* 343 (2–3): 327–330.

[6] Puttaswamy, N. G., N. M. Badiger and M. Raja Rao. 1991. "The General Purpose Scattering Chamber at the Nuclear Science Centre." DAE Symposium on Nuclear Physics. 34 B

[7] http://www.iuac.res.in/refac/np/gpsc/gpsc_main.html.

[8] Sharma, Vijay Raj. 2015. "Breakup fusion studies in heavy ion interactions at low energies." Ph.D. thesis, Aligarh Muslim University, Aligarh, India.

[9] SRIM06; http://www.srim.org/

[10] Kumar, BP Ajith. 2001. "CANDLE, Collection and Analysis of Nuclear Data using Linux Network." *DAE Symposium on Nuclear Physics,* Kolkata, India.

[11] Yadav, Abhishek. "Probing the incomplete fusion reaction dynamics using light heavy ions." Ph.D. Dissertation, Aligarh Muslim University, Aligarh, India. 2012.

[12] Brown, E., and R. B. Firestone. 1986. *Table of Radioactive isotopes.* New York: John Wiley and sons.

[13] Firestone, Richard B., Virgina S. Shirley, C. M. Baglin, SY Frank Chu, and J. Zipkin. 1996. *Table of Isotopes,* 8-th ed. New York: John Willey & Sons.

[14] Tuli, J. K. 2000. "Nuclear wallet card." National Nuclear data Centre, Brookhaven National Laboratory, Upton, New York, USA.

[15] Unnati. 2006. "Study of Angular and Recoil range Distributions for some residues produced in heavy ion interactions." Ph.D. dissertation, Aligarh Msulim University, Aligarh, India.

[16] Muralithar, S., B. Mukherjee, R. P. Singh, G. Mukherjee, P. Joshi, A. Punithan, B. K. Sahu et al. 2013. "A charged particle detector array for detection of light charged particles from nuclear reactions." *Nuclear Instruments and Methods in Physics Research Section A: Accelerators, Spectrometers, Detectors and Associated Equipment* 729: 849–855.

[17] Muralithar, S. 2014. "Nuclear structure at high spin using multidetector gamma array and ancillary detectors." *Pramana* 82 (4): 769–778.

Chapter 4

[1] Lunardon, M., C. Merigliano, G. Viesti, D. Fabris, G. Nebbia, M. Cinausero, G. de Angelis, E. Farnea, E. Fioretto, G. Prete, A. Brondi, G. La Rana, R. Moro, A. Principe, E. Vardaci, N. Gelli, F. Lucarelli, P. Pavan, D. R. Napoli, and G. Vedovato. 1999. "Alpha particle emission, incomplete fusion and population of high-spin states in the reaction 120 MeV ^{19}F + ^{181}Ta." *Nuclear Physics A* 652 (1): 3–16.

[2] Vergani, P., E. Gadioli, E. Vaciago, E. Fabrici, E. Gadioli Erba, M. Galmarini, G. Ciavola, and C. Marchetta. 1993. "Complete and incomplete fusion and emission of preequilibrium nucleons in the interaction of C^{12} with Au^{197} below 10 MeV/nucleon." *Physical Review C* 48 (4): 1815.

[3] Dasgupta, Mahananda, P. R. S. Gomes, D. J. Hinde, S. B. Moraes, R. M. Anjos, A. C. Berriman, Rachel D. Butt et al. 2004. "Effect of breakup on the fusion of Li^6, Li^7, and Be^9 with heavy nuclei." *Physical Review C* 70 (2): 024606.

[4] Gadioli, E., C. Birattari, M. Cavinato, E. Fabrici, E. Gadioli Erba, V. Allori, F. Cerutti, *et al.* 1998. "Angular distributions and forward recoil range distributions of residues created in the interaction of ^{12}C and ^{16}O ions with ^{103}Rh." *Nuclear Physics A* 641 (3): 271–296.

[5] Canto, L. F., R. Donangelo, Lia M. de Matos, M. S. Hussein, and P. Lotti. 1998. "Complete and incomplete fusion in heavy-ion collisions." *Physical Review C* 58 (2): 1107.

[6] Sharma, Manoj Kumar, B. K. Sharma, B. P. Singh, H. D. Bhardwaj, Rakesh Kumar, K. S. Golda, and R. Prasad. 2004. "Complete and incomplete fusion reactions in the ^{16}O + ^{169}Tm system: Excitation functions and recoil range distributions." *Physical Review C* 70 (4): 044606.

[7] Kumar Sharma, Manoj, B. P. Singh, Sunita Gupta, M. M. Musthafa, H. D. Bhardwaj, R. Prasad, and A. K. Sinha. 2003. "Complete and Incomplete Fusion: Measurement and Analysis of Excitation

Functions in ^{12}C + ^{128}Te System at Energies near and above the Coulomb Barrier." *Journal of the Physical Society of Japan* 72 (8): 1917–1925.

[8]   Yadav, Abhishek, Vijay R. Sharma, Pushpendra P. Singh, Devendra P. Singh, R. Kumar, M. K. Sharma, B. P. Singh, R. Prasad, and R. K. Bhowmik. 2012. "Effect of entrance-channel parameters on incomplete fusion reactions." *Physical Review C* 85 (6): 064617.

[9]   Yadav, Abhishek, Vijay R. Sharma, Pushpendra P. Singh, R. Kumar, Devendra P. Singh, M. K. Sharma, B. P. Singh, and R. Prasad. 2012. "Effect of α-Q value on incomplete fusion." *Physical Review C* 86 (1): 014603.

[10] Shuaib, Mohd, Vijay R. Sharma, Abhishek Yadav, Pushpendra P. Singh, Manoj Kumar Sharma, Devendra P. Singh, R. Kumar et al. 2016. "Incomplete fusion studies in the F 19+ Tb 159 system at low energies and its correlation with various systematics." *Physical Review C* 94 (1): 014613.

[11] Shuaib, Mohd, Vijay R. Sharma, Abhishek Yadav, Manoj Kumar Sharma, Pushpendra P. Singh, Devendra P. Singh, R. Kumar et al. 2017. "Influence of incomplete fusion on complete fusion at energies above the Coulomb barrier." *Journal of Physics G: Nuclear and Particle Physics* 44 (10): 105108.

[12] Brown, E., and R. B. Firestone. 1986. *Table of Radioactive isotopes (ed. VS Shirley)* New York: John Wiley and sons.

[13] Pühlhofer, F. 1977. "On the interpretation of evaporation residue mass distributions in heavy-ion induced fusion reactions." *Nuclear Physics A* 280 (1): 267–284.

[14] Blann, M. 1991. *NEA Data Bank*, Gif-sur-Yveette, France, Report No. *PSR-146.*

[15] Hauser, Walter, and Herman Feshbach. 1952. "The inelastic scattering of neutrons." *Physical review* 87 (2): 366.

[16] Becchetti Jr, F. D., and G. W. Greenlees. 1969. "Nucleon-nucleus optical-model parameters, A> 40, E< 50 MeV." *Physical Review* 182 (4): 1190.

[17] Satchler, G. R. 1965. "Analysis of the scattering of 28 MeV alpha particles." *Nuclear Physics* 70 (1): 177–195.

[18] Weisskopf, V. F., and D. H. Ewing. 1940. "Erratum: On the Yield of Nuclear Reactions with Heavy Elements (Phys. Rev. 57, 472 (1940))." *Physical Review* 57 (10). 935.

[19] Blann, Marshall. 1971. "Hybrid model for pre-equilibrium decay in nuclear reactions." *Physical Review Letters* 27 (6): 337.

[20] Bodansky, David. 1962. "Compound Statistical Features in Nuclear Reactions." *Annual review of nuclear science* 12 (1): 79–122.

[21] Evans, R.D. 1955. "The Atomic Nucleus." McGraw-Hill: New York.

[22] Gupta, Sunita, B. P. Singh, M. M. Musthafa, H. D. Bhardwaj, and R. Prasad. 2000. "Complete and incomplete fusion of ^{12}C with ^{165}Ho below 7 MeV/nucleon: Measurements and analysis of excitation functions." *Physical Review-Section C-Nuclear Physics* 61 (6) : 64613–64613.

[23] Gavron, A. 1980. "Statistical model calculations in heavy ion reactions." *Physical Review C* 21 (1): 230.

[24] Bass, Rainer. 1974. "Fusion of heavy nuclei in a classical model." *Nuclear Physics A* 231 (1): 45–63.

[25] Cavinato, M., E. Fabrici, E. Gadioli, E. Gadioli Erba, P. Vergani, M. Crippa, G. Colombo, I. Redaelli, and M. Ripamonti. 1995. "Study of the reactions occurring in the fusion of C^{12} and O^{16} with heavy nuclei at incident energies below 10 MeV/nucleon." *Physical Review C* 52 (5): 2577.

[26] Lestone, J. P. 1996. "Analysis of α-particle emission from F^{19} + Ta^{181} reactions leading to residues." *Physical Review C* 53 (4): 2014.

[27] Sharma, Vijay R., Abhishek Yadav, Pushpendra P. Singh, Devendra P. Singh, Sunita Gupta, M. K. Sharma, Indu Bala et al. 2014. "Influence of a one-neutron-excess projectile on low-energy incomplete fusion." *Physical Review C* 89 (2): 024608.

[28] Wapstra, A. H., and N. B. Gove. 1971. "Part I. Atomic mass table." *Atomic Data and Nuclear Data Tables* 9 (4–5): 267–301.

[29] Myers, William D., and Wladyslaw J. Swiatecki. 1966. "Nuclear masses and deformations." *Nuclear Physics* 81 (1): 1–60.

[30] Ladenbauer-Bellis, Inge-Maria, Ivor L. Preiss, and C. E. Anderson. 1962. "Excitation Functions for Heavy-Ion-Induced Reactions on Aluminum-27." *Physical Review* 125 (2): 606.

[31] Sharma, Manoj Kumar, B. P. Singh, Rakesh Kumar, K. S. Golda, H. D. Bhardwaj, and R. Prasad. 2006. "A study of the reactions occurring in ^{16}O + ^{159}Tb system: Measurement of excitation functions and recoil range distributions." *Nuclear Physics A* 776 (3–4): 83–104.

[32] Gupta, Unnati, Pushpendra P. Singh, Devendra P. Singh, Manoj Kumar Sharma, Abhishek Yadav, Rakesh Kumar, B. P. Singh, and R. Prasad. 2008. "Observation of large incomplete fusion in ^{16}O + ^{103}Rh system at$\approx$ 3–5 MeV/nucleon." *Nuclear Physics A* 811 (1–2): 77–92.

[33] Singh, Devendra P., Pushpendra P. Singh, Abhishek Yadav, Manoj Kumar Sharma, B. P. Singh, K. S. Golda, Rakesh Kumar, A. K. Sinha, and R. Prasad. 2009. "Investigation of the role of break-up processes on the fusion of O^{16} induced reactions." *Physical Review C* 80 (1): 014601.

[34] Singh, Devendra P., Pushpendra P. Singh, Abhishek Yadav, Manoj Kumar Sharma, B. P. Singh, K. S. Golda, Rakesh Kumar, A. K. Sinha, and R. Prasad. 2010. "Energy dependence of incomplete fusion processes in the O^{16} + Ta^{181} system: Measurement and analysis of forward-recoil–range distributions at E lab $\approx$ 7 MeV/nucleon." *Physical Review C* 81 (5): 054607.

[35] Gupta, Unnati, Pushpendra P. Singh, Devendra P. Singh, Manoj Kumar Sharma, Abhishek Yadav, Rakesh Kumar, S. Gupta, H. D. Bhardwaj, B. P. Singh, and R. Prasad. 2009. "Disentangling full and partial linear momentum transfer events in the ^{16}O + ^{169}Tm system at E_{proj} $\approx$ 5. 4 MeV/nucleon." *Physical Review C* 80 (2): 024613.

[36] Sharma, Manoj Kumar, Devendra P. Singh, Pushpendra P. Singh, B. P. Singh, H. D. Bhardwaj, and R. Prasad. 2007. "Reaction mechanism in the ^{16}O + ^{27}Al system: Measurements and analysis of excitation functions and angular distributions." *Physical Review C* 75 (6): 064608.

[37] Singh, Pushpendra P., B. P. Singh, M. K. Sharma, Unnati Gupta, Rakesh Kumar, D. Singh, R. P. Singh et al. 2009. "Probing of incomplete fusion dynamics by spin-distribution measurement." *Physics Letters B* 671 (1): 20–24.

[38] RadWare. Level Scheme Directory. Available at http://radware.phy.ornl.gov/agsdir1.html.

[39] Singh, Pushpendra P., Abhishek Yadav, Devendra P. Singh, Unnati Gupta, Manoj K. Sharma, R. Kumar, D. Singh et al. 2009. "Role of high ℓ values in the onset of incomplete fusion." *Physical Review C* 80 (6): 064603.

[40] Bhowmik Ranjan, S. Muralithar and R. P. Singh. 2001. "INGASORT, A New Program for the Analysis of Multiclover Array." DAE-BRNS Symposium on Nuclear Physics. 44 B: 422–423.

Singh R. P., Lagy T. Baby, R. K. Bhowmik. 2001. "Mote-carlo simulation of the performance of a clover detector." DAE-BRNS Symposium on Nuclear Physics. 44B:382–382.

[41] Yadav, Abhishek, Pushpendra P. Singh, Vijay R. Sharma, Unnati Gupta, Devendra P. Singh, Manoj K. Sharma, B. P. Singh et al. 2011. "Signature of pre-equilibrium-emission in forward-to-backward yield ratio measurement." *International Journal of Modern Physics E* 20 (10): 2133–2142.

[42] Sharma, Manoj Kumar, Pushpendra P. Singh, Vijay Raj Sharma, Mohd Shuaib, Devendra P. Singh, Abhishek Yadav, R. Kumar, B. P. Singh, and R. Prasad. 2016. "Precompound emission in low-energy heavy-ion interactions from recoil range and spin distributions of heavy residues: A new experimental method." *Physical Review C* 94 (4): 044617.

Chapter 5

[1] Diaz-Torres, A., and I. J. Thompson. 2002. "Effect of continuum couplings in fusion of halo ^{11}Be on ^{208}Pb around the Coulomb barrier." *Physical Review C* 65 (2): 024606.

[2] Diaz-Torres, Alexis, D. J. Hinde, Jeffrey Allan Tostevin, Mahananda Dasgupta, and L. R. Gasques. 2007. "Relating breakup and incomplete fusion of weakly bound nuclei through a classical trajectory model with stochastic breakup." *Physical review letters* 98 (15): 152701.

[3] Buthelezi, E. Z., E. Gadioli, G. F. Steyn, F. Albertini, C. Birattari, M. Cavinato, F. Cerutti et al. 2004. "Incomplete fusion of projectile fragments in the interaction of ^{12}C with ^{103}Rh up to 33 MeV per nucleon." *Nuclear Physics A* 734: 553–556.

[4] Gomes, P. R. S., I. Padron, E. Crema, O. A. Capurro, JO Fernández Niello, A. Arazi, G. V. Marti et al. 2006. "Comprehensive study of reaction mechanisms for the ^{9}Be + ^{144}Sm system at near-and sub-barrier energies." *Physical Review C* 73 (6): 064606.

[5] Gomes, P. R. S., I. Padron, M. D. Rodríguez, G. V. Marti, R. M. Anjos, J. Lubian, R. Veiga et al. 2004. "Fusion, reaction and break-up cross sections of weakly bound projectiles on ^{64}Zn." *Physics Letters B* 601 (1): 20–26.

[6] Dasgupta, Mahananda, D. J. Hinde, A. Mukherjee, and J. O. Newton. 2007. "New challenges in understanding heavy ion fusion." *Nuclear Physics A* 787 (1–4): 144–149.

[7] Singh, Pushpendra P., Abhishek Yadav, Devendra P. Singh, Unnati Gupta, Manoj K. Sharma, R. Kumar, D. Singh et al. 2009. "Role of high ℓ values in the onset of incomplete fusion." *Physical Review C* 80 (6): 064603.

[8] Singh, Pushpendra P., B. P. Singh, Manoj Kumar Sharma, R. Kumar, K. S. Golda, D. Singh, R. P. Singh et al. 2008. "Spin-distribution measurement: A sensitive probe for incomplete fusion dynamics." *Physical Review C* 78 (1): 017602.

[9] Singh, Pushpendra P., B. P. Singh, Manoj Kumar Sharma, Devendra P. Singh, R. Prasad, Rakesh Kumar, and K. S. Golda et al. 2008. "Influence of incomplete fusion on complete fusion: Observation of a large incomplete fusion fraction at E≈ 5–7 MeV/nucleon." *Physical Review C* 77 (1): 014607.

[10] Britt, Harold C., and Arthur R. Quinton. 1961. "Alpha Particles and Protons Emitted in the Bombardment of ^{197}Au and ^{209}Bi by ^{12}C, ^{14}N, and ^{16}O Projectiles." *Physical Review* 124 (3): 877.

[11] Kaufmann, Richard, and Richard Wolfgang. 1961. "Single-Nucleon Transfer Reactions of F^{19}, O^{16}, N^{14}, and C^{12}." *Physical Review* 121 (1): 206.

[12] Inamura, T., M. Ishihara, T. Fukuda, T. Shimoda, and H. Hiruta. 1977. "Gamma-rays from an incomplete fusion reaction induced by 95 MeV ^{14}N." *Physics Letters B* 68 (1): 51–54.

[13] Inamura, T., A. C. Kahler, D. R. Zolnowski, U. Garg, T. T. Sugihara, and M. Wakai. 1985. "Gamma-ray multiplicity distribution associated with massive transfer." *Physical Review C* 32 (5): 1539.

[14] Zolnowski, D. R., H. Yamada, S. E. Cala, A. C. Kahler, and T. T. Sugihara. 1978. "Evidence for" Massive Transfer" in Heavy-Ion Reactions on Rare-Earth Targets." *Physical Review Letters* 41 (2): 92.

[15] Geoffroy, K. Ao, D. G. Sarantites, M. L. Halbert, D. C. Hensley, R. A. Dayras, and J. H. Barker. 1979. "Angular momentum transfer in incomplete-fusion reactions." *Physical Review Letters* 43 (18): 1303.

[16] Trautmann, W., Ole Hansen, H. Tricoire, W. Hering, R. Ritzka, and W. Trombik. 1984. "Dynamics of Incomplete Fusion Reactions from γ-Ray Circular-Polarization Measurements." *Physical Review Letters* 53 (17): 1630.

[17] Inamura, T., T. Kojima, T. Nomura, T. Sugitate, and H. Utsunomiya. 1979. "Multiplicity of γ-rays following fast alpha;-particle emission in the 95 MeV ^{14}N + ^{159}Tb reaction." *Physics Letters B* 84 (1): 71–74.

[18] Singh, Pushpendra P., B. P. Singh, M. K. Sharma, Unnati Gupta, Rakesh Kumar, D. Singh, R. P. Singh et al. 2009. "Probing of incomplete fusion dynamics by spin-distribution measurement." *Physics Letters B* 671 (1): 20–24.

[19] Sharma, Vijay R., Abhishek Yadav, Pushpendra P. Singh, Indu Bala, Devendra P. Singh, Sunita Gupta, M. K. Sharma, R. Kumar, S. Muralithar, R. P. Singh, B. P. Singh, R. K. Bhowmik, and R. Prasad. 2015. "Spin distribution measurements in ^{16}O + ^{159}Tb system: incomplete fusion reactions." *Journal of Physics G: Nuclear and Particle* 42 (5): 055113.

[20] Sharma, Vijay R., Pushpendra P. Singh, Mohd Shuaib, Abhishek Yadav, Indu Bala, Manoj K. Sharma, S. Gupta et al. 2016. "Incomplete fusion in ^{16}O + ^{159}Tb." *Nuclear Physics A* 946: 182–193.

[21] Udagawa, T., and T. T. Tamura. 1980. "Breakup-fusion description of massive transfer reactions with emission of fast light particles." *Physical Review Letters* 45 (16): 1311.

[22] Udagawa, T., D. Price, and T. Tamura. 1982. "Exact-finite-range DWBA calculations of massive transfer reactions treated as breakup-fusion reactions." *Physics Letters B* 116 (5): 311–314.

[23] Wilczynski, J., K. Siwek-Wilczynska, J. Van Driel, S. Gonggrijp, D. C. J. M. Hageman, R. V. F. Janssens, J. Łukasiak, Rh H. Siemssen, and S. Y. Van der Werf. 1982. "Binary l-matched reactions in ^{14}N + ^{159}Tb collisions." *Nuclear Physics A* 373 (1): 109–140.

[24] Siwek-Wilczyńska, K., EH du Marchie Van Voorthuysen, J. Van Popta, R. H. Siemssen, and J. Wilczyński. 1979. "Incomplete Fusion in ^{12}C + ^{160}Gd Collisions Interpreted in Terms of a Generalized Concept of Critical Angular Momentum." *Physical Review Letters* 42 (24): 1599.

[25] Takada, E., T. Shimoda, N. Takahashi, T. Yamaya, K. Nagatani, T. Udagawa, and T. Tamura. 1981. "Projectile breakup reaction and evidence of a breakup-fusion mechanism." *Physical Review C* 23 (2): 772.

[26] Bondorf, Jacob P., J. N. De, G. Fai, A. O. T. Karvinen, B. Jakobsson, and J. Randrup. 1980. "Promptly emitted particles in nuclear collisions." *Nuclear Physics A* 333 (2): 285–301.

[27] Gupta, Unnati, Pushpendra P. Singh, Devendra P. Singh, Manoj Kumar Sharma, Abhishek Yadav, Rakesh Kumar, S. Gupta, H. D. Bhardwaj, B. P. Singh, and R. Prasad. 2009. "Disentangling full and partial linear momentum transfer events in the ^{16}O + ^{169}Tm system at $E_{proj} \approx 5.4$ MeV/nucleon." *Physical Review C* 80 (2): 024613.

[28] Singh, Devendra P., Pushpendra P. Singh, Abhishek Yadav, Manoj Kumar Sharma, B. P. Singh, K. S. Golda, Rakesh Kumar, A. K. Sinha, and R. Prasad. 2010. "Energy dependence of incomplete fusion processes in the ^{16}O + ^{181}Ta system: Measurement and analysis of forward-recoil–range distributions at E lab ≈ 7 MeV/nucleon." *Physical Review C* 81 (5): 054607.

[29] Yadav, Abhishek, Pushpendra P. Singh, Mohd Shuaib, Vijay R. Sharma, Indu Bala, Sunita Gupta, D. P. Singh et al. 2017. "Systematic study of low-energy incomplete fusion: Role of entrance channel parameters." *Physical Review C* 96 (4): 044614.

[30] Morgenstern, H., W. Bohen, W. Galster, K. Grabisch, and A. Kyanowski. 1984. "Influence of the Mass Asymmetry on the Onset of Incomplete and the Limit to Complete Fusion." *Physical Review Letters* 52 (13): 1104.

[31] Shuaib, Mohd, Vijay R. Sharma, Abhishek Yadav, Manoj Kumar Sharma, Pushpendra P. Singh, Devendra P. Singh, R. Kumar et al. 2017. "Influence of incomplete fusion on complete fusion at energies above the Coulomb barrier." *Journal of Physics G: Nuclear and Particle Physics* 44 (10): 105108.

[32] Wilczyński, J., K. Siwek-Wilczyńska, J. Van Driel, S. Gonggrijp, D. C. J. M. Hageman, R. V. F. Janssens, J. Łukasiak, and R. H. Siemssen. 1980. "Incomplete-Fusion Reactions in the N 14+ Tb 159 System and a" Sum-Rule Model" for Fusion and Incomplete-Fusion Reactions." *Physical Review Letters* 45 (8): 606.

[33] Hagino, K., N. Rowley, and A. T. Kruppa. 1999. "A program for coupled-channel calculations with all order couplings for heavy-ion fusion reactions." *Computer Physics Communications* 123 (1–3): 143–152.

[34] Bohr, Niels. 1936. "Neutron capture and nuclear constitution." *Nature* 137: 344–348.

[35] Sharma, Manoj Kumar, Pushpendra P. Singh, Vijay Raj Sharma, Mohd Shuaib, Devendra P. Singh, Abhishek Yadav, R. Kumar, B. P. Singh, and R. Prasad. 2016. "Precompound emission in low-energy heavy-ion interactions from recoil range and spin distributions of heavy residues: A new experimental method." *Physical Review C* 94 (4): 044617.

[36] Yadav, Abhishek, Pushpendra P. Singh, Vijay R. Sharma, Unnati Gupta, Devendra P. Singh, Manoj K. Sharma, B. P. Singh et al. 2011. "Signature of pre-equilibrium-emission in forward-to-backward yield ratio measurement." *International Journal of Modern Physics E* 20 (10): 2133–2142.

[37] Zolnowski, D. R., H. Yamada, S. E. Cala, A. C. Kahler, and T. T. Sugihara. 1978. "Evidence for" Massive Transfer" in Heavy-Ion Reactions on Rare-Earth Targets." *Physical Review Letters* 41 (2): 92.

[38] Dracoulis, G. D., A. P. Byrne, T. Kibédi, T. R. McGoram, and S. M. Mullins. 1997. "Incomplete fusion as a spectroscopic tool." *Journal of Physics G: Nuclear and Particle Physics* 23 (10): 1191.

[39] Görgen, A., H. Hübel, D. Ward, S. Chmel, R. M. Clark, M. Cromaz, R. M. Diamond et al. 1999. "Spectroscopy of ^{200}Hg after incomplete fusion reaction." *The European Physical Journal A-Hadrons and Nuclei* 6 (2): 141–147.

[40] Baktash, C. ed. 1999. *Proceedings of the Nuclear Structure 98 Conference, Gatlinburg. August 1998.* (AIP Conference Proceedings).

[41] Nasirov Avazbek, Kyungil Kim, Giuseppe Mandaglio, Giorgio Giardina, Akhtam Muminov, and Youngman Kim. 2013. "Main restrictions in the synthesis of new superheavy elements: Quasifission and/or fusion fission." *The European Physical Journal A* 49: 147

Index